21st Century Canon of Solar Eclipses

BLACK & WHITE EDITION

Fred Espenak

Edition 1.0
September 2016

21st Century Canon of Solar Eclipses – Black & White Edition

Astropixels Publishing
P.O. Box 16197
Portal, AZ 85632

www.astropixels.com/pubs

This book may be ordered at: www.astropixels.com/pubs/21CCSE.html

More about solar eclipses of the 21st Century can be found at:

www.eclipsewise.com/solar/SEcatalog/21CCSEcat.html

Printed in the United States of America

ISBN 978-1-941983-11-9

First Edition (Version 1.0a)

Front Cover: A time sequence shows the partial phases and totality during the total solar eclipse of 2001 Jun 21 from Chisamba, Zambia. Photo copyright ©2001 by Fred Espenak.

Back Cover: Portrait of Fred Espenak, Copyright © 2016 by Patricia Totten Espenak.

Table of Contents

SECTION 1: SOLAR ECLIPSE FUNDAMENTALS..5
    1.1 Introduction...5
    1.2 Classification of Solar Eclipses..5
    1.3 Visual Appearance of Partial Solar Eclipses...6
    1.4 Visual Appearance of Annular Solar Eclipses...6
    1.5 Visual Appearance of Total Solar Eclipse 7
    1.6 Safely Observing Solar Eclipses 8
    1.7 Central Line and Duration of Totality...9

SECTION 2: SOLAR ECLIPSE PREDICTIONS 11
    2.1 Solar Eclipse Contacts... 11
    2.2 Mean Lunar Radius 11
    2.3 Solar and Lunar Coordinates 12
    2.4 Measurement of Time 12
    2.5 ΔT (Delta T) 13
    2.6 Date Format 13

SECTION 3: SOLAR ECLIPSE STATISTICS.. 14
    3.1 Statistical Distribution of Eclipse Types 14
    3.2 Eclipse Frequency and the Calendar Year 15
    3.3 Extremes in Greatest Central Duration and Eclipse Magnitude 16
    3.4 Eclipse Seasons 16
    3.5 Quincena 17

SECTION 4: EXPLANATION OF SOLAR ECLIPSE CATALOG IN APPENDIX A............................ 18
    4.1 Introduction... 18
    4.2 Cat Num (Catalog Number)... 18
    4.3 Calendar Date 18
    4.4 TD of Greatest Eclipse (Terrestrial Dynamical Time of Greatest Eclipse) 18
    4.5 ΔT (Delta T) 18
    4.6 Luna Num (Lunation Number)... 18
    4.7 Saros Num (Saros Series Number) 19
    4.8 Ecl Type (Solar Eclipse Type) 19
    4.9 QLE (Quincena Lunar Eclipse Parameter)... 19
    4.10 Gamma 20
    4.11 Ecl Mag (Eclipse Magnitude).. 20
    4.12 Lat Long (Latitude and Longitude).. 20
    4.13 Sun Alt (Altitude of Sun) 20
    4.14 Sun Azm (Azimuth of Sun) 20
    4.15 Path Width 20
    4.16 Central Line Dur (Central Line Duration) 20
    4.17 EclipseWise.com and the 21st Century Canon 20

SECTION 5: EXPLANATION OF SOLAR ECLIPSE MAPS IN APPENDICES B, C AND D 21
    5.1 Explanation of Small Global Eclipse Maps in Appendix B 21
    5.2 Explanation of Large Global Eclipse Maps in Appendix C 23
    5.3 Explanation of Central Solar Eclipse Maps in Appendix D 25

REFERENCES... 26

APPENDIX A.. 27
    Solar Eclipse Catalog: 2001 to 2100 27
    Key to Solar Eclipse Catalog 28

APPENDIX B.. 33

SMALL GLOBAL SOLAR ECLIPSE MAPS: 2001 TO 2100 33
KEY TO SMALL GLOBAL SOLAR ECLIPSE MAPS 34
APPENDIX C **55**
LARGE GLOBAL SOLAR ECLIPSE MAPS: 2017 TO 2066 55
KEY TO LARGE GLOBAL SOLAR ECLIPSE MAPS 56
APPENDIX D **171**
CENTRAL SOLAR ECLIPSE MAPS: 2001 TO 2100 171
KEY TO CENTRAL SOLAR ECLIPSE MAPS 172

Section 1: Solar Eclipse Fundamentals

1.1 Introduction

The Moon's orbit is tilted about 5.1° to Earth's orbit around the Sun. The points where the two orbits appear to cross are called the nodes. When the New Moon occurs near one of these nodes, the Moon's shadow can sometimes fall on some portion of Earth and a solar eclipse takes place.

The Moon's shadow is composed of three cone-shaped components. The outer or penumbral shadow is a zone where the Sun's rays are partially blocked. Nested within the penumbra is the umbral shadow — a region where direct rays from the Sun are completely blocked. The conical umbra tapers to a point beyond which extends an expanding cone called the antumbra. From within this third shadow, the Moon appears smaller than the Sun and is seen in silhouette against the solar disk.

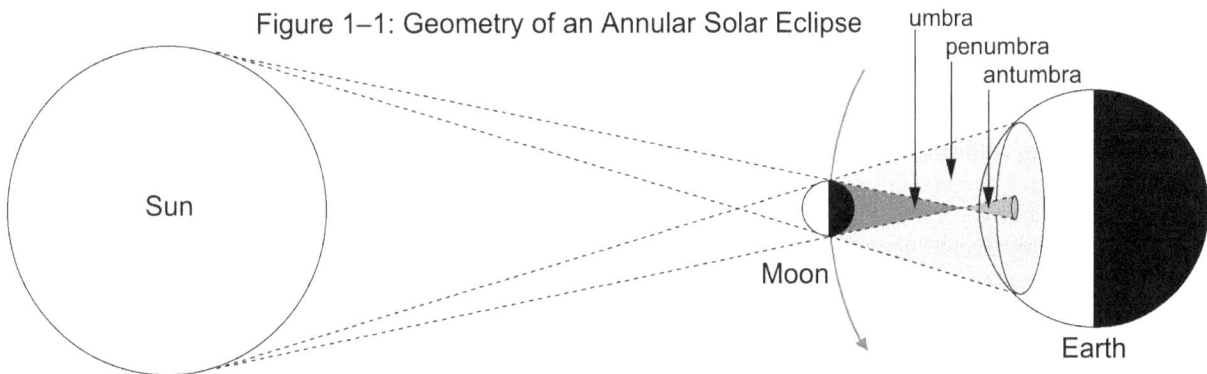

Figure 1–1: Geometry of an Annular Solar Eclipse

Figure 1–1 illustrates the geometry of an annular solar eclipse. A partial eclipse is visible from within the large penumbral shadow, while the annular eclipse is confined to the much smaller antumbral shadow.

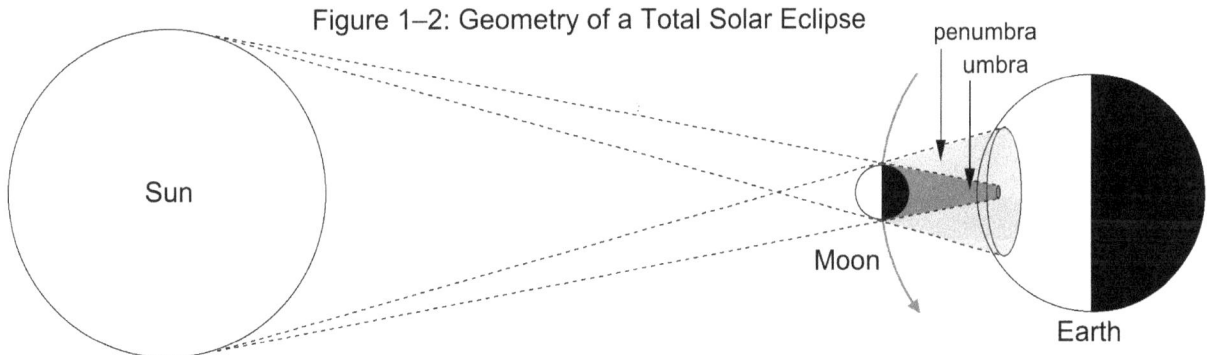

Figure 1–2: Geometry of a Total Solar Eclipse

Figure 1–2 illustrates the geometry of a total solar eclipse. A partial eclipse is visible from within the large penumbral shadow, while the total eclipse is only seen from the much smaller umbral shadow.

1.2 Classification of Solar Eclipses

There are four basic types of solar eclipses:

1. **Partial Solar Eclipse** — The Moon's penumbral shadow traverses Earth while the umbral and antumbral shadows miss Earth. A portion of the Sun's disk is obscured from within the penumbra.
2. **Annular Solar Eclipse** — The Moon's penumbral and antumbral shadows traverse Earth. The Moon's umbral shadow completely misses Earth. The Moon's disk appears smaller than the Sun so a

bright ring surrounds the Moon when viewed from within the antumbral shadow. A partial eclipse is seen within the penumbral shadow (Figure 1–1).

3. **Total Solar Eclipse** — The Moon's penumbral and umbral shadows traverse Earth. The Moon's antumbral shadow extends beyond Earth's surface. The Moon's disk appears larger than the Sun and completely covers the solar disk when viewed from within the umbral shadow. A partial eclipse is seen within the penumbral shadow (Figure 1–2).
4. **Hybrid Solar Eclipse** — The Moon's penumbral, umbral and antumbral shadows all traverse parts of Earth. The curvature of Earth's surface brings some regions into the umbra and others into the antumbra. The eclipse is total within the umbra and annular within the antumbra. A partial eclipse is seen within the penumbral shadow. Hybrid eclipses are also known as annular-total eclipses.

Annular, total and hybrid eclipses are sometimes referred to as *central* eclipses[1]. However, on rare occasions it is possible to have an annular or total eclipse that is non-central (Sect. 3.1).

1.3 Visual Appearance of Partial Solar Eclipses

When the Moon's penumbral shadow strikes Earth, a partial eclipse of the Sun is visible from that region. The Moon's apparent motion with respect to the Sun is relatively slow — the partial phases can last up to three hours or more. During this time, the Moon's dark limb slowly creeps across the Sun's disk.

Partial eclipses are dangerous to look at with the unprotected eye because the Sun is still extremely bright. Special techniques are needed to safely view the eclipse (Sect. 1.6). Even when a partial eclipse reaches its maximum phase, the sky and landscape remain bright. But careful inspection of dappled sunlight beneath a leafy tree will reveal multiple images of the eclipse. The gaps between the tree leaves act like pinhole cameras and project images of the eclipse onto the ground below. If the eclipse occurs on a mostly cloudy day, brief views of the partial phases may be possible as the Sun passes though the more translucent clouds.

The Moon's penumbral shadow is 6700 to 7300 kilometers in diameter and can cover a significant fraction of the daytime hemisphere of Earth. Consequently, partial eclipses may be visible from large geographic areas as the penumbra sweeps across Earth's surface.

Photo 1–1 shows various phases of the Annular solar eclipse of 2005 Oct 03. ©2005 F. Espenak

1.4 Visual Appearance of Annular Solar Eclipses

During an annular eclipse, the Moon's penumbral and antumbral shadows sweep across Earth. Compared to the penumbra, the antumbra is much smaller and has a maximum diameter of 374 kilometers. Because of this, the antumbra covers a tiny fraction of Earth's surface.

A partial eclipse is visible within the penumbral shadow, but only observers located in the much narrower track of the antumbra will see an annular eclipse. For this reason, the antumbra's trajectory across Earth is called the path of annularity.

---

[1] A central eclipse is one in which the central axis of the Moon's umbral/antumbral shadow intersects with Earth's surface.

All annular eclipses begin with a series of partial phases lasting about an hour. At the peak of the eclipse, the Moon's disk can be seen in complete silhouette against the Sun. The remaining solar photosphere appears as a ring of intensely bright light surrounding the Moon. The annular phase can last a maximum of 12 ½ minutes, but is more typically 3 to 6 minutes in length. After annularity, another series of partial phases occurs as the Moon gradually uncovers the Sun.

Special precautions must be used to watch the eclipse (Sect. 1.6). Even during the annular phase, the Sun is dangerously bright and cannot be viewed without a solar filter. The landscape and sky remain bright throughout the eclipse, giving little indication of the celestial event in progress.

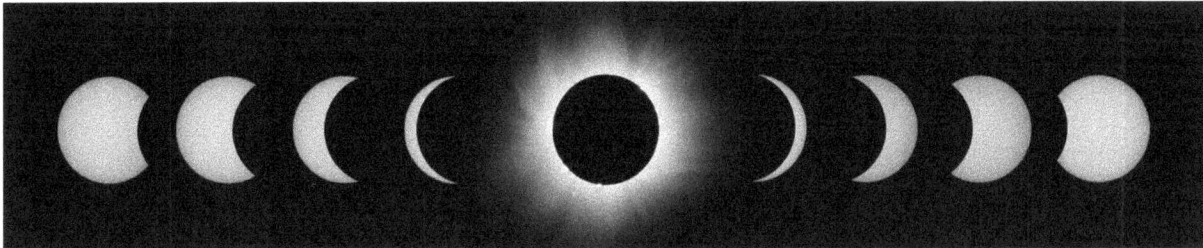

Photo 1–2 shows various phases of the total solar eclipse of 1999 Aug 11. ©1999 F. Espenak

1.5 Visual Appearance of Total Solar Eclipse

During a total eclipse, the Moon's penumbral and umbral shadows fall upon Earth. The umbra has a maximum diameter of 273 kilometers. The narrow track traced out as the umbra sweeps across Earth's surface is called the path of totality. The Sun is completely obscured by the Moon from within this zone. The total phase can last up to 7 ½ minutes, but is more typically 2 to 3 minutes in length.

Total eclipses all begin and end with a series of partial phases lasting about an hour. But this is where the resemblance between partial and annular eclipses ends — the total phase is the most spectacular astronomical event visible to the naked eye. During totality, the Sun's outer atmosphere — the solar corona — appears as a gossamer halo surrounding the Moon, and bright stars and planets are visible.

The eclipse begins to take on a unique character about five minutes before the total phase commences. Sunlight takes on a foreboding quality and casts abnormally sharp shadows. The approaching lunar umbra darkens the western sky and the air temperature is noticeably cooler. A minute before totality, ghostly shadow bands[2] ripple across the ground.

The ambient light grows feeble even though the crescent Sun is still too bright to see. In the final seconds, the Sun's corona emerges from the glare as the solar crescent shrinks to a brilliant jewel. This celestial diamond ring[3] lingers for a moment before the sunlight is extinguished and totality begins.

The Sun's glorious corona is now displayed to full advantage in the darkened sky. Standing within the Moon's umbra affords the rare and unprecedented opportunity to gaze directly at the glowing million-degree plasma surrounding our star. Twisted, tortured, and constrained by the Sun's enormous magnetic fields, the solar corona is revealed to the naked eye only during the brief seconds when the Moon completely blocks the brilliant disk of the Sun. An eerie twilight bathes the landscape and the colors of dusk surround the horizon.

Minutes race by like seconds. Suddenly, a sparkling bead of sunlight reappears along one edge of the Moon and quickly grows to blindingly bright proportions. Daylight returns as the corona fades and the total phase ends. Another hour of partial phases remains before the eclipse ends, it is anticlimactic after totality.

---

[2] Shadow bands are caused by the thin slit-like solar crescent illuminating Earth's atmosphere moments before and after a total solar eclipse. They appear as thin parallel undulating lines of alternating light and dark that race across the ground. This motion is caused by atmospheric winds.

[3] This phenomenon is commonly known as the *diamond ring effect*.

While filters are required for viewing the partial phases, they must be removed for totality. The total phase is the only time it is completely safe to look directly at the Sun without protection. In fact, the total phase is not even visible through solar filters because the Sun's corona is a million times fainter than the photosphere[4].

Figure 1–3: Path of Totality

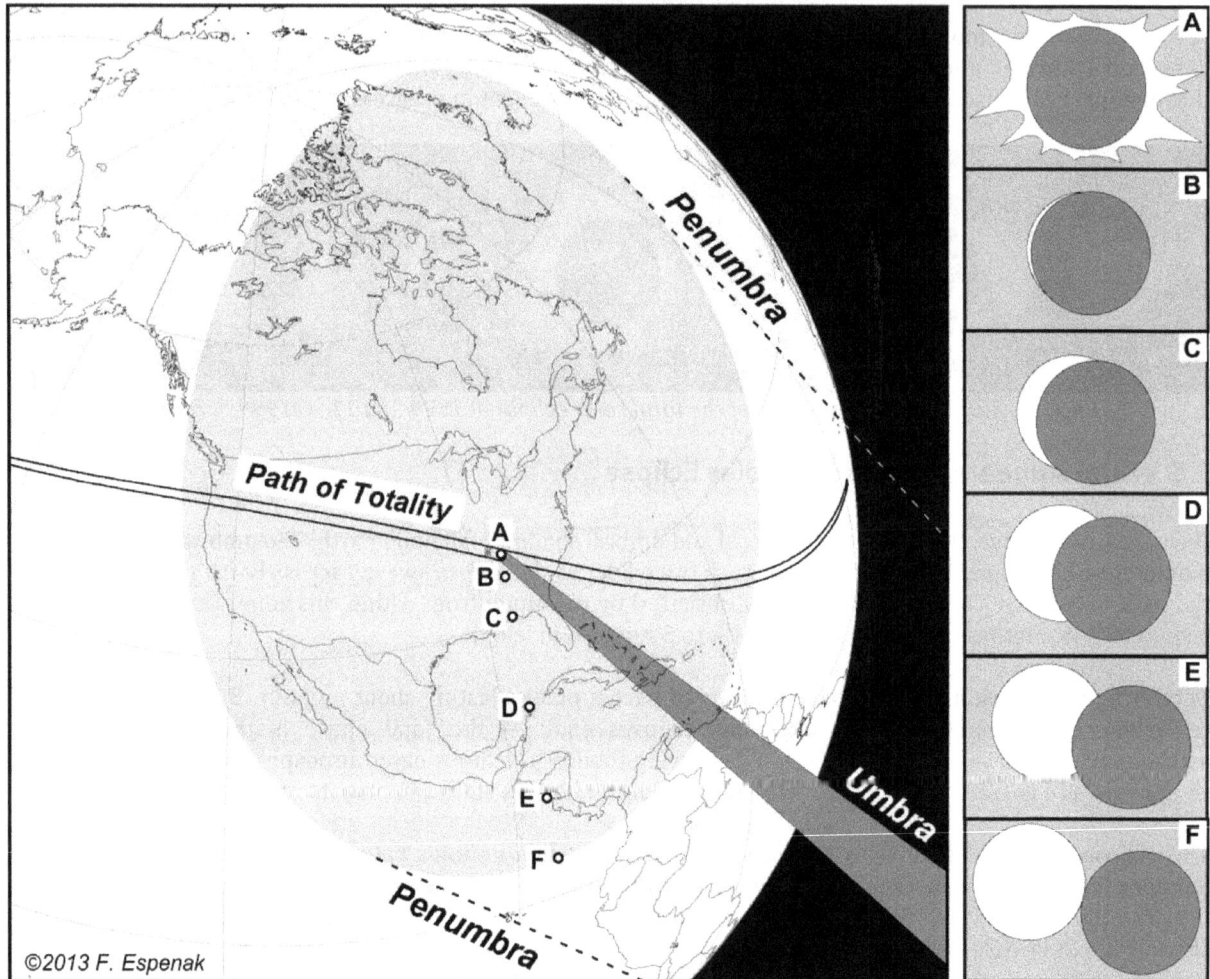

Figure 1–3 illustrates the path of totality for the total solar eclipse of 2017 Aug 21. The appearance of maximum eclipse from six different geographic locations on the map (labeled A through F) is shown in the side bar. The total eclipse is visible from position A because it lies in the path of totality.

1.6 Safely Observing Solar Eclipses

Partial eclipses, annular eclipses, and the partial phases of total eclipses are never safe to watch without special precautions. Even when 99% of the Sun's surface is obscured during the partial phases, the remaining photospheric crescent is intensely bright and cannot be viewed safely without eye protection (Chou, 1981; Marsh, 1982). Do not attempt to observe the partial or annular phases of any solar eclipse with the naked eye. Failure to use appropriate filtration may result in permanent eye damage or blindness. The only time it is safe to view the Sun directly with the naked eye is during the brief period of the total eclipse phase when the Sun's disk is completely covered by the Moon.

---

[4] The photosphere is the visible surface of the Sun's disk.

The same equipment and techniques used to observe the Sun outside of eclipse can be used for viewing partial phases and annular eclipses (Littmann, Espenak & Willcox, 2008, Reynolds & Sweetsir, 1995; Pasachoff & Covington, 1993). The most inexpensive of these methods is by projection — a pinhole or small opening is used to project the Sun's image onto a screen placed two or more feet beyond the opening. Projected images of the Sun can even be seen on the ground by creating small openings between interlaced fingers, or in the dappled sunlight beneath a leafy tree.

Observing the Sun directly is possible when using filters specifically designed for this purpose. Such filters usually have a thin layer of metal deposited on their surfaces that attenuates both visible and infrared energy. Another widely available filter is a number 14 welder's glass, obtained from welding supply outlets. Black polymer has recently become a popular, inexpensive alternative. This material can be cut with scissors and adapted to any kind of viewing device. No filter is safe to use with an optical device (i.e., telescope, binoculars, etc.) unless it has been specifically designed for that purpose. Sources for filters can be found at:

eclipsewise.com/extra/equipment.html#Solar_Filters

Local science museums, planetariums and amateur astronomy clubs are good sources for additional information. Remember, only the total phase of an eclipse can be safely viewed without a filter.

1.7 Central Line and Duration of Totality

The axis of the Moon's shadow determines the central line of the path of totality (and annularity). For the purposes of this discussion, it is assumed that the central line lies in the middle of the eclipse path.[5]

The duration of totality is longest on the central line, so great effort is often made to get as close to it as possible. However, the duration actually drops off quite slowly with distance from the central line. For example, a location 20% from the center to the edge of the path still has a totality lasting 98% of the central line duration. Even if one travels half way to the path limit, the duration is still 87% of the central line value.

Although figure 1–4 plots data for the 2017 total solar eclipse, the same relationship holds for all total eclipse paths. The solid curve is calculated assuming a smooth lunar limb. In reality, the Moon's limb profile has mountains that extend the duration and valleys that shorten it. These topographic features typically change the duration by a second or two for most of the path — their effects become significantly larger near the umbral path limits. The dotted line in figure 1–4 has been calculated using the Moon's true profile.

The following expression is useful for calculating the duration of totality for any location at a perpendicular distance of δ kilometers from the central line. It assumes a smooth profile for the Moon's limb.

$$d = D \times [\, 1 - (\, \delta/Z\,)^2\,]^{1/2} \text{ seconds} \qquad (1\text{--}2)$$

Where: d = duration of totality at point of interest (seconds)
D = duration of totality on the central line (seconds)
δ = perpendicular distance from the central line to position of interest (kilometers)
Z = perpendicular distance from the central line to the path edge (kilometers)

The Moon's limb profile varies with the lunar libration, which changes from eclipse to eclipse. Consequently, the effects of the limb profile are different at each eclipse.

---

[5] The position of the central line is actually offset from the center. The difference is greatest in cases where the Sun's altitude is low. Curvature of Earth's surface is responsible for this shift.

Figure 1–4: Duration vs. Distance from the Central Line

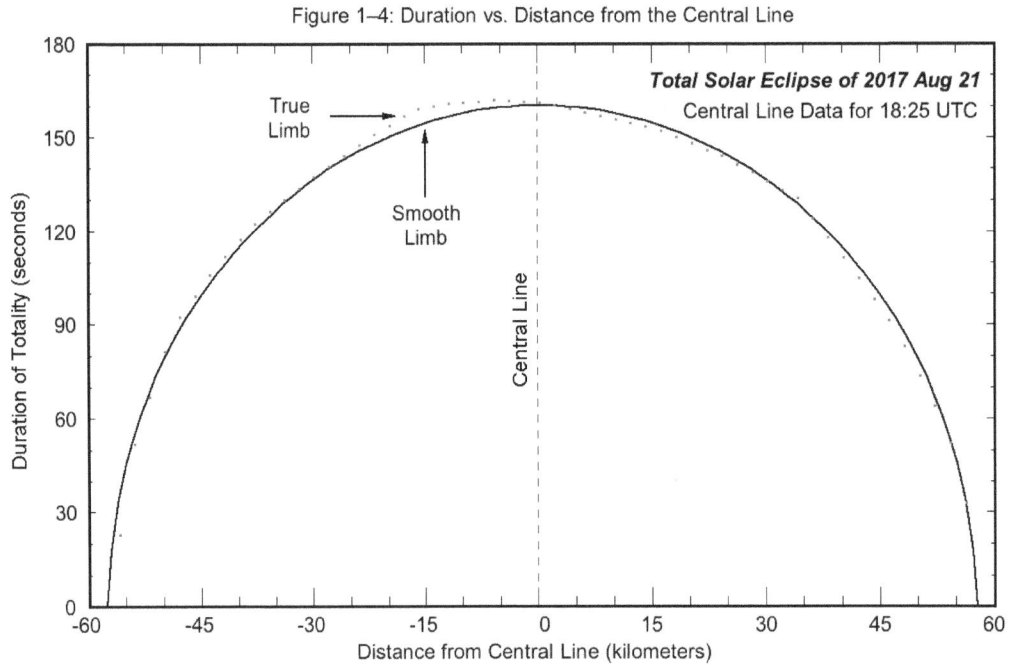

Figure 1–4 shows how the duration of totality changes with distance from the central line. The solid curve represents the duration using the Moon's mean limb. At a distance ~87% from the central line to the path limit, the duration drops to 50%. It is only when the distance from the central line to the path limit reaches ~90% that the duration takes a precipitous drop. But even at a distance of 96%, the duration is still 28% of the central line value. The dotted curve uses the Moon's true limb to calculate the duration using Kaguya/Herald limb data.

Photo 1–5 *Pat and Fred Espenak gaze at the Sun's corona during the total solar eclipse of 2008 August 01 from Jinta, China. ©2008 F. Espenak*

Section 2: Solar Eclipse Predictions

2.1 Solar Eclipse Contacts

During the course of a solar eclipse, the instants when the Moon's disk becomes tangent to the Sun's disk are known as eclipse contacts. They mark various stages or phases of a solar eclipse.

Partial solar eclipses have two primary contacts.

First Contact (C1) — Instant of first exterior tangency of the Moon with the Sun (Partial Eclipse Begins)
Fourth Contact (C4) — Instant of last exterior tangency of the Moon with the Sun (Partial Eclipse Ends)

Central solar eclipses (total, annular or hybrid) have four primary contacts. Contacts C2 and C3 mark the instants when the Moon's disk is first and last internally tangent to the Sun. These are the times when the annular or total phase of the eclipse begins and ends, respectively.

First Contact (C1) — Instant of first exterior tangency of the Moon with the Sun (Partial Eclipse Begins)
Second Contact (C2) — Instant of first interior tangency of the Moon with the Sun
    (Annular or Total Eclipse Begins)
Third Contact (C3) — Instant of last interior tangency of the Moon with the Sun
    (Annular or Total Eclipse Ends)
Fourth Contact (C4) — Instant of last exterior tangency of the Moon with the Sun (Partial Eclipse Ends)

Figure 2–1: Solar Eclipse Contacts

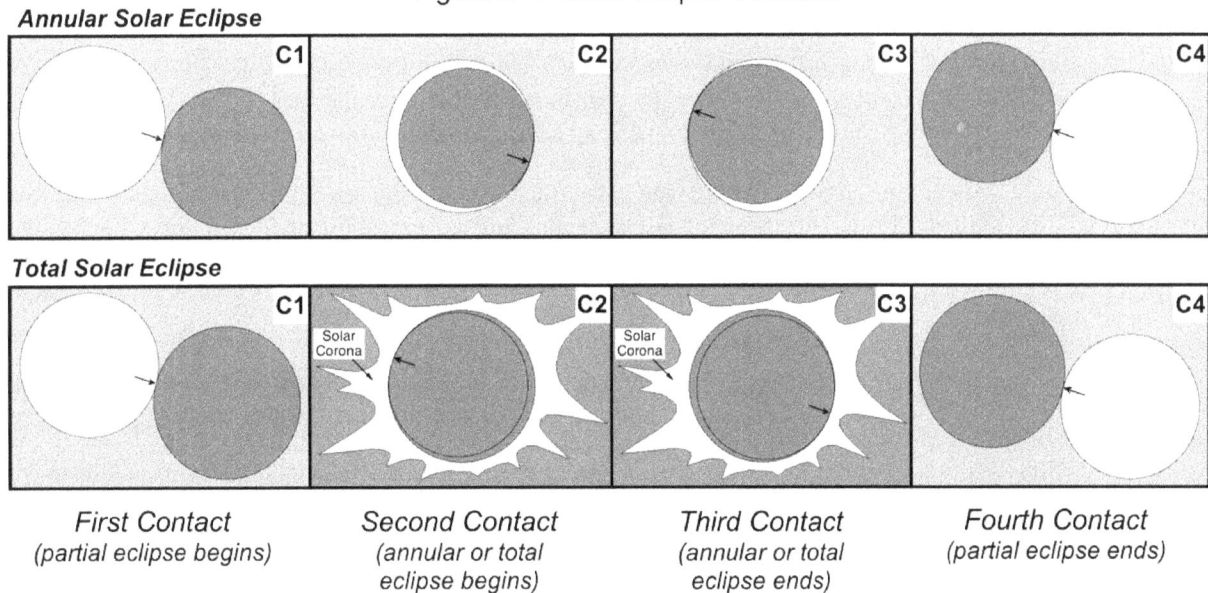

Annular Solar Eclipse

Total Solar Eclipse

First Contact	Second Contact	Third Contact	Fourth Contact
(partial eclipse begins)	(annular or total eclipse begins)	(annular or total eclipse ends)	(partial eclipse ends)

Figure 2–1 illustrates the four contacts for annular and total solar eclipses. The arrows indicate the contact point of the Sun's limb in each diagram.

2.2 Mean Lunar Radius

The International Astronomical Union (IAU) has adopted a value of k=0.2725076 for the mean lunar radius. This value is currently used by the Nautical Almanac Office for all solar eclipse predictions (Fiala and Lukac, 1983) and is believed to be the best mean radius, averaging mountain peaks and low valleys along the Moon's rugged limb. However, this introduces a problem in predicting the character and duration of central eclipses, particularly total eclipses. A total eclipse can be defined as an eclipse in which the Sun's disk is completely occulted by the Moon. This cannot occur so long as any photospheric rays are visible through deep valleys along the Moon's limb (Meeus, Grosjean and Vanderleen, 1966).

This work uses the IAU's accepted value of k (k=0.2725076) for all penumbral (exterior) contacts. In order to avoid eclipse type misidentification and to predict central durations, which are closer to the actual durations observed at total eclipses, we depart from convention by adopting the smaller value for k (k=0.272281) for all central (interior) contacts. This is consistent with predictions published in *Thousand Year Canon of Solar Eclipses* (Espenak 2014). Consequently, the smaller k produces shorter central durations and narrower paths for total eclipses when compared with calculations using the IAU value for k. Similarly, the smaller k predicts longer central durations and wider paths for annular eclipses.

2.3 Solar and Lunar Coordinates

The coordinates of the Sun and the Moon used in the eclipse predictions presented here have been calculated with the JPL DE405 (Jet Propulsion Laboratory Developmental Ephemeris 405). The DE405 is based upon the International Celestial Reference Frame (ICRF), the adopted reference frame of the IAU. The DE405 includes both nutation or libration and has an absolute accuracy of several kilometers for planetary positions,. In most cases this corresponds to a small fraction of an arc-second.

The Moon's center of figure does not coincide with its center of mass. To compensate, an empirical correction is sometimes added to the Moon's center of mass position. Unfortunately, the large variation in lunar libration from one eclipse to the next minimizes the effectiveness of this empirical correction. Because of this, no correction has been made to the Moon's center of mass position in the *21st Century Canon*.

2.4 Measurement of Time

The most natural form of time measurement is the solar day (usually measured from solar noon to solar noon). The length of the solar day varies during the year because of the eccentricity of Earth's orbit around the Sun. Mean solar time resolves this problem by using an average to define the mean solar day.

In 1884, Greenwich Mean Time (GMT) — the mean solar time on the Greenwich Meridian (0° longitude) — was adopted as the standard reference time for clocks around the world. A fundamental basis of GMT is the assumption that Earth's rotation on its axis is constant. It wasn't until the mid-twentieth century that astronomers realized the rotation period is gradually increasing. Earth is slowing down because of tidal friction with the Moon.

For purposes of orbital calculations, time using Earth's rotation was abandoned for a more uniform time scale based on Earth's orbit about the Sun. In 1952, Ephemeris Time was introduced to address the problem. The ephemeris second was defined as a fraction of the tropical year[6] for 1900 Jan 01 as calculated from Newcomb's *Tables of the Sun* (1895). Ephemeris Time was used for Solar System ephemeris calculations until 1979.

Terrestrial Dynamical Time (TD) is the modern replacement for Ephemeris Time and is used in theories of planetary motion in the Solar System. TD is based on International Atomic Time (TAI), which is a high-precision standard using several hundred atomic clocks worldwide. To ensure continuity with Ephemeris Time, TD was defined to match ET for the date 1977 Jan 01. In 1991, the IAU refined the definition of TD to make it more precise. It was also renamed Terrestrial Time (TT) although the author prefers to use the older name Terrestrial Dynamical Time.

---

[6] The tropical year is the length of time that the Sun takes to return to the same position in the cycle of seasons, as seen from Earth (e.g., the time from vernal equinox to vernal equinox).

Civilian time used throughout the world is still based on mean solar time, although indirectly. While Greenwich Mean Time was determined though observations of the Sun, its modern day replacement, Universal Time (actually UT1) is based on Earth's rotation using observations of distant quasars. UT1 is a nonuniform time because Earth is gradually slowing down at an irregular rate. At present (2016), the accumulated error in the rotation of Earth in the course of one year is ~0.5 seconds.

Coordinated Universal Time (UTC) is derived from International Atomic Time (TAI). The length of the UTC second is defined in terms of an atomic transition of cesium and is accurate to approximately one nanosecond (billionth of a second) per day. UTC was defined to closely parallel UT1. However, the two time systems are intrinsically incompatible since UTC is uniform while UT1 is based on Earth's rotation, which is gradually slowing. In order to keep the two times within 0.9 seconds of each other, a leap second is added to UTC as needed (currently once every few years).

Today, UTC is the time standard used to define time zones around the world. It is the time reference for GPS satellites and aviation, and is used to synchronize the clocks of computers across the Internet.

2.5 ΔT (Delta T)

The orbital positions of the Sun and the Moon, required by eclipse predictions, are calculated using Terrestrial Dynamical Time (TD) because it is a uniform time scale. However, world time zones and daily life are based on Universal Time[7] (UT1). In order to convert eclipse predictions from TD to UT1, the difference between these two time scales must be known. The parameter ΔT (Delta T) is the arithmetic difference, in seconds, between the two as:

$$\Delta T = TD - UT1 \tag{2-1}$$

Past values of ΔT can be deduced from historical records. In spite of their relatively low precision, these data represent the only evidence for the value of ΔT prior to 1600. In the centuries following the introduction of the telescope (circa 1609), thousands of high quality observations have been made of lunar occultations of stars, affording valuable data with increased accuracy in the determination of ΔT.

In modern times, the determination of ΔT is made using atomic clocks and radio observations of quasars. From 1955 to 2010, the average 1-year change in ΔT ranges from 0.18 seconds to 1.06 seconds. Future changes in ΔT are unknown since theoretical models of the physical causes are imprecise. Extrapolations from the table weighted by the long period trend from tidal braking of the Moon offer estimates of +71 seconds in 2024, +85 seconds in 2050, and +127 seconds in 2100.

Polynomial expressions for ΔT based on this data can be found at: *eclipsewise.com/help/deltatpoly2014.html*

2.6 Date Format

There are a number of ways to write the calendar date through variations in the order of day, month, and year. The International Organization for Standardization's (ISO) 8601 advises a numeric date representation, which organizes the elements from the largest to the smallest. The exact format is YYYY–MM–DD where YYYY is the calendar year, MM is the month of the year between 01 (January) and 12 (December), and DD is the day of the month between 01 and 31. For example, the 27th day of April in the year 1943 would then be expressed as 1943-04-27. The ISO convention is adopted here, but the month number has been replaced with the three-letter English abbreviation of the month name for additional clarity. From the previous example, the date then is expressed as 1943 Apr 27.

---

[7] World time zones are actually based on Coordinated Universal Time (UTC). It is an atomic time synchronized and adjusted to stay within a second of astronomically determined Universal Time (UT1) through the addition of an occasional "leap second" to compensate for the gradual slowing of Earth's rotation.

Section 3: Solar Eclipse Statistics

3.1 Statistical Distribution of Eclipse Types

Eclipses of the Sun can only occur during the New Moon phase. It is then possible for the Moon's penumbral, umbral and/or antumbral shadows to sweep across Earth's surface thereby producing an eclipse. There are four types of solar eclipses:

1. **Partial** — Moon's penumbral shadow traverses Earth (umbral and antumbral shadows miss Earth)
2. **Annular** — Moon's penumbral and antumbral shadows traverse Earth (Moon is too far from Earth to completely cover the Sun)
3. **Total** — Moon's penumbral and umbral shadows traverse Earth (Moon is close enough to Earth to completely cover the Sun)
4. **Hybrid** — Moon's penumbral, umbral and antumbral shadows traverse Earth (eclipse is annular or total along different sections of its path). Hybrid eclipses are also known as annular-total eclipses.

During the 100-year period from 2001 to 2100, Earth experiences 224 eclipses of the Sun. The statistical distribution of the four eclipse types over this interval is shown in Table 3–1.

Table 3–1: Distribution of Basic Eclipse Types

Eclipse Type	Abbreviation	Number	Percent
All Eclipses	—	224	100.0%
Partial	P	77	34.4%
Annular	A	72	32.1%
Total	T	68	30.4%
Hybrid	H	7	3.1%

All partial eclipses are events in which some portion of the Moon's penumbral shadow passes across Earth's surface. In comparison all annular, total and hybrid eclipses can be characterized as events in which some portion of the Moon's umbral and/or antumbral shadow crosses Earth.

In the case of umbral or antumbral eclipses (annular, total or hybrid), they can be further categorized as:

1. **Central (two limits)** — The central axis of the Moon's umbral or antumbral shadow traverses Earth thereby producing a central line in the eclipse track. The umbra or antumbra falls entirely upon Earth producing a ground track with both a northern and southern limit.
2. **Central (one limit)** — The central axis of the Moon's umbral or antumbral shadow traverses Earth. However, a portion of the umbra or antumbra misses Earth throughout the eclipse thereby producing a ground track with just one limit.
3. **Non-Central** — The central axis of the Moon's umbral or antumbral shadow misses Earth. However, one edge of the shadow grazes Earth producing a ground track with one limit and no central line.

Using these categories, the distribution of the 72 annular eclipses appears in Table 3–2.

Table 3–2: Statistics of Annular Eclipses

Annular Eclipses	Number	Percent
All Annular Eclipses	72	100.0%
Central (two limits)	68	94.4%
Central (one limit)	2	2.8%
Non-Central (one limit)	2	2.8%

Central annular eclipses with one limit include: 2003 May 31, and 2044 Feb 28. Non-central annular eclipses include: 2014 Apr 29, and 2043 Oct 03.

Similarly, the distribution of the 68 total eclipses is shown in Table 3–3.

Table 3–3: Statistics of Total Eclipses

Total Eclipses	Number	Percent
All Total Eclipses	68	100.0%
Central (two limits)	67	98.5%
Central (one limit)	0	0.0%
Non-Central (one limit)	1	1.5%

There are no central total eclipses with a single limit during the 21st Century. However, there is one non-central total eclipse (one limit) on 2043 Apr 09.

All 7 hybrid eclipses are central with two limits. Hybrid eclipses with a single limit (both central and non-central) are exceedingly rare. An estimate of the mean frequency of non-central hybrid eclipses is one out of every 600 million eclipses or once every 250 million years (Meeus, 2002).

3.2 Eclipse Frequency and the Calendar Year

There are 2 to 5 solar eclipses in every calendar year. Table 3–4 shows the distribution in the number of eclipses per year for the 100 years covered in the *21st Century Canon*.

Table 3–4: Number of Eclipses per Year

Number of Eclipses per Year	Number of Years	Percent
2	82	73.2%
3	12	16.1%
4	6	10.7%
5	0	0.0%

When two eclipses occur in one calendar year, they can be any combination of P, A, T or H (partial, annular, total or hybrid) with the one exception — they can not both be T.

When three eclipses occur in one calendar year, there are 14 possible combinations of P, A, T or H. In most cases all three eclipses are partial.

Years in which four eclipses take place are: 2011, 2029, 2047, 2065, 2076, and 2094.

The maximum number of five solar eclipses in one calendar year is quite rare. No year during the 21st Century has five solar eclipses. The last time this occurred was in 1935 and the next instance is in 2206.

3.3 Extremes in Greatest Central Duration and Eclipse Magnitude

The longest and shortest central eclipses of the century as well as largest and smallest partial eclipses appear in Table 3–5.

Table 3-05: Extremes in Central Duration and Eclipse Magnitude

Extrema Type	Date (Dynamical Time)	Central Line Duration	Eclipse Magnitude
Longest Annular	2010 Jan 15	11m08s	—
Shortest Annular	2085 Dec 16	00m19s	—
Longest Total	2009 Jul 22	06m39s	—
Shortest Total	2068 May 31	01m06s	—
Longest Hybrid	2013 Nov 03	01m40s	—
Shortest Hybrid	2067 Dec 06	00m08s	—
Largest Partial	2051 Apr 11	—	0.9849
Smallest Partial	2098 Oct 24	—	0.0057

*Central line duration at the instant of greatest eclipse

3.4 Eclipse Seasons

Because of its ~5.1° inclination the Moon's orbit crosses the ecliptic at two points or nodes. If New Moon takes place within approximately 17° of a node[8], a solar eclipse will be visible from some location on Earth.

The Sun makes one complete circuit of the ecliptic in 365.24 days, so its average angular velocity is 0.99° per day. At this rate, it takes 34.5 days for the Sun to cross the 34° wide eclipse zone centered on each node. Because the Moon's orbit with respect to the Sun has a mean duration of 29.53 days, there will always be one and possibly two solar eclipses during each 34.5-day interval when the Sun passes through the nodal eclipse zones. These time periods are called eclipse seasons.

The mid-point of each eclipse season is separated by 173.3 days because this is the mean time for the Sun to travel from one node to the next. The period is a little less that half a calendar year because the lunar nodes slowly regress westward by 19.3° per year.

Figure 3–1: Eclipse Seasons and Orbital Nodes

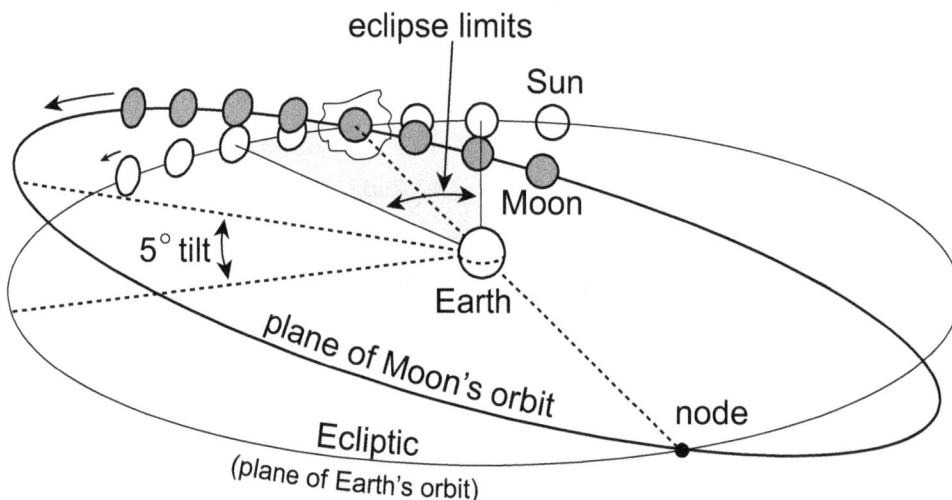

---

[8] The exact angular distance from the node depends on the distances of the Sun and the Moon from Earth, which determine their angular diameters.

21ST CENTURY CANON OF SOLAR ECLIPSES

3.5 Quincena

The mean time interval between New Moon and Full Moon is 14.77 days. This is less than half the duration of a 34.5-day eclipse season. As a consequence, the same Sun–node alignment geometry responsible for producing a solar eclipse always results in a complementary lunar eclipse within a fortnight. The lunar eclipse may precede or succeed the solar eclipse. In either case, the pair of eclipses is referred to here as a quincena[9]. The QLE (Quincena Lunar Eclipse parameter) identifies the type of the lunar eclipse and whether it precedes or succeeds a particular solar eclipse. There are three basic types of lunar eclipses:

1. penumbral lunar eclipse (n) — Moon passes partly or completely within Earth's penumbral shadow
2. partial lunar eclipse (p) — Moon passes partly within Earth's umbral shadow
3. total lunar eclipse (t) — Moon passes completely within Earth's umbral shadow

The QLE is a two character string consisting of one or more of the above lunar eclipse types. The first character in the QLE identifies the type of lunar eclipse preceding a solar eclipse. The second character identifies the type of lunar eclipse succeeding a solar eclipse. In most instances, one of the two characters is "–" indicating no lunar eclipse occurs. For example, a QLE of "–p" means that no lunar eclipse precedes a solar eclipse, but a partial lunar eclipse follows the solar eclipse 15 days later.

On rare occasions, a double quincena occurs in which a solar eclipse is both preceded and succeeded by a lunar eclipse.

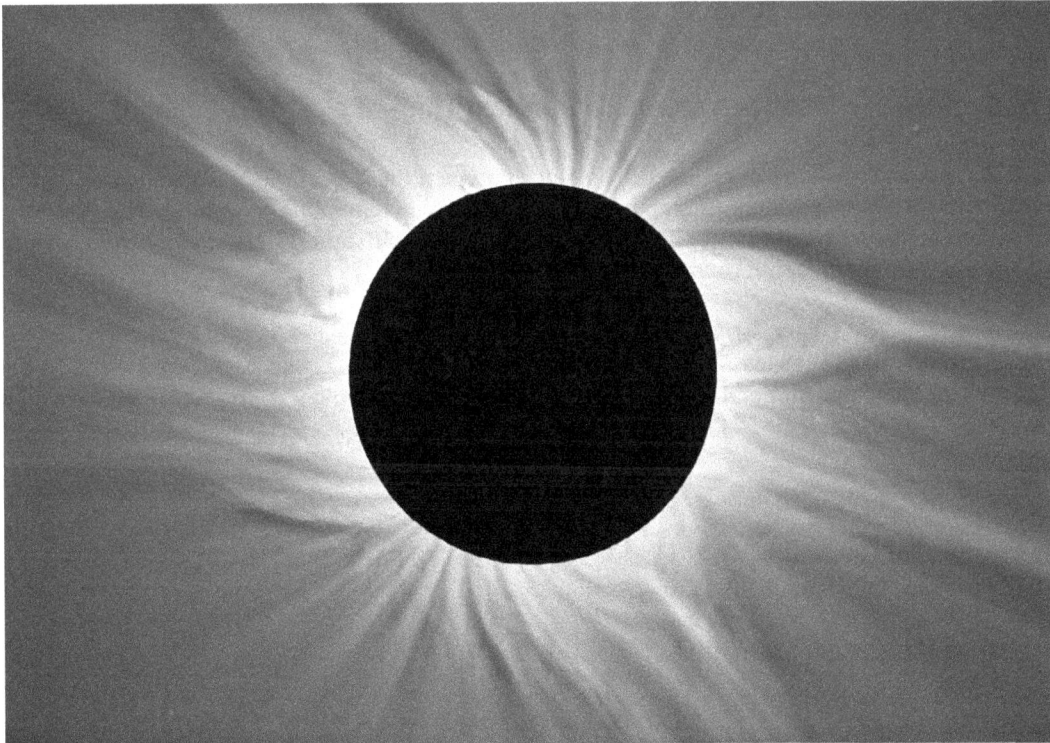

Photo 4–2 Fine details in the solar corona are captured during the total solar eclipse of 2006 March 29 from Jalu LIBYA.
© 2006 F. Espenak, www.MrEclipse.com

[9] Quincena is a Spanish word meaning *a period of fifteen days*. This also happens to be the time interval, rounded to the nearest day, between New Moon and Full Moon, or Full Moon and New Moon. So *quincena* is a convenient and appropriate term for describing a pair of eclipses (one solar and one lunar) separated by this period.

Section 4: Explanation of Solar Eclipse Catalog in Appendix A

4.1 Introduction

Earth experiences 224 eclipses of the Sun during the 21st Century. The catalog in *Appendix A* summarizes the principal characteristics of each eclipse and complements the maps in *Appendices B, C* and *D*.

Each line in the catalog corresponds to a single eclipse and provides concise parameters to characterize the eclipse. The calendar date and time of the instant of greatest eclipse are given, along with the value of Delta T . The lunation number and Saros series are listed along with the eclipse type. Gamma is the distance of the shadow axis from Earth's center at greatest eclipse, while the eclipse magnitude is the fraction of the Sun's diameter obscured at that instant. The geographic latitude and longitude of the umbral or antumbral axis are given for greatest eclipse, along with the Sun's altitude and azimuth, the width of the path, and the central line duration of totality or annularity. For both partial and non-central total or annular eclipses, the latitude and longitude correspond to the point closest to the shadow cone axis at greatest eclipse. The Sun's altitude is always 0° at this location. Detailed descriptions of each field in the catalog appear in the following sections.

4.2 Cat Num (Catalog Number)

The catalog number is the sequential number assigned to each eclipse from 1 to 224.

4.3 Calendar Date

Gregorian calendar date at greatest eclipse is given in the ISO order of YEAR-MONTH-DAY.

4.4 TD of Greatest Eclipse (Terrestrial Dynamical Time of Greatest Eclipse)

The instant of greatest eclipse occurs when the distance between the axis of the Moon's shadow cone and the center of Earth reaches a minimum. For partial eclipses, this instant differs slightly from the instant of greatest magnitude due to Earth's flattening. For total eclipses, this instant differs slightly from the instant of greatest duration, although the differences are relatively small.

Greatest eclipse is given in Terrestrial Dynamical Time or TD (Sect. 2.4), which is a time system based on International Atomic Time. As such, TD is the modern equivalent to its predecessor Ephemeris Time and is used in theories of planetary motion in the Solar System. To determine the geographic visibility of an eclipse, TD is converted to Universal Time (UT1) using the parameter Delta T (Sect. 2.5).

4.5 ΔT (Delta T)

ΔT (Delta T) is the arithmetic difference, in seconds, between Terrestrial Dynamical Time (TD) and Universal Time (UT1). For more information on ΔT, see Section 2.5.

4.6 Luna Num (Lunation Number)

The lunation number is the number of synodic months or lunations since New Moon on 2000 Jan 06. It can be converted to the Brown Lunation Number[10] by adding 953.

---

[10] The *Brown Lunation Number* defines lunation 1 as beginning at the first New Moon of 1923, the year when Ernest W. Brown's lunar theory was introduced in the major national astronomical almanacs.

4.7 Saros Num (Saros Series Number)

Each eclipse belongs to a Saros series using a numbering system first introduced by van den Bergh (1955). The eclipses with an odd Saros number take place at the ascending node of the Moon's orbit; those with an even Saros number take place at the descending node. This relationship is reversed for *lunar* eclipses.

The Saros is a period of 223 synodic months (~ 18 years, 11 days, and 8 hours). Eclipses separated by this period belong to the same Saros series and share similar geometry and characteristics.

4.8 Ecl Type (Solar Eclipse Type)

The first value in this 2-character parameter gives the eclipse type. The four basic types of solar eclipses are:

1. **Partial Solar Eclipse (P)** — The Moon's penumbral shadow traverses Earth; the Moon's umbral and antumbral shadows completely miss Earth
2. **Annular Solar Eclipse (A)** — The Moon's penumbral and antumbral shadows traverse Earth; the Moon is too far from Earth to completely cover the Sun
3. **Total Solar Eclipse (T)** — The Moon's penumbral and umbral shadow traverse Earth; the Moon is close enough to Earth to completely cover the Sun
4. **Hybrid Solar Eclipse (H)** — The Moon's penumbral, umbral and antumbral shadows traverse different parts of Earth; eclipse appears either total or annular along different sections of its path; hybrid eclipses are also known as annular-total eclipses

The second character of the eclipse type is a qualifier defined as follows.

1. m = Middle eclipse of Saros series
2. n = Central eclipse with no northern limit
3. s = Central eclipse with no southern limit
4. + = Non-central eclipse with no northern limit
5. − = Non-central eclipse with no southern limit
6. 2 = Hybrid eclipse path begins total and ends annular
7. 3 = Hybrid eclipse path begins annular and ends total
8. b = Saros series begins (first eclipse in a Saros series)
9. e = Saros series ends (last eclipse in a Saros series)

Qualifiers 1 through 5 are used with annular, total or hybrid eclipses. Qualifiers 6 and 7 apply only to special classes of hybrid eclipses while qualifiers 8 and 9 are used exclusively with partial eclipses.

4.9 QLE (Quincena Lunar Eclipse Parameter)

A lunar eclipse always occurs within 15 days of a solar eclipse. The Quincena Lunar Eclipse parameter (QLE) identifies the type of the lunar eclipse and whether it precedes or succeeds a particular solar eclipse. There are three basic types of lunar eclipses:

1. **Penumbral Lunar Eclipse (n)** — The Moon passes partly or completely within Earth's penumbra
2. **Partial Lunar Eclipse (p)** — The Moon passes partly within Earth's umbra
3. **Total Lunar Eclipse (t)** — The Moon passes completely within Earth's umbra

The QLE consists of a two-character string. The characters identify the type of lunar eclipse preceding and succeeding a solar eclipse, respectively. In most instances, one of the two characters in the QLE is "–" indicating no lunar eclipse occurs. On some occasions, a double quincena occurs in which a solar eclipse is both preceded and succeeded by a lunar eclipse. The QLE then consists of two characters identifying the types of the two lunar eclipses. (Section 3.5).

4.10 Gamma

Gamma is the minimum distance from the lunar shadow axis to the center of Earth, in units of Earth's equatorial radius. This distance is positive or negative, depending on whether the axis of the shadow cone passes north or south of Earth's center. If gamma is between +0.997 and –0.997, the eclipse is a central one (either total, annular, or hybrid). The limiting value 0.997 differs from unity because of the flattening of Earth.

4.11 Ecl Mag (Eclipse Magnitude)

The eclipse magnitude is defined as the fraction of the Sun's diameter occulted by the Moon. For partial eclipses, the eclipse magnitude at the instant of greatest eclipse is given for the geographic position closest to the Moon's shadow axis. For total, annular, and hybrid eclipses, the eclipse magnitude is replaced by the ratio of the topocentric apparent diameters of the Moon and the Sun at greatest eclipse. The eclipse magnitude is always < 1.0 for partial and annular eclipses, but ≥ 1.0 for total and hybrid eclipses.

4.12 Lat Long (Latitude and Longitude)

The latitude and longitude correspond to the position intersected by the lunar shadow axis at greatest eclipse.

4.13 Sun Alt (Altitude of Sun)

The Sun's altitude at the geographic position intersected by the lunar shadow axis is given at the instant of greatest eclipse. For partial eclipses, the Sun's altitude is always 0° because the shadow axis misses Earth. In this case, the geographic position corresponds to the point closest to the shadow axis.

4.14 Sun Azm (Azimuth of Sun)

The Sun's azimuth at the geographic position intersected by the lunar shadow axis is given at the instant of greatest eclipse. The values 0°, 90°, 180°, and 270° correspond to the north, east, south and west, respectively.

4.15 Path Width

For total, annular, and hybrid eclipses, the width of the path of totality or annularity (kilometers) is given at the geographic position intersected by the lunar shadow axis at the instant of greatest eclipse.

4.16 Central Line Dur (Central Line Duration)

For total, annular, or hybrid eclipses, the central line duration of the total or annular phase (minutes and seconds) is given at the geographic position intersected by the lunar shadow axis at greatest eclipse.

For total and hybrid eclipses, this duration is close to the maximum of the total phase along the entire path. For annular eclipses, the duration at greatest eclipse may be near either the minimum or maximum duration of the annular phase along the path. If the annular phase duration exceeds approximately 2.3 min, then it is close to the maximum duration along the central line track. If the annular phase duration is less, however, then it corresponds to a minimum and the annular duration increases towards the ends of the path.

4.17 EclipseWise.com and the 21ˢᵗ Century Canon

Much of the content of the *21ˢᵗ Century Canon of Solar Eclipses* is based on the eclipse predictions website *www.EclipseWise.com*. A plain text file containing the entire solar eclipse catalog in *Appendix A* can be downloaded at: *www.EclipseWise.com/solar/SEpubs/21CCSE.txt.*

A web based version of the catalog is available at: *www.eclipsewise.com/solar/SEcatalog/21CCSEcat.html*

Section 5: Explanation of Solar Eclipse Maps in Appendices B, C and D

5.1 Explanation of Small Global Eclipse Maps in Appendix B

Earth experiences 224 eclipses of the Sun during the 21st Century. An individual global map for each eclipse appears in *Appendix B*.

The geographic visibility of each eclipse is illustrated with an orthographic projection map of Earth showing the path of the Moon's penumbral (partial) and umbral/antumbral (total, hybrid, or annular) shadows with respect to the continental coastlines, political boundaries (circa 2016) and the Equator. North is to the top and the daylight terminator is drawn for the instant of greatest eclipse. An 'x' symbol marks the sub-solar point where the Sun appears directly overhead at that time. The salient features of the eclipse maps are identified in Figure 5–1, which serves as a key.

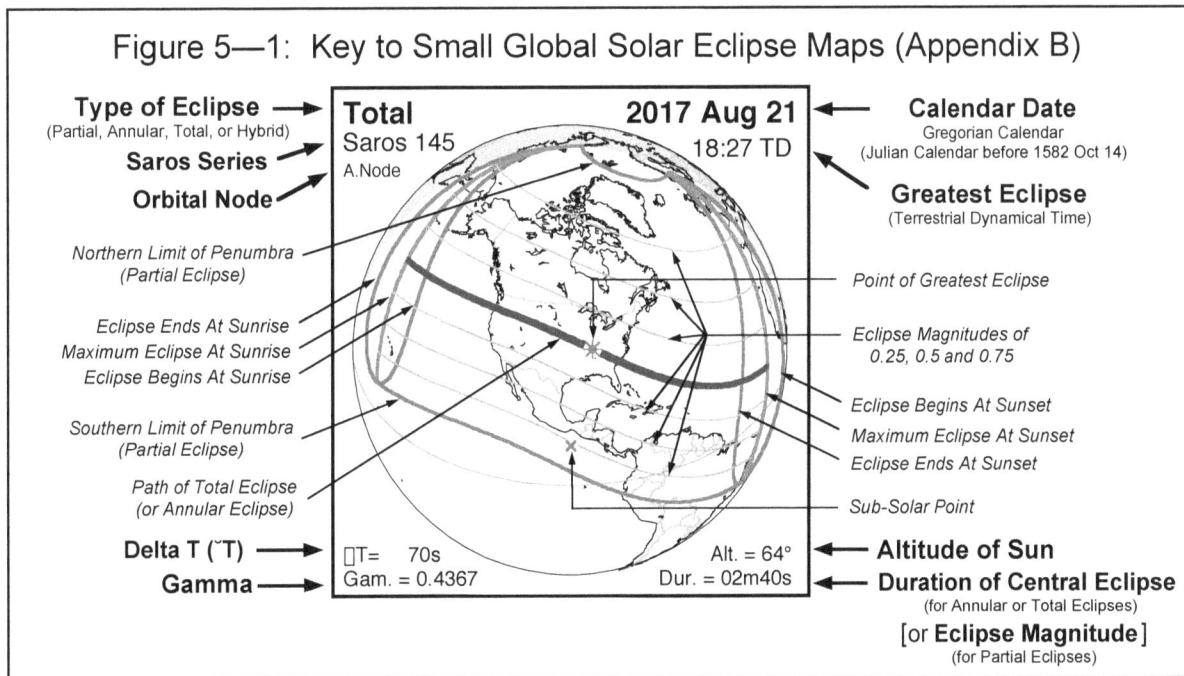

Figure 5—1: Key to Small Global Solar Eclipse Maps (Appendix B)

The limits of the Moon's penumbral shadow delineate the region of visibility of a partial solar eclipse. This irregular or saddle shaped region often covers more than half the daylight hemisphere of Earth and consists of several distinct zones or limits. At the northern and/or southern boundaries lie the limits of the penumbra's path. Partial eclipses have only one of these limits, as do central eclipses when the Moon's shadow axis falls no closer than about 0.45 radii from Earth's center. Great loops at the western and eastern extremes of the penumbra's path identify the areas where the eclipse begins/ends at sunrise and sunset, respectively. If the penumbra has both a northern and southern limit, the rising and setting curves form two separate, closed loops (e.g., 2017 Aug 21). Otherwise, the curves are connected in a distorted figure eight (e.g., 2019 Jul 02). Bisecting the *eclipse begins/ends at sunrise and sunset* loops is the curve of maximum eclipse at sunrise (western loop) and sunset (eastern loop).

The eclipse magnitude is defined as the fraction of the Sun's diameter occulted by the Moon. A curve of constant eclipse magnitude delineates the locus of all points where the local magnitude at maximum eclipse is equal to a constant value. The maps include *curves of constant eclipse magnitude* for values of 0.25, 0.5 and 0.75. These curves run exclusively between the curves of maximum eclipse at sunrise and sunset. They are approximately parallel to the northern/southern penumbral limits and the umbral/antumbral paths of central eclipses. The northern and southern limits of the penumbra may be thought of as curves of eclipse magnitude of 0.0. For total eclipses, the northern and southern limits of the umbra are curves of eclipse magnitude of 1.0.

Greatest eclipse is the instant when the axis of the Moon's shadow cone passes closest to Earth's center. Although greatest eclipse differs slightly from the instants of greatest magnitude and greatest duration (for total eclipses), the differences are relatively small. The point on Earth's surface intersected by the axis of the Moon's shadow cone at greatest eclipse is marked by an asterisk symbol '*'. For partial eclipses, the shadow axis misses Earth entirely, so the point of greatest eclipse lies on the day/night terminator and the Sun appears on the horizon.

Data relevant to an eclipse appear in the corners of each map. In the top left corner are the eclipse type (total, hybrid, annular, or partial), the Saros series of the eclipse, and the node of the Moon's orbit where the eclipse occurs. To the top right are the Gregorian calendar date, the time of greatest eclipse (Terrestrial Dynamical Time), and the value of Delta T (ΔT).

The bottom left corner lists gamma (the minimum distance of the Moon's shadow cone axis from Earth's center, in Earth equatorial radii). The Sun's altitude at the geographic position of greatest eclipse is found to the lower right. The content of the final datum in the bottom right corner depends on the type of eclipse. If the eclipse is partial then the eclipse magnitude is given. If the eclipse is total, hybrid or annular, then the duration of the total or annular phase is given at the instant of greatest eclipse.

The list below briefly summarizes the parameters appearing in each global eclipse map.

Solar Eclipse Type – One of four basic types of solar eclipses: Partial, Annular Total or Hybrid (See Sect. 4.1).

Saros Series – The Saros series that the eclipse belongs to. Eclipses with an odd number take place at the Moon's ascending node, while those with an even number are at the descending node (See Sect. 4.7).

Node – The orbital node near which the eclipse takes place. The ascending node (A. Node) is the point where the Moon travels from south to north through Earth's orbital plane. Similarly, the descending node (D. Node) is the point where the Moon travels from north to south.

Calendar Date – Gregorian calendar date at greatest eclipse is given in the ISO order of YEAR-MONTH-DAY.

Greatest Eclipse – The instant (Terrestrial Dynamical Time) when the distance between the axis of the Moon's shadow cone and the center of Earth reaches a minimum (See Sect. 4.4).

ΔT (Delta T) – the arithmetic difference, in seconds, between Terrestrial Dynamical Time (TD) and Universal Time (UT1), (See Sect. 2.5).

Gamma – The minimum distance from the lunar shadow axis to the center of Earth, in units of Earth's equatorial radius (See Sect. 4.10).

Altitude of Sun – The Sun's altitude at the geographic position intersected by the lunar shadow axis is given at the instant of greatest eclipse (See Sect. 4.13).

Duration of Central Eclipse – The central line duration of the total or annular phase (in minutes and seconds) is given at the instant of greatest eclipse (See Sect. 4.16).

Eclipse Magnitude – The fraction of the Sun's diameter occulted by the Moon at the instant of greatest eclipse (See Sect. 4.11).

5.2 Explanation of Large Global Eclipse Maps in Appendix C

The heart of the *21st Century Canon of Solar Eclipses* is the set of 113 full page global maps appearing in Appendix C — one map for every solar eclipse from 2017 through 2066. This feature is similar to the *Fifty Year Canon of Solar Eclipses: 1986 – 2035* (Espenak 1986) and makes *21st Century Canon* its modern successor.

The salient features of the large global maps are identified in Figure 5–2, which serves as a key.

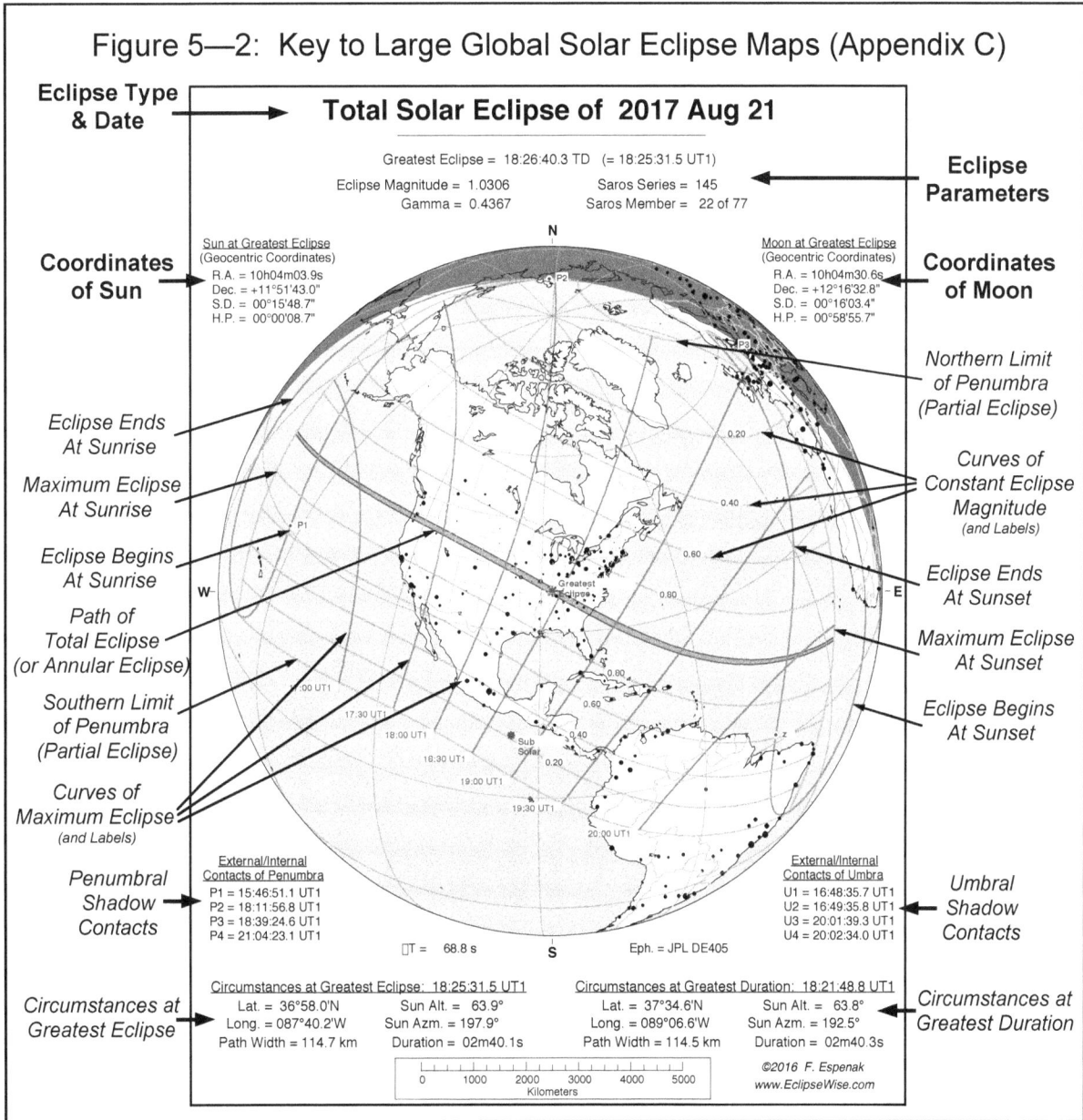

Figure 5—2: Key to Large Global Solar Eclipse Maps (Appendix C)

Each orthographic projection map shows the entire path of penumbral (partial) and umbral (total) or antumbral (annular) eclipse with respect to the continental coastlines, political boundaries (circa 2016), and major cities. North is to the top and the daylight terminator is plotted for the instant of greatest eclipse. An asterisk "*" indicates the sub-solar point on Earth (i.e., Sun in the Zenith).

23

The limits of the Moon's penumbral shadow delineate the region of visibility of the partial solar eclipse. This irregular or saddle shaped region often covers more than half of the daylight hemisphere of Earth and consists of several distinct zones or limits. At the northern and/or southern boundaries lie the limits of the penumbra's path. Partial eclipses have only one of these limits, as do central eclipses when the Moon's shadow axis falls no closer than about 0.45 radii from Earth's center.

Great loops at the western and eastern extremes of the penumbra's path identify the areas where the eclipse begins/ends at sunrise and sunset, respectively. If the penumbra has both a northern and southern limit, the rising and setting curves form two separate, closed loops. Otherwise, the curves are connected in a distorted figure eight. Bisecting the 'eclipse begins/ends at sunrise and sunset' loops is the curve of maximum eclipse at sunrise (western loop) and sunset (eastern loop. The points *P1* and *P4* mark the coordinates where the penumbral shadow first contacts (partial eclipse begins) and last contacts (partial eclipse ends) Earth's surface. If the penumbral path has both a northern and southern limit, then points *P2* and *P3* are also plotted. These correspond to the coordinates where the penumbral shadow cone is internally tangent to Earth's disk.

A curve of maximum eclipse is the locus of all points where the eclipse is at maximum at a given time. Curves of maximum eclipse are plotted at each half-hour Universal Time and are labeled with the time. They generally run between the penumbral limits in the north/south direction, or from the maximum eclipse at sunrise and sunset curves to one of the limits.

The curves of constant eclipse magnitude delineate the locus of all points where the magnitude at maximum eclipse is constant. These curves run between the curves of maximum eclipse at sunrise and sunset . They are roughly parallel to the northern/southern penumbral limits and the umbral paths of central eclipses (total, annular, and hybrid). In fact, the northern and southern limits of the penumbra are curves of constant magnitude of 0.0. The adjacent curves are for magnitudes of 0.2, 0.4, 0.6 and 0.8 and are labeled as such.

For total eclipses, the northern and southern limits of the umbra are curves of constant magnitude of 1.0. Antumbral path limits for annular eclipses are curves of maximum eclipse magnitude. The path of each total eclipse is darkly shaded. Similarly, the path of each annular eclipse is lightly shaded.

Greatest eclipse is defined as the instant when the axis of the Moon's shadow passes closest to Earth's center. Although greatest eclipse differs slightly from the instants of greatest magnitude and greatest duration (for total eclipses), the differences are small. An asterisk '*' marks the point on Earth's surface intersected by the lunar shadow axis at greatest eclipse. For partial eclipses, the shadow axis misses Earth entirely, so the point of greatest eclipse lies on the day/night terminator and the Sun appears on the horizon.

Data pertinent to the eclipse appear with each map. At the top is listed the instant of greatest eclipse, expressed in both Terrestrial Dynamical Time and Universal Time. The eclipse magnitude is defined as the fraction of the Sun's diameter obscured by the Moon at greatest eclipse. For central eclipses, the magnitude listed is actually the geocentric ratio of diameters of the Moon and the Sun. Gamma is the minimum distance (in Earth radii) of the Moon's shadow axis from Earth's center at greatest eclipse. The Saros series of the eclipse is listed, followed by the member position. The first number identifies the sequence position of the eclipse in the Saros, while the second number is the number of eclipses in the entire series.

In the upper left and right corners are the geocentric coordinates of the Sun and the Moon, respectively, at the instant of greatest eclipse. They are:

R.A. –   Right Ascension
Dec. –   Declination
S.D. –   Apparent Semi-Diameter
H.P. –   Horizontal Parallax

To the lower left are exterior/interior contact times of the Moon's penumbra with Earth, which are defined:

P1 – Instant of first exterior tangency of Penumbra with Earth's limb. (Partial Eclipse Begins)
P2 – Instant of first interior tangency of Penumbra with Earth's limb.
P3 – Instant of last interior tangency of Penumbra with Earth's limb.
P4 – Instant of last exterior tangency of Penumbra with Earth's limb. (Partial Eclipse Ends)

Not all eclipses have *P2* and *P3* penumbral contacts. They are only present in cases where the penumbral shadow falls completely within Earth's disk. For central eclipses, the lower right corner lists exterior/interior contact times of the Moon's umbral shadow with Earth's limb which are defined as follows:

U1 – Instant of first exterior tangency of Umbra with Earth's limb.
(Umbral [Total/Annular] Eclipse Begins)
U2 – Instant of first interior tangency of Umbra with Earth's limb.
U3 – Instant of last interior tangency of Umbra with Earth's limb.
U4 – Instant of last exterior tangency of Umbra with Earth's limb.
(Umbral [Total/Annular] Eclipse Ends)

At bottom left are the geographic coordinates of the position of greatest eclipse along with the local circumstances at that location (i.e., Sun altitude, Sun azimuth, path width and duration of totality/annularity). For central eclipses, similar data are also given at the bottom right for the position of greatest duration.

5.3 Explanation of Central Solar Eclipse Maps in Appendix D

The final series of 12 maps in Appendix D show the path of every central solar eclipse (total, annular and hybrid) with respect to continental coastlines, political boundaries (circa 2016), and major cities. The maps employ a transverse Mercator projection and are centered on three geographic regions:

1. North and South America
2. Europe and Africa
3. Asia and Australia

Furthermore, each map covers a 25-year period:

1. 2001–2025
2. 2026–2050
3. 2051–2075
4. 2076–2100

Each eclipse is identified by its date at the instant of greatest eclipse. An asterisk symbol marks the location of greatest eclipse.

The paths of total, annular, and hybrid eclipses can be identified as follows.

Total Eclipses – the path is darkly shaded;
the date appears in bold and is preceded by the letter **T**

Annular Eclipses – the path is lightly shaded;
the date appears in italics and is preceded by the letter **A**

Hybrid Eclipses – the path is darkly shaded along the total sections and
lightly shaded along the annular sections;
the date appears in bold-italics and is preceded by the letter **H**

References

Astronomical Almanac for 1986, Washington: US Government Printing Office; London: HM Stationery Office (1985).

Astronomical Almanac for 2011, Washington: US Government Printing Office; London: HM Stationery Office (2010).

Espenak, F., *Fifty Year Canon of Solar Eclipses: 1986–2035*, Sky Publishing Corp., Cambridge, Massuchusetts (1987).

Espenak, F., *Thousand Year Canon of Solar Eclipses: 1501 to 2500*, Astropixels Publishing, Portal, Arizona (2014).

Espenak, F., *Thousand Year Canon of Lunar Eclipses: 1501 to 2500*, Astropixels Publishing, Portal, Arizona (2014).

Espenak, F., and Meeus, J., *Five Millennium Canon of Solar Eclipses: –1999 to +3000 (2000 BCE to 3000 CE)*, NASA Tech. Pub. 2006–214141, NASA Goddard Space Flight Center, Greenbelt, Maryland (2006).

Espenak, F., and Meeus, J., *Five Millennium Canon of Lunar Eclipses: –1999 to +3000 (2000 BCE to 3000 CE)*, NASA Tech. Pub. 2006–214172, NASA Goddard Space Flight Center, Greenbelt, Maryland (2009a).

Espenak, F., and Meeus, J., *Five Millennium Catalog of Lunar Eclipses: –1999 to +3000 (2000 BCE to 3000 CE)*, NASA Tech. Pub. 2006–214173, NASA Goddard Space Flight Center, Greenbelt, Maryland (2009b).

Espenak, F., and Meeus, J., *Five Millennium Catalog of Solar Eclipses: –1999 to +3000 (2000 BCE to 3000 CE)*, NASA Tech. Pub. 2006–214174, NASA Goddard Space Flight Center, Greenbelt, Maryland (2009c).

Explanatory Supplement to the Ephemeris, H.M. Almanac Office, London (1974).

Littmann, M., Espenak, F., and Willcox, K., *Totality—Eclipses of the Sun*, 3rd Ed., Oxford University Press, New York (2008).

Meeus, J., *Mathematical Astronomy Morsels*, Willmann-Bell, pp. 56–62 (1997).

——, *More Mathematical Astronomy Morsels*, Willmann-Bell, pp. 70–72 (2002).

——, Grosjean, C.C., and Vanderleen, W., *Canon of Solar Eclipses*, Pergamon Press, Oxford, United Kingdom (1966).

Mucke, H., and Meeus, J., *Canon of Solar Eclipses: –2003 to +2526*, Astronomisches Büro, Vienna (1983).

Newcomb, S., "Tables of the Motion of the Earth on its Axis Around the Sun," *Astron. Papers Amer. Eph.*, Vol. 6, Part I (1895).

van den Bergh, *Periodicity and Variation of Solar (and Lunar) Eclipses*, Tjeenk Willink, and Haarlem, Netherlands (1955).

van der Sluys, M. , *http://hemel.waarnemen.com/Computing/deltat.html* (2010).

von Oppolzer, T.R., *Canon der Finsternisse*, Wien, (1887); Gingerich, O., (Translator) *Canon of Eclipses*, Dover Publications, New York (1962).

Appendix A

Solar Eclipse Catalog: 2001 to 2100

Key to Solar Eclipse Catalog

Cat Num — sequential Catalog Number assigned to each eclipse from 1 to 2,389

Calendar Date — Gregorian date of Greatest Eclipse (Julian date prior to 1582 Oct 04)

TD of Greatest Eclipse — Terrestrial Dynamical Time of Greatest Eclipse

ΔT — arithmetic difference between Terrestrial Dynamical Time Universal Time (seconds)

Luna Num — number of synodic months, or lunations, since New Moon on 2000 Jan 06

Saros Num — Saros Series Number of eclipse

Ecl Type — Solar Eclipse Type

> P = Partial Solar Eclipse
> A = Annular Solar Eclipse
> T = Total Solar Eclipse
> H = Hybrid Solar Eclipse
> m = Middle eclipse of Saros series
> n = Central eclipse of with no northern limit
> s = Central eclipse of with no southern limit
> + = Non-central eclipse of with no northern limit
> – = Non-central eclipse of with no southern limit
> 2 = Hybrid eclipse path begins total and ends annular
> 3 = Hybrid eclipse path begins annular and ends total
> b = Saros series begins (first eclipse in a Saros series)
> e = Saros series ends (last eclipse in a Saros series)

QLE — Quincena Lunar Eclipse Parameter

> n = Penumbral Lunar Eclipse
> p = Partial Lunar Eclipse
> t = Total Lunar Eclipse

Gamma — minimum distance from the axis of the lunar shadow to the center of Earth

Ecl Mag — Eclipse Magnitude; fraction of the Sun's diameter obscured by the Moon

Lat & Lng — latitude and longitude where the Sun appears in zenith at greatest eclipse

Sun Alt & Sun Azm — altitude and azimuth of the Sun at greatest eclipse

Path Width — width of the central path (km) at greatest eclipse (total, annular & hybrid eclipses)

Central Line Dur — Central Line Duration (minutes. seconds) at greatest eclipse

Cat Num	Calendar Date	TD of Greatest Eclipse	ΔT s	Luna Num	Saros Num	Ecl Type	QLE	Gamma	Ecl Mag	Lat °	Long °	Sun Alt °	Sun Azm °	Path Width km	Central Line Dur
001	2001 Jun 21	12:04:46	64	18	127	T	-p	-0.5701	1.0495	11.3S	2.7E	55	355	200	04m57s
002	2001 Dec 14	20:53:01	64	24	132	A	-n	0.4089	0.9681	0.6N	130.7W	66	188	126	03m53s
003	2002 Jun 10	23:45:22	64	30	137	A	nn	0.1993	0.9962	34.5N	178.6W	78	169	13	00m23s
004	2002 Dec 04	07:32:16	64	36	142	T	n-	-0.3020	1.0244	39.5S	59.6E	72	16	87	02m04s
005	2003 May 31	04:09:23	64	42	147	An	t-	0.9960	0.9384	66.6N	24.5W	3	35	−	03m37s
006	2003 Nov 23	22:50:22	65	48	152	T	t-	-0.9638	1.0379	72.7S	88.4E	15	111	496	01m57s
007	2004 Apr 19	13:35:05	65	53	119	P	-t	-1.1335	0.7367	61.6S	44.3E	0	295		
008	2004 Oct 14	03:00:23	65	59	124	P	-t	1.0348	0.9283	61.2N	153.7W	0	253		
009	2005 Apr 08	20:36:51	65	65	129	H	-n	-0.3473	1.0074	10.6S	119.0W	70	332	27	00m42s
010	2005 Oct 03	10:32:47	65	71	134	A	-p	0.3306	0.9576	12.9N	28.7E	71	209	162	04m32s
011	2006 Mar 29	10:12:23	65	77	139	T	n-	0.3843	1.0515	23.2N	16.7E	67	149	184	04m07s
012	2006 Sep 22	11:41:16	65	83	144	A	p-	-0.4062	0.9352	20.6S	9.1W	66	31	261	07m09s
013	2007 Mar 19	02:32:58	65	89	149	P	t-	1.0728	0.8756	61.0N	55.5E	0	92		
014	2007 Sep 11	12:32:24	65	95	154	P	t-	-1.1255	0.7507	61.0S	90.2W	0	80		
015	2008 Feb 07	03:56:10	65	100	121	A	-t	-0.9570	0.9650	67.6S	150.5W	16	269	444	02m12s
016	2008 Aug 01	10:22:12	66	106	126	T	-p	0.8307	1.0394	65.7N	72.3E	34	235	237	02m27s
017	2009 Jan 26	07:59:45	66	112	131	A	-n	-0.2820	0.9282	34.1S	70.2E	73	337	280	07m54s
018	2009 Jul 22	02:36:25	66	118	136	T	nn	0.0698	1.0799	24.2N	144.1E	86	198	258	06m39s
019	2010 Jan 15	07:07:39	66	124	141	A	p-	0.4002	0.9190	1.6N	69.3E	66	165	333	11m08s
020	2010 Jul 11	19:34:38	66	130	146	T	p-	-0.6788	1.0580	19.7S	121.9W	47	14	259	05m20s
021	2011 Jan 04	08:51:42	66	136	151	P	t-	1.0626	0.8576	64.7N	20.8E	0	155		
022	2011 Jun 01	21:17:18	66	141	118	P	-t	1.2130	0.6011	67.8N	46.8E	0	6		
023	2011 Jul 01	08:39:30	66	142	156	Pb	t-	-1.4917	0.0971	65.2S	28.6E	0	21		
024	2011 Nov 25	06:21:25	67	147	123	P	-t	-1.0536	0.9047	68.6S	82.4W	0	165		
025	2012 May 20	23:53:54	67	153	128	A	-p	0.4828	0.9439	49.1N	176.3E	61	171	237	05m46s
026	2012 Nov 13	22:12:55	67	159	133	T	-n	-0.3719	1.0500	40.0S	161.3W	68	11	179	04m02s
027	2013 May 10	00:26:20	67	165	138	A	pn	-0.2694	0.9544	2.2N	175.5E	74	350	173	06m03s
028	2013 Nov 03	12:47:36	67	171	143	H3	n-	0.3272	1.0159	3.5N	11.7W	71	192	58	01m40s
029	2014 Apr 29	06:04:33	67	177	148	A-	t-	-1.0000	0.9868	70.6S	131.3E	0	319	−	−
030	2014 Oct 23	21:45:39	68	183	153	P	t-	1.0908	0.8114	71.2N	97.2W	0	231		
031	2015 Mar 20	09:46:47	68	188	120	T	-t	0.9454	1.0446	64.4N	6.6W	18	135	463	02m47s
032	2015 Sep 13	06:55:19	68	194	125	P	-t	-1.1004	0.7875	72.1S	2.3W	0	77		
033	2016 Mar 09	01:58:19	68	200	130	T	-n	-0.2609	1.0450	10.1N	148.8E	75	162	155	04m09s
034	2016 Sep 01	09:08:02	68	206	135	A	-n	-0.3330	0.9736	10.7S	37.8E	70	16	100	03m06s
035	2017 Feb 26	14:54:33	69	212	140	A	n-	-0.4578	0.9922	34.7S	31.2W	63	340	31	00m44s
036	2017 Aug 21	18:26:40	69	218	145	T	p-	0.4367	1.0306	37.0N	87.7W	64	198	115	02m40s
037	2018 Feb 15	20:52:33	69	224	150	P	t-	-1.2116	0.5991	71.0S	0.6E	0	228		
038	2018 Jul 13	03:02:16	69	229	117	P	-t	-1.3542	0.3365	67.9S	127.4E	0	8		
039	2018 Aug 11	09:47:28	69	230	155	P	t-	1.1476	0.7368	70.4N	174.5E	0	321		
040	2019 Jan 06	01:42:38	69	235	122	P	-t	1.1417	0.7146	67.4N	153.6E	0	178		
041	2019 Jul 02	19:24:07	70	241	127	T	-p	-0.6466	1.0459	17.4S	109.0W	50	359	201	04m33s
042	2019 Dec 26	05:18:53	70	247	132	A	-n	0.4135	0.9701	1.0N	102.2E	66	184	118	03m39s
043	2020 Jun 21	06:41:15	70	253	137	Am	nn	0.1209	0.9940	30.5N	79.7E	83	174	21	00m38s
044	2020 Dec 14	16:14:39	70	259	142	T	n-	-0.2939	1.0254	40.3S	68.0W	73	10	90	02m10s
045	2021 Jun 10	10:43:07	70	265	147	A	t-	0.9152	0.9435	80.8N	66.8W	23	90	527	03m51s
046	2021 Dec 04	07:34:38	71	271	152	T	p-	-0.9526	1.0367	76.8S	46.2W	17	115	419	01m54s
047	2022 Apr 30	20:42:37	71	276	119	P	-t	-1.1901	0.6396	62.1S	71.5W	0	304		
048	2022 Oct 25	11:01:20	71	282	124	P	-t	1.0701	0.8619	61.6N	77.3E	0	244		
049	2023 Apr 20	04:17:56	71	288	129	H	-n	-0.3952	1.0132	9.6S	125.8E	67	334	49	01m16s
050	2023 Oct 14	18:00:41	71	294	134	A	-p	0.3753	0.9520	11.4N	83.1W	68	208	187	05m17s
051	2024 Apr 08	18:18:29	71	300	139	T	n-	0.3431	1.0566	25.3N	104.1W	70	149	198	04m28s
052	2024 Oct 02	18:46:13	72	306	144	A	p-	-0.3509	0.9326	22.0S	114.5W	69	31	266	07m25s
053	2025 Mar 29	10:48:36	72	312	149	P	t-	1.0405	0.9376	61.1N	77.1W	0	83		
054	2025 Sep 21	19:43:04	72	318	154	P	t-	-1.0651	0.8550	60.9S	153.5E	0	89		
055	2026 Feb 17	12:13:06	72	323	121	A	-t	-0.9743	0.9630	64.7S	86.7E	12	268	616	02m20s
056	2026 Aug 12	17:47:06	72	329	126	T	-p	0.8977	1.0386	65.2N	25.2W	26	248	294	02m18s
057	2027 Feb 06	16:00:48	73	335	131	A	-n	-0.2952	0.9281	31.3S	48.5W	73	334	282	07m51s
058	2027 Aug 02	10:07:50	73	341	136	T	nn	0.1421	1.0790	25.5N	33.2E	82	202	258	06m23s
059	2028 Jan 26	15:08:59	73	347	141	A	p-	0.3901	0.9208	3.0N	51.6W	67	161	323	10m27s
060	2028 Jul 22	02:56:40	73	353	146	T	p-	-0.6056	1.0560	15.6S	126.7E	53	17	230	05m10s

Cat Num	Calendar Date	TD of Greatest Eclipse	ΔT s	Luna Num	Saros Num	Ecl Type	QLE	Gamma	Ecl Mag	Lat °	Long °	Sun Alt °	Sun Azm °	Path Width km	Central Line Dur
061	2029 Jan 14	17:13:48	73	359	151	P	t-	1.0553	0.8714	63.7N	114.2W	0	145		
062	2029 Jun 12	04:06:13	74	364	118	P	-t	1.2943	0.4576	66.8N	66.2W	0	355		
063	2029 Jul 11	15:37:19	74	365	156	P	t-	-1.4191	0.2303	64.3S	85.6W	0	30		
064	2029 Dec 05	15:03:58	74	370	123	P	-t	-1.0609	0.8911	67.5S	135.6E	0	177		
065	2030 Jun 01	06:29:13	74	376	128	A	-p	0.5626	0.9443	56.5N	80.1E	55	176	250	05m21s
066	2030 Nov 25	06:51:37	74	382	133	T	-n	-0.3867	1.0468	43.6S	71.2E	67	7	169	03m44s
067	2031 May 21	07:16:04	74	388	138	A	nn	-0.1970	0.9589	8.9N	71.7E	79	354	152	05m26s
068	2031 Nov 14	21:07:31	75	394	143	H	n-	0.3078	1.0106	0.6S	137.6W	72	189	38	01m08s
069	2032 May 09	13:26:42	75	400	148	A	t-	-0.9375	0.9957	51.3S	7.1W	20	345	44	00m22s
070	2032 Nov 03	05:34:13	75	406	153	P	t-	1.0643	0.8554	70.4N	132.6E	0	218		
071	2033 Mar 30	18:02:36	75	411	120	T	-t	0.9778	1.0462	71.3N	155.8W	11	111	781	02m37s
072	2033 Sep 23	13:54:31	76	417	125	P	-t	-1.1583	0.6890	72.2S	121.3W	0	91		
073	2034 Mar 20	10:18:45	76	423	130	T	-n	0.2894	1.0458	16.1N	22.2E	73	162	159	04m09s
074	2034 Sep 12	16:19:28	76	429	135	A	-p	-0.3936	0.9736	18.2S	72.6W	67	18	102	02m58s
075	2035 Mar 09	23:05:54	76	435	140	A	n-	-0.4368	0.9919	29.0S	155.0W	64	340	31	00m48s
076	2035 Sep 02	01:56:46	76	441	145	T	p-	0.3727	1.0320	29.1N	158.0E	68	199	116	02m54s
077	2036 Feb 27	04:46:49	77	447	150	P	t-	-1.1942	0.6286	71.6S	131.5W	0	242		
078	2036 Jul 23	10:32:06	77	452	117	P	-t	-1.4250	0.1992	68.9S	3.5E	0	19		
079	2036 Aug 21	17:25:45	77	453	155	P	t-	1.0825	0.8622	71.1N	47.0E	0	309		
080	2037 Jan 16	09:48:55	77	458	122	P	-t	1.1477	0.7049	68.5N	20.8E	0	166		
081	2037 Jul 13	02:40:36	77	464	127	T	-p	-0.7246	1.0413	24.8S	139.1E	43	3	201	03m58s
082	2038 Jan 05	13:47:11	78	470	132	A	-n	0.4169	0.9728	2.1N	25.5W	65	179	107	03m18s
083	2038 Jul 02	13:32:55	78	476	137	A	nn	0.0398	0.9911	25.4N	21.9W	88	179	31	01m00s
084	2038 Dec 26	01:00:10	78	482	142	T	n-	-0.2881	1.0269	40.3S	163.9E	73	5	95	02m18s
085	2039 Jun 21	17:12:54	78	488	147	A	p-	0.8312	0.9454	78.9N	102.1W	33	153	365	04m05s
086	2039 Dec 15	16:23:46	79	494	152	T	p-	-0.9458	1.0356	80.9S	172.7E	18	123	380	01m51s
087	2040 May 11	03:43:02	79	499	119	P	-t	-1.2529	0.5306	62.8S	174.4E	0	313		
088	2040 Nov 04	19:09:02	79	505	124	P	-t	1.0993	0.8074	62.2N	53.4W	0	234		
089	2041 Apr 30	11:52:21	79	511	129	T	-p	-0.4492	1.0189	9.6S	12.2E	63	337	72	01m51s
090	2041 Oct 25	01:36:22	80	517	134	A	-p	0.4133	0.9467	9.9N	162.8E	66	206	213	06m07s
091	2042 Apr 20	02:17:30	80	523	139	T	n-	0.2956	1.0614	27.0N	137.3E	73	151	210	04m51s
092	2042 Oct 14	02:00:42	80	529	144	A	n-	-0.3030	0.9301	23.7S	137.8E	72	30	273	07m44s
093	2043 Apr 09	10:57:49	80	535	149	T+	t-	1.0031	1.0096	61.3N	151.9E	0	74	–	–
094	2043 Oct 03	03:01:49	81	541	154	A-	t-	-1.0102	0.9497	61.0S	35.2E	0	98	–	–
095	2044 Feb 28	20:24:40	81	546	121	As	-t	-0.9954	0.9600	62.2S	25.6W	4	260	–	02m27s
096	2044 Aug 23	01:17:02	81	552	126	T	-t	0.9613	1.0364	64.3N	120.5W	15	264	453	02m04s
097	2045 Feb 16	23:56:07	81	558	131	A	-n	-0.3125	0.9285	28.3S	166.2W	72	331	281	07m47s
098	2045 Aug 12	17:42:39	82	564	136	T	-n	0.2116	1.0774	25.9N	78.6W	78	206	256	06m06s
099	2046 Feb 05	23:06:26	82	570	141	A	p-	0.3765	0.9232	4.8N	171.4W	68	157	310	09m42s
100	2046 Aug 02	10:21:13	82	576	146	T	p-	-0.5350	1.0531	12.7S	15.1E	58	21	206	04m51s
101	2047 Jan 26	01:33:18	82	582	151	P	t-	1.0450	0.8908	62.9N	111.7E	0	135		
102	2047 Jun 23	10:52:31	83	587	118	P	-t	1.3766	0.3129	65.8N	178.0W	0	346		
103	2047 Jul 22	22:36:17	83	588	156	P	t-	-1.3477	0.3605	63.4S	160.1E	0	40		
104	2047 Dec 16	23:50:12	83	593	123	P	-t	-1.0661	0.8817	66.4S	6.6W	0	188		
105	2048 Jun 11	12:58:53	83	599	128	A	-p	0.6468	0.9441	63.7N	11.5W	49	184	271	04m58s
106	2048 Dec 05	15:35:27	83	605	133	T	-n	-0.3973	1.0440	46.1S	56.4W	66	1	160	03m28s
107	2049 May 31	13:59:59	84	611	138	A	nn	-0.1187	0.9631	15.3N	29.9W	83	358	134	04m45s
108	2049 Nov 25	05:33:48	84	617	143	H	n-	0.2943	1.0057	3.8S	95.2E	73	185	21	00m38s
109	2050 May 20	20:42:50	84	623	148	H	t-	-0.8688	1.0038	40.1S	123.8W	29	352	27	00m21s
110	2050 Nov 14	13:30:53	85	629	153	P	t-	1.0447	0.8874	69.5N	1.0E	0	206		
111	2051 Apr 11	02:10:39	85	634	120	P	-t	1.0169	0.9849	71.6N	32.1E	0	63		
112	2051 Oct 04	21:02:15	85	640	125	P	-t	-1.2094	0.6024	72.0S	117.7E	0	105		
113	2052 Mar 30	18:31:53	85	646	130	T	-n	0.3239	1.0466	22.4N	102.6W	71	161	164	04m08s
114	2052 Sep 22	23:39:10	86	652	135	A	-p	-0.4480	0.9734	25.7S	174.9E	63	20	106	02m51s
115	2053 Mar 20	07:08:19	86	658	140	A	n-	-0.4089	0.9919	23.0S	82.9E	66	341	31	00m50s
116	2053 Sep 12	09:34:09	86	664	145	T	n-	0.3140	1.0328	21.5N	41.7E	72	199	116	03m04s
117	2054 Mar 09	12:33:40	87	670	150	P	t-	-1.1711	0.6678	72.0S	97.9E	0	256		
118	2054 Aug 03	18:04:02	87	675	117	Pe	-t	-1.4941	0.0656	69.8S	121.4W	0	31		
119	2054 Sep 02	01:09:34	87	676	155	P	t-	1.0215	0.9793	71.7N	82.4W	0	296		
120	2055 Jan 27	17:54:05	87	681	122	P	-t	1.1550	0.6932	69.5N	112.3W	0	154		

Cat Num	Calendar Date	TD of Greatest Eclipse	ΔT s	Luna Num	Saros Num	Ecl Type	QLE	Gamma	Ecl Mag	Lat °	Long °	Sun Alt °	Sun Azm °	Path Width km	Central Line Dur
121	2055 Jul 24	09:57:50	88	687	127	T	-p	-0.8012	1.0359	33.3S	25.7E	37	8	202	03m17s
122	2056 Jan 16	22:16:45	88	693	132	A	-n	0.4199	0.9760	3.9N	153.6W	65	175	95	02m52s
123	2056 Jul 12	20:21:59	88	699	137	A	nn	-0.0426	0.9878	19.4N	123.8W	88	3	43	01m26s
124	2057 Jan 05	09:47:52	88	705	142	T	n-	-0.2837	1.0287	39.2S	35.1E	73	359	102	02m29s
125	2057 Jul 01	23:40:15	89	711	147	A	p-	0.7455	0.9464	71.5N	176.3W	41	177	298	04m22s
126	2057 Dec 26	01:14:35	89	717	152	T	p-	-0.9405	1.0348	84.9S	21.7E	19	141	355	01m50s
127	2058 May 22	10:39:26	89	722	119	P	-t	-1.3194	0.4141	63.5S	61.1E	0	322		
128	2058 Jun 21	00:19:35	89	723	157	Pb	t-	1.4869	0.1261	65.9N	9.8E	0	13		
129	2058 Nov 16	03:23:07	90	728	124	P	-t	1.1224	0.7644	62.9N	174.1E	0	225		
130	2059 May 11	19:22:16	90	734	129	T	-p	-0.5080	1.0242	10.7S	100.5W	59	340	95	02m23s
131	2059 Nov 05	09:18:15	90	740	134	A	-p	0.4454	0.9417	8.7N	47.0E	63	203	238	07m00s
132	2060 Apr 30	10:10:00	91	746	139	T	n-	0.2422	1.0660	28.0N	20.8E	76	154	222	05m15s
133	2060 Oct 24	09:24:10	91	752	144	A	nn	-0.2625	0.9277	25.8S	28.0E	75	28	281	08m06s
134	2061 Apr 20	02:56:49	91	758	149	T	t-	0.9578	1.0476	64.5N	59.1E	16	97	559	02m37s
135	2061 Oct 13	10:32:10	92	764	154	A	t-	-0.9639	0.9469	62.1S	54.5W	15	79	743	03m41s
136	2062 Mar 11	04:26:16	92	769	121	P	-t	-1.0238	0.9331	61.0S	147.2W	0	263		
137	2062 Sep 03	08:54:27	92	775	126	P	-t	1.0192	0.9749	61.3N	150.2E	0	286		
138	2063 Feb 28	07:43:30	93	781	131	A	-p	-0.3360	0.9293	25.2S	77.6E	70	329	280	07m41s
139	2063 Aug 24	01:22:11	93	787	136	T	-n	0.2771	1.0750	25.6N	168.3E	74	209	252	05m49s
140	2064 Feb 17	07:00:23	93	793	141	A	p-	0.3597	0.9262	7.0N	69.6E	69	154	295	08m56s
141	2064 Aug 12	17:46:06	94	799	146	T	p-	-0.4652	1.0495	10.9S	96.1W	62	24	184	04m28s
142	2065 Feb 05	09:52:26	94	805	151	P	t-	1.0336	0.9123	62.2N	22.0W	0	125		
143	2065 Jul 03	17:33:52	94	810	118	P	-t	1.4619	0.1639	64.8N	71.7E	0	336		
144	2065 Aug 02	05:34:17	94	811	156	P	t-	-1.2758	0.4903	62.7S	46.4E	0	49		
145	2065 Dec 27	08:39:56	95	816	123	P	-t	-1.0688	0.8769	65.4S	149.3W	0	198		
146	2066 Jun 22	19:25:48	95	822	128	A	-p	0.7330	0.9435	70.1N	96.5W	43	198	309	04m40s
147	2066 Dec 17	00:23:40	95	828	133	T	-n	-0.4043	1.0416	47.4S	175.6E	66	355	152	03m14s
148	2067 Jun 11	20:42:26	96	834	138	A	nn	-0.0387	0.9670	21.0N	130.3W	88	2	119	04m05s
149	2067 Dec 06	14:03:43	96	840	143	H	n-	0.2845	1.0011	6.0S	32.5W	74	181	4	00m08s
150	2068 May 31	03:56:39	96	846	148	T	p-	-0.7970	1.0110	31.0S	123.1E	37	357	63	01m06s
151	2068 Nov 24	21:32:30	97	852	153	P	t-	1.0299	0.9109	68.5N	131.2W	0	194		
152	2069 Apr 21	10:11:09	97	857	120	P	-t	1.0624	0.8992	71.0N	101.4W	0	50		
153	2069 May 20	17:53:18	97	858	158	Pb	t-	-1.4852	0.0879	68.8S	70.1W	0	342		
154	2069 Oct 15	04:19:56	97	863	125	P	-t	-1.2524	0.5298	71.6S	5.6W	0	119		
155	2070 Apr 11	02:36:09	98	869	130	T	-n	0.3652	1.0472	29.1N	134.9E	68	162	168	04m04s
156	2070 Oct 04	07:08:57	98	875	135	A	-p	-0.4950	0.9731	32.8S	60.2E	60	21	110	02m44s
157	2071 Mar 31	15:01:06	98	881	140	A	n-	-0.3739	0.9919	16.7S	37.2W	68	342	31	00m52s
158	2071 Sep 23	17:20:28	99	887	145	T	n-	0.2620	1.0333	14.2N	76.9W	75	198	116	03m11s
159	2072 Mar 19	20:10:31	99	893	150	P	t-	-1.1405	0.7199	72.2S	30.5W	0	270		
160	2072 Sep 12	08:59:20	100	899	155	T	t-	0.9655	1.0558	69.8N	101.8E	14	240	732	03m13s
161	2073 Feb 07	01:55:59	100	904	122	P	-t	1.1651	0.6768	70.5N	114.7E	0	141		
162	2073 Aug 03	17:15:23	100	910	127	T	-t	-0.8763	1.0294	43.2S	89.6W	28	14	206	02m29s
163	2074 Jan 27	06:44:15	101	916	132	A	-n	0.4251	0.9798	6.6N	78.6E	65	171	79	02m21s
164	2074 Jul 24	03:10:32	101	922	137	A	nn	-0.1242	0.9838	12.8N	133.5E	83	7	58	01m57s
165	2075 Jan 16	18:36:04	101	928	142	T	n-	-0.2799	1.0311	37.2S	94.3W	74	354	110	02m42s
166	2075 Jul 13	06:05:44	102	934	147	A	p-	0.6583	0.9467	63.1N	95.0E	49	186	262	04m45s
167	2076 Jan 06	10:07:28	102	940	152	T	p-	-0.9373	1.0342	87.2S	173.9W	20	203	340	01m49s
168	2076 Jun 01	17:31:22	102	945	119	P	-t	-1.3897	0.2897	64.4S	51.4W	0	331		
169	2076 Jul 01	06:50:43	103	946	157	P	t-	1.4005	0.2746	67.0N	98.3W	0	3		
170	2076 Nov 26	11:43:01	103	951	124	P	-t	1.1401	0.7315	63.7N	39.9E	0	215		
171	2077 May 22	02:46:05	103	957	129	T	-p	-0.5725	1.0290	13.1S	148.1E	55	343	119	02m54s
172	2077 Nov 15	17:07:56	104	963	134	A	-p	0.4705	0.9371	7.8N	71.0W	62	199	262	07m54s
173	2078 May 11	17:56:55	104	969	139	T	n-	0.1838	1.0701	28.1N	93.9W	79	158	232	05m40s
174	2078 Nov 04	16:55:44	104	975	144	A	nn	-0.2285	0.9255	27.8S	83.5W	77	25	287	08m29s
175	2079 May 01	10:50:13	105	981	149	T	p-	0.9081	1.0512	66.2N	46.5W	24	108	406	02m55s
176	2079 Oct 24	18:11:21	105	987	154	A	t-	-0.9243	0.9484	63.4S	160.8W	22	72	495	03m39s
177	2080 Mar 21	12:20:15	106	992	121	P	-t	-1.0578	0.8734	60.9S	85.7E	0	271		
178	2080 Sep 13	16:38:09	106	998	126	P	-t	1.0724	0.8743	61.1N	25.6E	0	277		
179	2081 Mar 10	15:23:31	106	1004	131	A	-p	-0.3653	0.9304	22.4S	36.9W	68	329	277	07m36s
180	2081 Sep 03	09:07:31	107	1010	136	T	-n	0.3379	1.0720	24.6N	53.4E	70	211	247	05m33s

31

Cat Num	Calendar Date	TD of Greatest Eclipse	ΔT s	Luna Num	Saros Num	Ecl Type	QLE	Gamma	Ecl Mag	Lat °	Long °	Sun Alt °	Sun Azm °	Path Width km	Central Line Dur
181	2082 Feb 27	14:47:00	107	1016	141	A	p-	0.3361	0.9298	9.4N	47.3W	70	152	277	08m12s
182	2082 Aug 24	01:16:21	108	1022	146	T	n-	-0.4004	1.0452	10.3S	151.5E	66	26	163	04m01s
183	2083 Feb 16	18:06:36	108	1028	151	P	t-	1.0170	0.9433	61.6N	154.3W	0	116		
184	2083 Jul 15	00:14:23	108	1033	118	Pe	-t	1.5464	0.0169	64.0N	37.9W	0	327		
185	2083 Aug 13	12:34:41	108	1034	156	P	t-	-1.2064	0.6146	62.1S	67.7W	0	58		
186	2084 Jan 07	17:30:24	109	1039	123	P	-t	-1.0715	0.8723	64.4S	68.3E	0	209		
187	2084 Jul 03	01:50:26	109	1045	128	A	-p	0.8208	0.9421	75.0N	169.3W	35	222	377	04m25s
188	2084 Dec 27	09:13:48	110	1051	133	T	-n	-0.4094	1.0396	47.3S	47.4E	66	349	146	03m04s
189	2085 Jun 22	03:21:16	110	1057	138	A	nn	0.0453	0.9704	26.2N	131.0E	87	186	106	03m29s
190	2085 Dec 16	22:37:48	111	1063	143	A	n-	0.2786	0.9971	7.3S	161.1W	74	176	10	00m19s
191	2086 Jun 11	11:07:14	111	1069	148	T	p-	-0.7215	1.0174	23.2S	12.2E	44	2	86	01m48s
192	2086 Dec 06	05:38:55	111	1075	153	P	p-	1.0194	0.9271	67.4N	96.0E	0	182		
193	2087 May 02	18:04:42	112	1080	120	P	-t	1.1139	0.8011	70.3N	127.3E	0	37		
194	2087 Jun 01	01:27:14	112	1081	158	P	t-	-1.4186	0.2146	67.8S	165.1E	0	354		
195	2087 Oct 26	11:46:57	112	1086	125	P	-t	-1.2882	0.4696	71.0S	130.8W	0	132		
196	2088 Apr 21	10:31:49	113	1092	130	T	-p	0.4135	1.0474	36.0N	14.8E	65	163	173	03m58s
197	2088 Oct 14	14:48:05	113	1098	135	A	-p	-0.5349	0.9727	39.7S	56.3W	57	21	115	02m38s
198	2089 Apr 10	22:44:42	113	1104	140	A	n-	-0.3319	0.9919	10.2S	155.0W	71	344	30	00m53s
199	2089 Oct 04	01:15:23	114	1110	145	T	n-	0.2167	1.0333	7.4N	162.5E	77	197	115	03m14s
200	2090 Mar 31	03:38:08	114	1116	150	P	t-	-1.1028	0.7843	72.1S	156.6W	0	284		
201	2090 Sep 23	16:56:36	115	1122	155	T	t-	0.9157	1.0562	60.7N	40.8W	23	218	463	03m36s
202	2091 Feb 18	09:54:40	115	1127	122	P	-t	1.1779	0.6558	71.2N	18.0W	0	128		
203	2091 Aug 15	00:34:43	116	1133	127	T	-t	-0.9490	1.0216	55.6S	150.2E	18	23	237	01m38s
204	2092 Feb 07	15:10:20	116	1139	132	A	-n	0.4322	0.9840	9.9N	49.0W	64	168	62	01m48s
205	2092 Aug 03	09:59:33	116	1145	137	A	nn	-0.2044	0.9794	5.6N	30.0E	78	10	75	02m31s
206	2093 Jan 27	03:22:16	117	1151	142	T	n-	-0.2737	1.0340	34.1S	136.1E	74	350	119	02m58s
207	2093 Jul 23	12:32:04	117	1157	147	A	p-	0.5717	0.9463	54.6N	1.0E	55	191	241	05m11s
208	2094 Jan 16	18:59:03	118	1163	152	T	p-	-0.9333	1.0342	84.8S	10.9W	21	267	329	01m52s
209	2094 Jun 13	00:22:11	118	1168	119	P	-t	-1.4613	0.1618	65.3S	163.9W	0	341		
210	2094 Jul 12	13:24:35	118	1169	157	P	t-	1.3149	0.4225	68.0N	152.5E	0	352		
211	2094 Dec 07	20:05:56	119	1174	124	P	-t	1.1547	0.7046	64.7N	95.3W	0	205		
212	2095 Jun 02	10:07:40	119	1180	129	T	-p	-0.6396	1.0332	16.7S	36.9E	50	347	145	03m18s
213	2095 Nov 27	01:02:57	120	1186	134	A	-p	0.4903	0.9330	7.2N	169.5E	61	195	285	08m47s
214	2096 May 22	01:37:14	120	1192	139	T	nn	0.1196	1.0737	27.3N	153.1E	83	162	241	06m06s
215	2096 Nov 15	00:36:15	121	1198	144	A	nn	-0.2018	0.9237	29.7S	163.0E	78	22	294	08m53s
216	2097 May 11	18:34:31	121	1204	149	T	p-	0.8516	1.0538	67.4N	149.8W	31	121	339	03m10s
217	2097 Nov 04	02:01:25	121	1210	154	A	t-	-0.8926	0.9494	65.8S	86.5E	26	68	411	03m36s
218	2098 Apr 01	20:02:31	122	1215	121	P	-t	-1.1005	0.7984	61.0S	38.4W	0	280		
219	2098 Sep 25	00:31:16	122	1221	126	P	-t	1.1184	0.7871	61.1N	101.3W	0	268		
220	2098 Oct 24	10:36:11	122	1222	164	Pb	t-	-1.5407	0.0057	61.8S	95.8W	0	116		
221	2099 Mar 21	22:54:32	123	1227	131	A	-p	-0.4016	0.9318	20.0S	149.4W	66	329	275	07m32s
222	2099 Sep 14	16:57:53	123	1233	136	T	-n	0.3942	1.0684	23.4N	63.1W	67	211	241	05m18s
223	2100 Mar 10	22:28:11	124	1239	141	A	n-	0.3077	0.9338	12.0N	162.8W	72	151	257	07m29s
224	2100 Sep 04	08:49:20	124	1245	146	T	n-	-0.3384	1.0402	10.5S	38.7E	70	28	142	03m32s

Appendix B

Small Global Solar Eclipse Maps: 2001 to 2100

Key to Small Global Solar Eclipse Maps

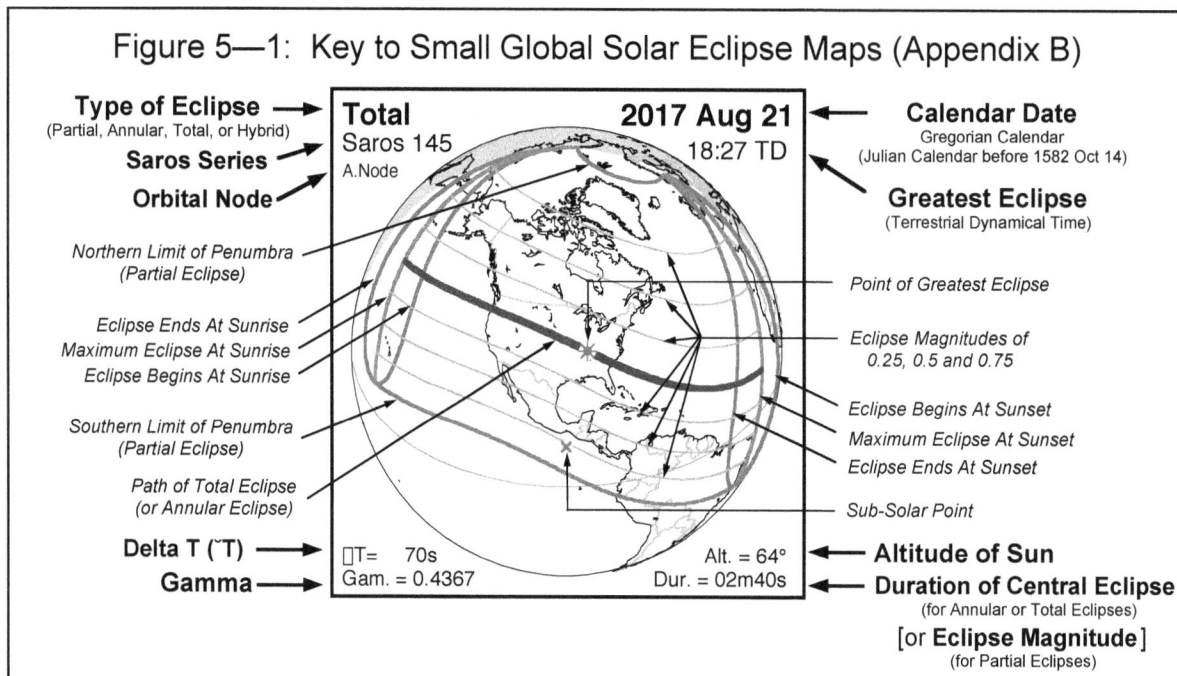

Figure 5—1: Key to Small Global Solar Eclipse Maps (Appendix B)

Solar Eclipse Type – One of four basic types of solar eclipses: Partial, Annular Total or Hybrid (See Sect. 4.1).

Saros Series – The Saros series that the eclipse belongs to. Eclipses with an odd number take place at the Moon's ascending node, while those with an even number are at the descending node (See Sect. 4.7).

Node – The orbital node near which the eclipse takes place. The ascending node (A. Node) is the point where the Moon travels from south to north through Earth's orbital plane. Similarly, the descending node (D. Node) is the point where the Moon travels from north to south.

Calendar Date – Gregorian calendar date at greatest eclipse is given in the ISO order of YEAR-MONTH-DAY.

Greatest Eclipse – The instant (Terrestrial Dynamical Time) when the distance between the axis of the Moon's shadow cone and the center of Earth reaches a minimum (See Sect. 4.4).

ΔT (Delta T) – the arithmetic difference, in seconds, between Terrestrial Dynamical Time (TD) and Universal Time (UT1), (See Sect. 2.5).

Gamma – The minimum distance from the lunar shadow axis to the center of Earth, in units of Earth's equatorial radius (See Sect. 4.10).

Altitude of Sun – The Sun's altitude at the geographic position intersected by the lunar shadow axis is given at the instant of greatest eclipse (See Sect. 4.13).

Duration of Central Eclipse – The central line duration of the total or annular phase (in minutes and seconds) is given at the instant of greatest eclipse (See Sect. 4.16).

Eclipse Magnitude – The fraction of the Sun's diameter occulted by the Moon at the instant of greatest eclipse (See Sect. 4.11).

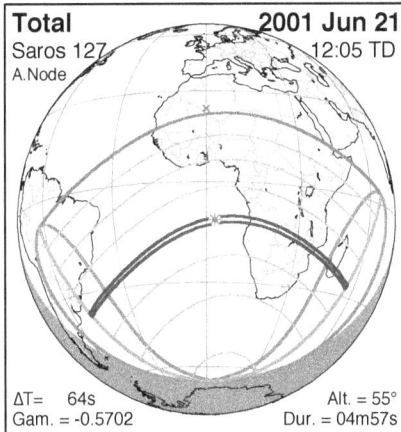

Total   **2001 Jun 21**
Saros 127   12:05 TD
A.Node

ΔT= 64s   Alt. = 55°
Gam. = -0.5702   Dur. = 04m57s

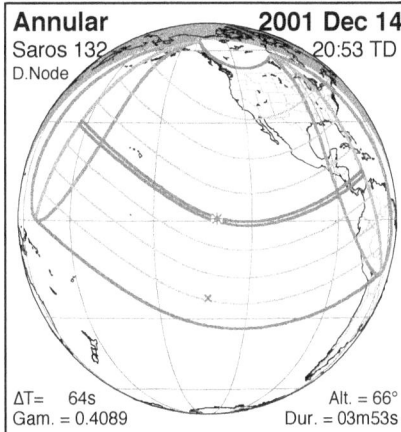

Annular   **2001 Dec 14**
Saros 132   20:53 TD
D.Node

ΔT= 64s   Alt. = 66°
Gam. = 0.4089   Dur. = 03m53s

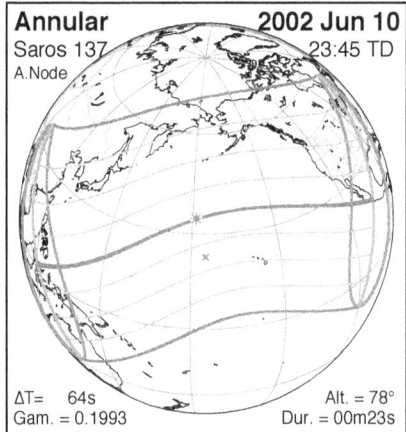

Annular   **2002 Jun 10**
Saros 137   23:45 TD
A.Node

ΔT= 64s   Alt. = 78°
Gam. = 0.1993   Dur. = 00m23s

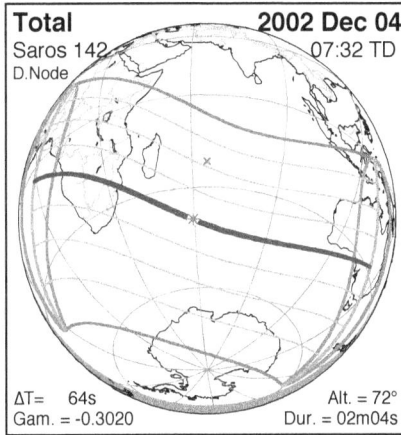

Total   **2002 Dec 04**
Saros 142   07:32 TD
D.Node

ΔT= 64s   Alt. = 72°
Gam. = -0.3020   Dur. = 02m04s

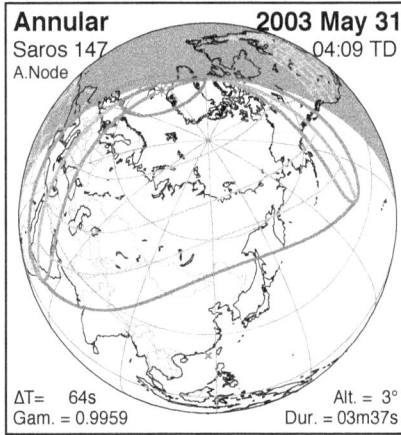

Annular   **2003 May 31**
Saros 147   04:09 TD
A.Node

ΔT= 64s   Alt. = 3°
Gam. = 0.9959   Dur. = 03m37s

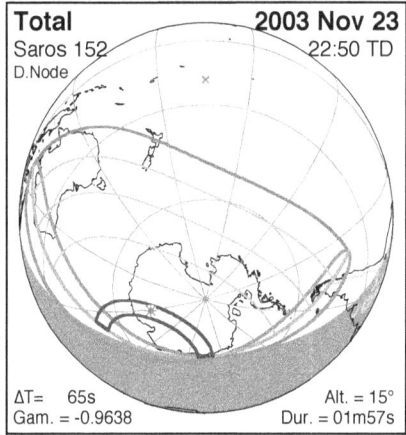

Total   **2003 Nov 23**
Saros 152   22:50 TD
D.Node

ΔT= 65s   Alt. = 15°
Gam. = -0.9638   Dur. = 01m57s

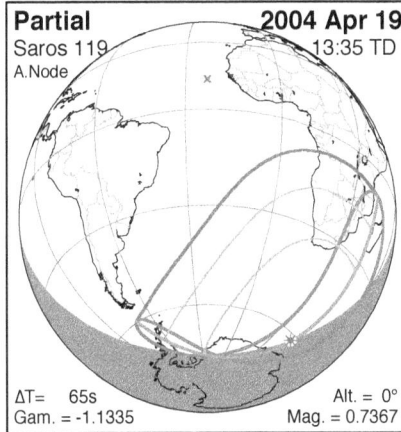

Partial   **2004 Apr 19**
Saros 119   13:35 TD
A.Node

ΔT= 65s   Alt. = 0°
Gam. = -1.1335   Mag. = 0.7367

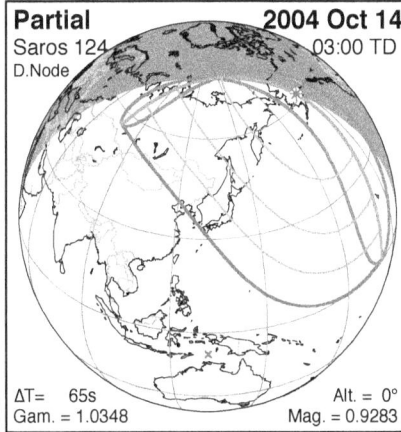

Partial   **2004 Oct 14**
Saros 124   03:00 TD
D.Node

ΔT= 65s   Alt. = 0°
Gam. = 1.0348   Mag. = 0.9283

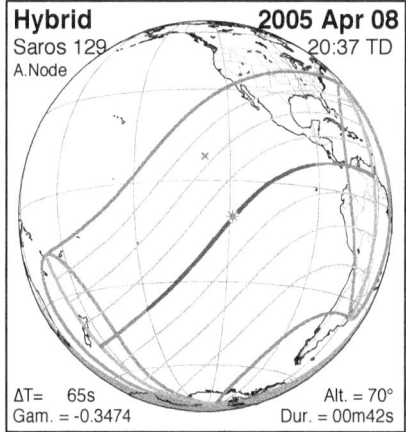

Hybrid   **2005 Apr 08**
Saros 129   20:37 TD
A.Node

ΔT= 65s   Alt. = 70°
Gam. = -0.3474   Dur. = 00m42s

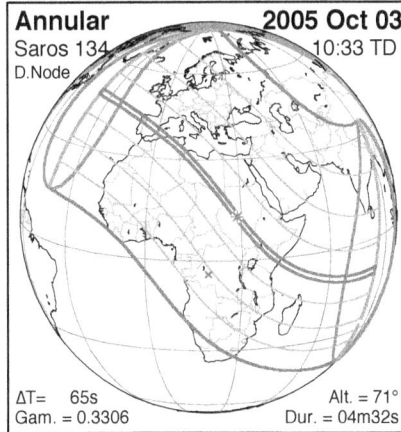

Annular   **2005 Oct 03**
Saros 134   10:33 TD
D.Node

ΔT= 65s   Alt. = 71°
Gam. = 0.3306   Dur. = 04m32s

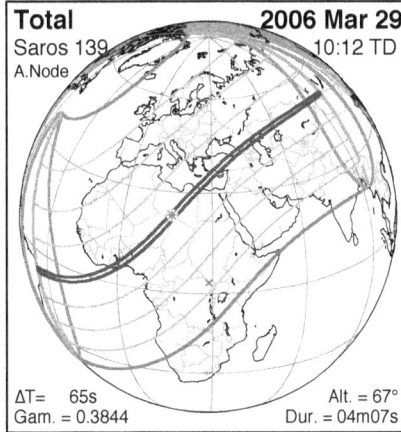

Total   **2006 Mar 29**
Saros 139   10:12 TD
A.Node

ΔT= 65s   Alt. = 67°
Gam. = 0.3844   Dur. = 04m07s

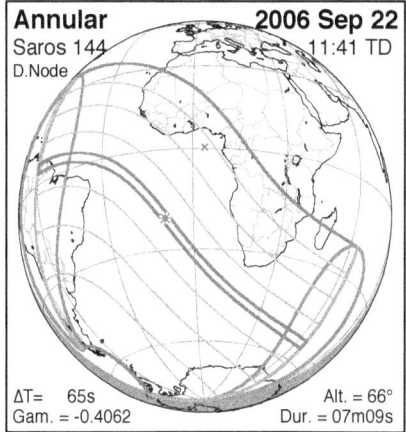

Annular   **2006 Sep 22**
Saros 144   11:41 TD
D.Node

ΔT= 65s   Alt. = 66°
Gam. = -0.4062   Dur. = 07m09s

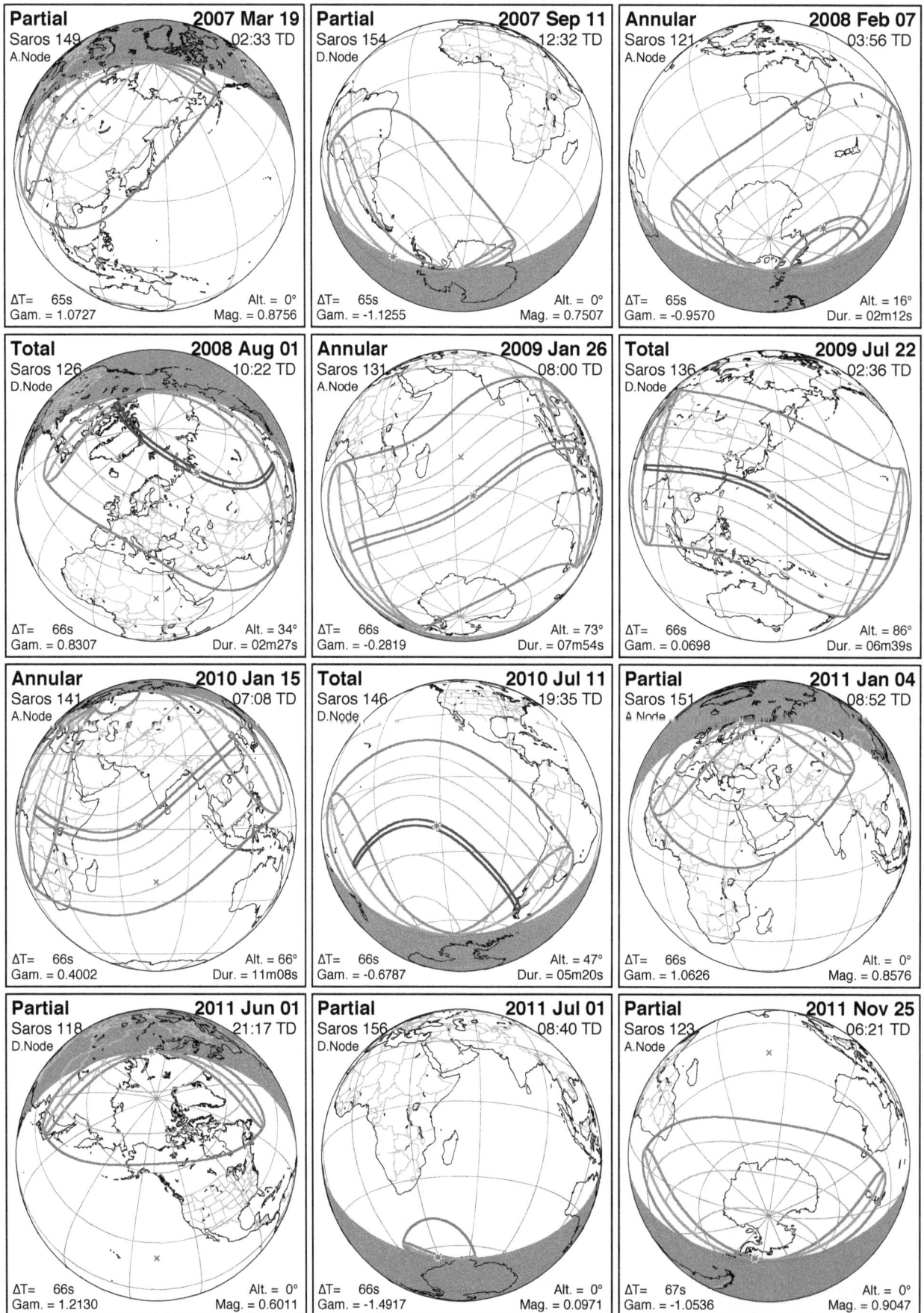

Partial	2007 Mar 19
Saros 149	02:33 TD
A.Node	
ΔT= 65s	Alt. = 0°
Gam. = 1.0727	Mag. = 0.8756

Partial	2007 Sep 11
Saros 154	12:32 TD
D.Node	
ΔT= 65s	Alt. = 0°
Gam. = -1.1255	Mag. = 0.7507

Annular	2008 Feb 07
Saros 121	03:56 TD
A.Node	
ΔT= 65s	Alt. = 16°
Gam. = -0.9570	Dur. = 02m12s

Total	2008 Aug 01
Saros 126	10:22 TD
D.Node	
ΔT= 66s	Alt. = 34°
Gam. = 0.8307	Dur. = 02m27s

Annular	2009 Jan 26
Saros 131	08:00 TD
A.Node	
ΔT= 66s	Alt. = 73°
Gam. = -0.2819	Dur. = 07m54s

Total	2009 Jul 22
Saros 136	02:36 TD
D.Node	
ΔT= 66s	Alt. = 86°
Gam. = 0.0698	Dur. = 06m39s

Annular	2010 Jan 15
Saros 141	07:08 TD
A.Node	
ΔT= 66s	Alt. = 66°
Gam. = 0.4002	Dur. = 11m08s

Total	2010 Jul 11
Saros 146	19:35 TD
D.Node	
ΔT= 66s	Alt. = 47°
Gam. = -0.6787	Dur. = 05m20s

Partial	2011 Jan 04
Saros 151	08:52 TD
A.Node	
ΔT= 66s	Alt. = 0°
Gam. = 1.0626	Mag. = 0.8576

Partial	2011 Jun 01
Saros 118	21:17 TD
D.Node	
ΔT= 66s	Alt. = 0°
Gam. = 1.2130	Mag. = 0.6011

Partial	2011 Jul 01
Saros 156	08:40 TD
D.Node	
ΔT= 66s	Alt. = 0°
Gam. = -1.4917	Mag. = 0.0971

Partial	2011 Nov 25
Saros 123	06:21 TD
A.Node	
ΔT= 67s	Alt. = 0°
Gam. = -1.0536	Mag. = 0.9047

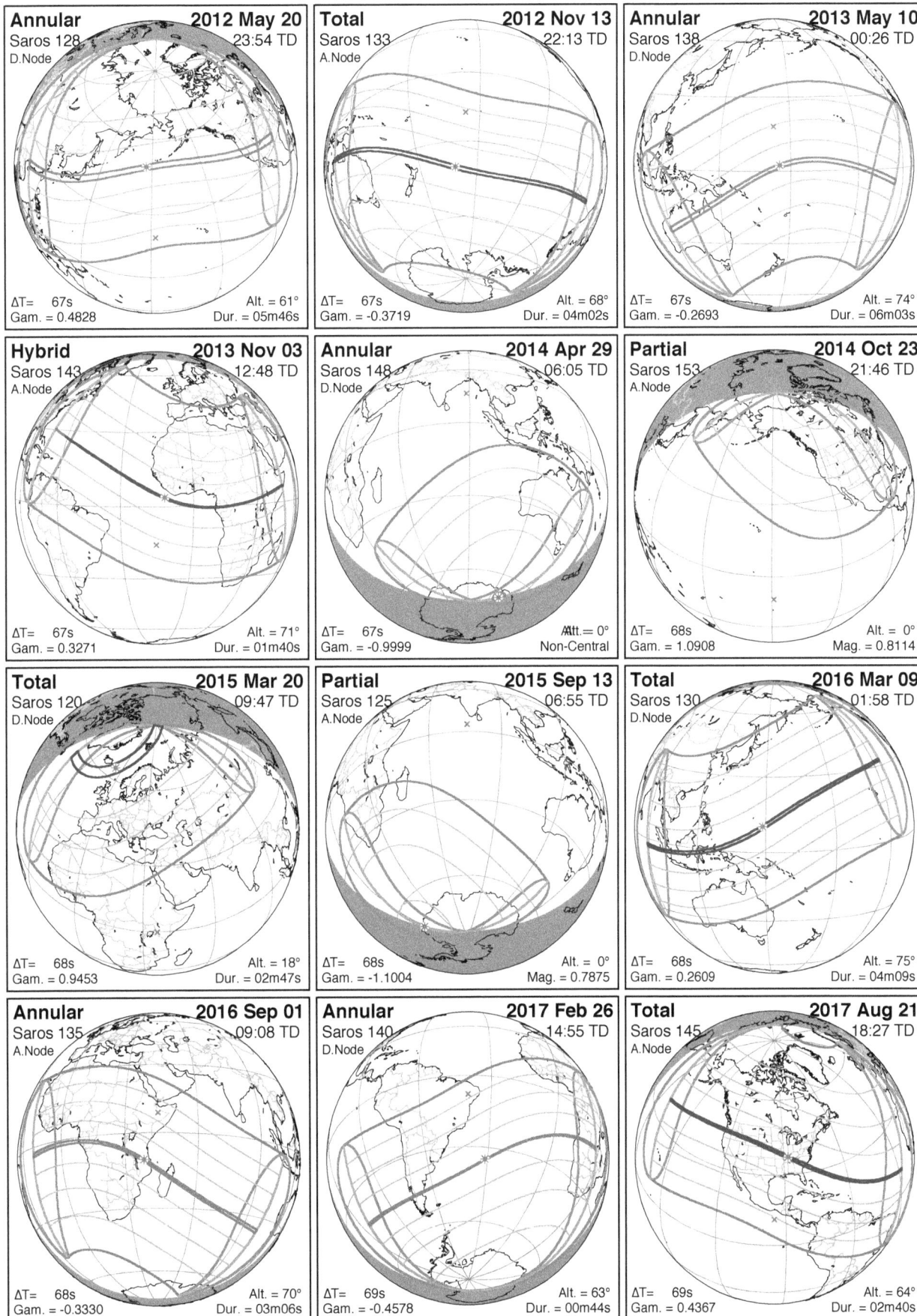

Annular	2012 May 20
Saros 128	23:54 TD
D.Node	

ΔT= 67s — Alt. = 61°
Gam. = 0.4828 — Dur. = 05m46s

Total	2012 Nov 13
Saros 133	22:13 TD
A.Node	

ΔT= 67s — Alt. = 68°
Gam. = -0.3719 — Dur. = 04m02s

Annular	2013 May 10
Saros 138	00:26 TD
D.Node	

ΔT= 67s — Alt. = 74°
Gam. = -0.2693 — Dur. = 06m03s

Hybrid	2013 Nov 03
Saros 143	12:48 TD
A.Node	

ΔT= 67s — Alt. = 71°
Gam. = 0.3271 — Dur. = 01m40s

Annular	2014 Apr 29
Saros 148	06:05 TD
D.Node	

ΔT= 67s — Alt. = 0°
Gam. = -0.9999 — Non-Central

Partial	2014 Oct 23
Saros 153	21:46 TD
A.Node	

ΔT= 68s — Alt. = 0°
Gam. = 1.0908 — Mag. = 0.8114

Total	2015 Mar 20
Saros 120	09:47 TD
D.Node	

ΔT= 68s — Alt. = 18°
Gam. = 0.9453 — Dur. = 02m47s

Partial	2015 Sep 13
Saros 125	06:55 TD
A.Node	

ΔT= 68s — Alt. = 0°
Gam. = -1.1004 — Mag. = 0.7875

Total	2016 Mar 09
Saros 130	01:58 TD
D.Node	

ΔT= 68s — Alt. = 75°
Gam. = 0.2609 — Dur. = 04m09s

Annular	2016 Sep 01
Saros 135	09:08 TD
A.Node	

ΔT= 68s — Alt. = 70°
Gam. = -0.3330 — Dur. = 03m06s

Annular	2017 Feb 26
Saros 140	14:55 TD
D.Node	

ΔT= 69s — Alt. = 63°
Gam. = -0.4578 — Dur. = 00m44s

Total	2017 Aug 21
Saros 145	18:27 TD
A.Node	

ΔT= 69s — Alt. = 64°
Gam. = 0.4367 — Dur. = 02m40s

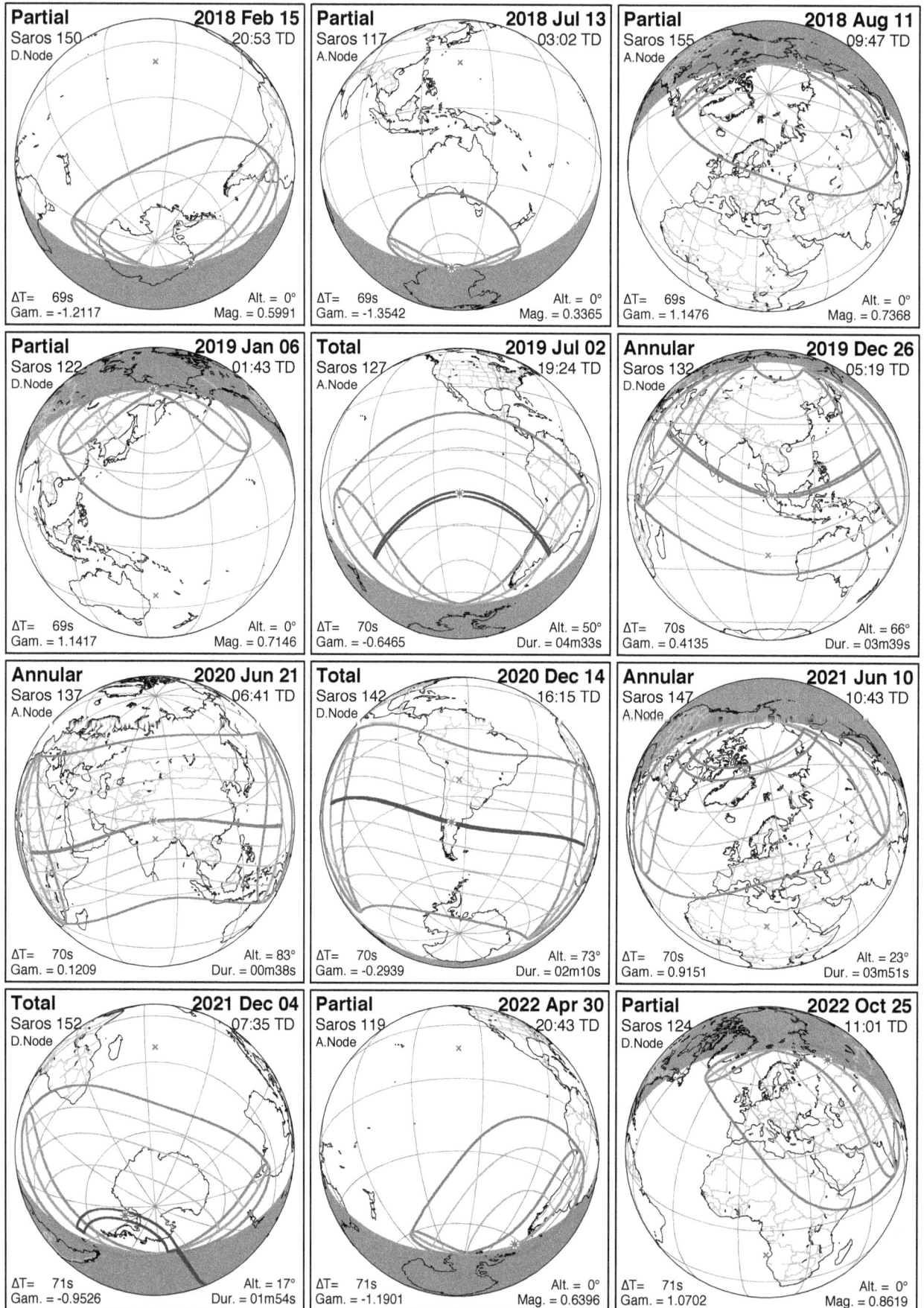

Partial	**2018 Feb 15**
Saros 150	20:53 TD
D.Node	
ΔT= 69s	Alt. = 0°
Gam. = -1.2117	Mag. = 0.5991

Partial	**2018 Jul 13**
Saros 117	03:02 TD
A.Node	
ΔT= 69s	Alt. = 0°
Gam. = -1.3542	Mag. = 0.3365

Partial	**2018 Aug 11**
Saros 155	09:47 TD
A.Node	
ΔT= 69s	Alt. = 0°
Gam. = 1.1476	Mag. = 0.7368

Partial	**2019 Jan 06**
Saros 122	01:43 TD
D.Node	
ΔT= 69s	Alt. = 0°
Gam. = 1.1417	Mag. = 0.7146

Total	**2019 Jul 02**
Saros 127	19:24 TD
A.Node	
ΔT= 70s	Alt. = 50°
Gam. = -0.6465	Dur. = 04m33s

Annular	**2019 Dec 26**
Saros 132	05:19 TD
D.Node	
ΔT= 70s	Alt. = 66°
Gam. = 0.4135	Dur. = 03m39s

Annular	**2020 Jun 21**
Saros 137	06:41 TD
A.Node	
ΔT= 70s	Alt. = 83°
Gam. = 0.1209	Dur. = 00m38s

Total	**2020 Dec 14**
Saros 142	16:15 TD
D.Node	
ΔT= 70s	Alt. = 73°
Gam. = -0.2939	Dur. = 02m10s

Annular	**2021 Jun 10**
Saros 147	10:43 TD
A.Node	
ΔT= 70s	Alt. = 23°
Gam. = 0.9151	Dur. = 03m51s

Total	**2021 Dec 04**
Saros 152	07:35 TD
D.Node	
ΔT= 71s	Alt. = 17°
Gam. = -0.9526	Dur. = 01m54s

Partial	**2022 Apr 30**
Saros 119	20:43 TD
A.Node	
ΔT= 71s	Alt. = 0°
Gam. = -1.1901	Mag. = 0.6396

Partial	**2022 Oct 25**
Saros 124	11:01 TD
D.Node	
ΔT= 71s	Alt. = 0°
Gam. = 1.0702	Mag. = 0.8619

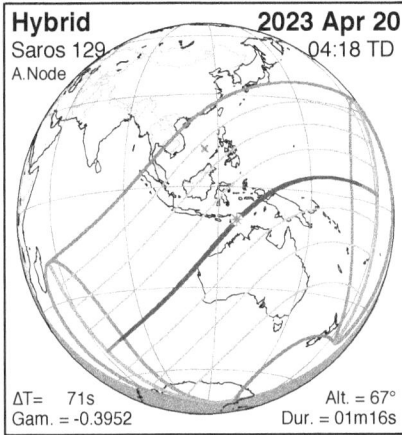

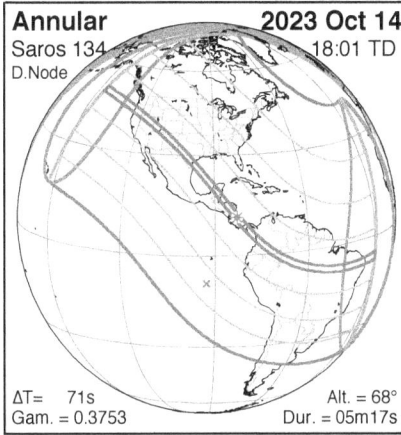

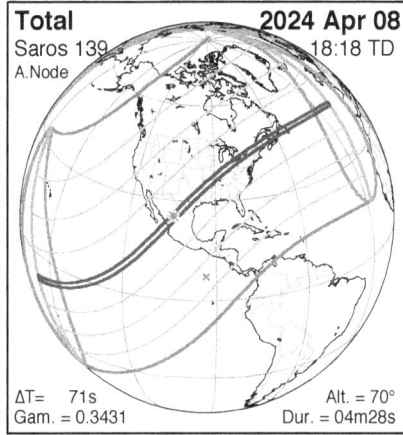

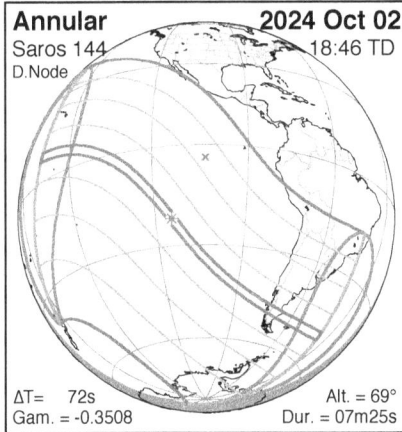

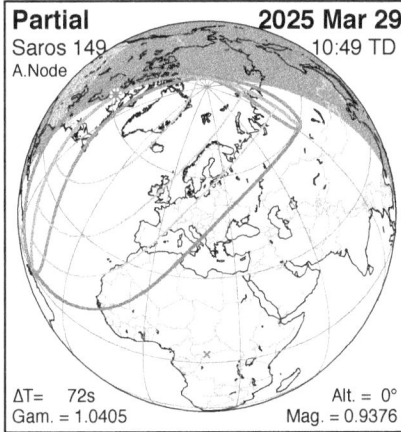

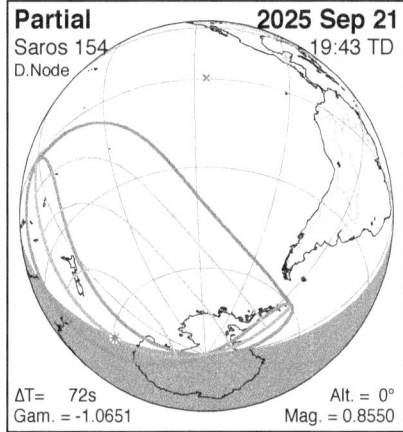

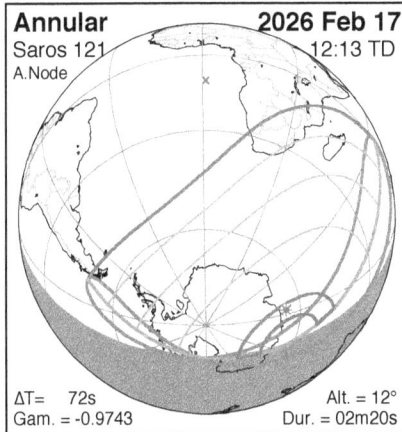

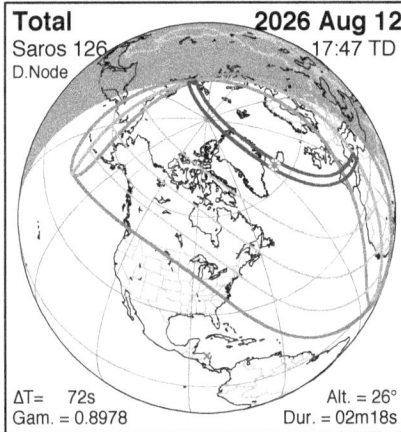

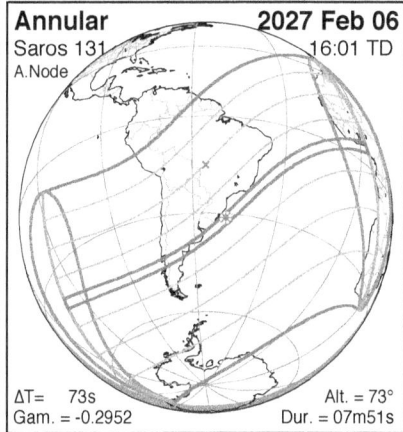

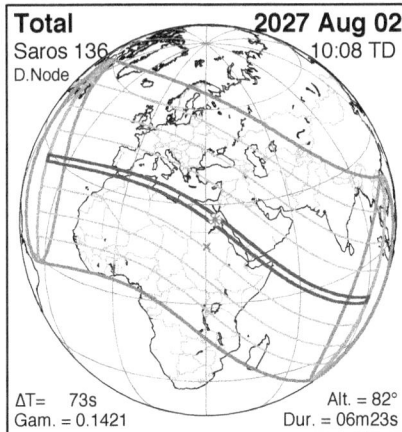

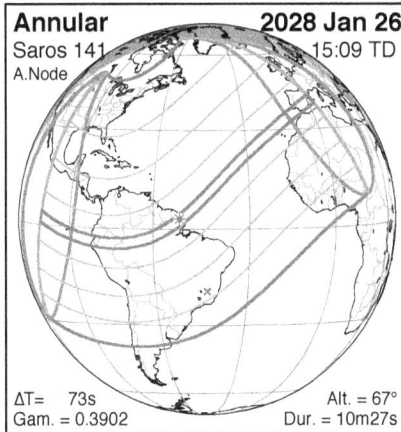

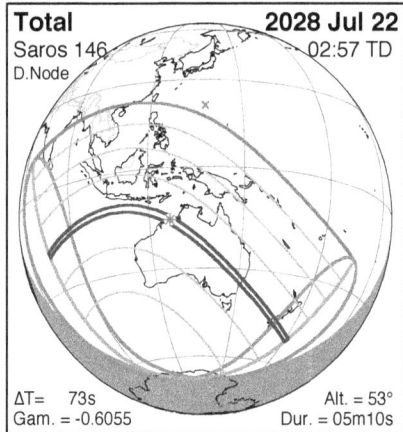

Hybrid	**2023 Apr 20**
Saros 129	04:18 TD
A.Node	
ΔT= 71s	Alt. = 67°
Gam. = -0.3952	Dur. = 01m16s

Annular	**2023 Oct 14**
Saros 134	18:01 TD
D.Node	
ΔT= 71s	Alt. = 68°
Gam. = 0.3753	Dur. = 05m17s

Total	**2024 Apr 08**
Saros 139	18:18 TD
A.Node	
ΔT= 71s	Alt. = 70°
Gam. = 0.3431	Dur. = 04m28s

Annular	**2024 Oct 02**
Saros 144	18:46 TD
D.Node	
ΔT= 72s	Alt. = 69°
Gam. = -0.3508	Dur. = 07m25s

Partial	**2025 Mar 29**
Saros 149	10:49 TD
A.Node	
ΔT= 72s	Alt. = 0°
Gam. = 1.0405	Mag. = 0.9376

Partial	**2025 Sep 21**
Saros 154	19:43 TD
D.Node	
ΔT= 72s	Alt. = 0°
Gam. = -1.0651	Mag. = 0.8550

Annular	**2026 Feb 17**
Saros 121	12:13 TD
A.Node	
ΔT= 72s	Alt. = 12°
Gam. = -0.9743	Dur. = 02m20s

Total	**2026 Aug 12**
Saros 126	17:47 TD
D.Node	
ΔT= 72s	Alt. = 26°
Gam. = 0.8978	Dur. = 02m18s

Annular	**2027 Feb 06**
Saros 131	16:01 TD
A.Node	
ΔT= 73s	Alt. = 73°
Gam. = -0.2952	Dur. = 07m51s

Total	**2027 Aug 02**
Saros 136	10:08 TD
D.Node	
ΔT= 73s	Alt. = 82°
Gam. = 0.1421	Dur. = 06m23s

Annular	**2028 Jan 26**
Saros 141	15:09 TD
A.Node	
ΔT= 73s	Alt. = 67°
Gam. = 0.3902	Dur. = 10m27s

Total	**2028 Jul 22**
Saros 146	02:57 TD
D.Node	
ΔT= 73s	Alt. = 53°
Gam. = -0.6055	Dur. = 05m10s

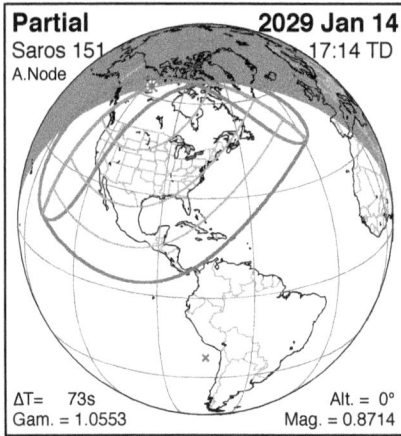

Partial   **2029 Jan 14**
Saros 151   17:14 TD
A.Node

ΔT= 73s   Alt. = 0°
Gam. = 1.0553   Mag. = 0.8714

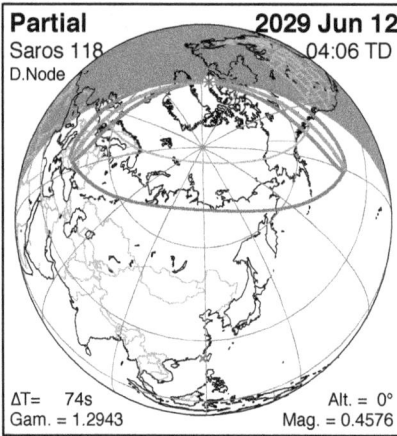

Partial   **2029 Jun 12**
Saros 118   04:06 TD
D.Node

ΔT= 74s   Alt. = 0°
Gam. = 1.2943   Mag. = 0.4576

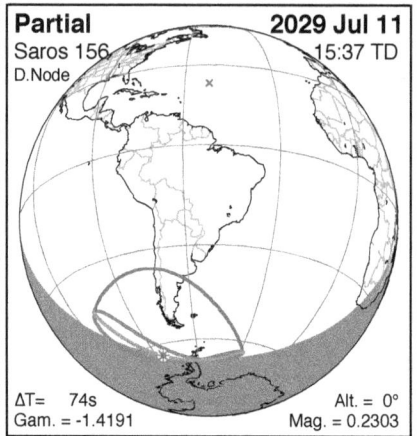

Partial   **2029 Jul 11**
Saros 156   15:37 TD
D.Node

ΔT= 74s   Alt. = 0°
Gam. = -1.4191   Mag. = 0.2303

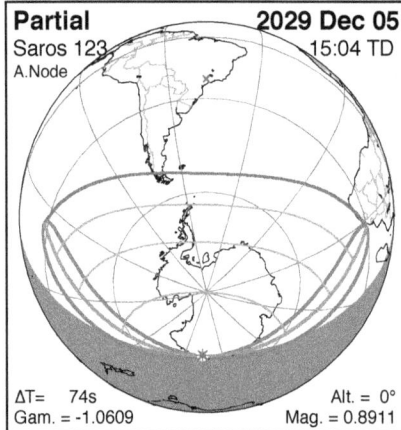

Partial   **2029 Dec 05**
Saros 123   15:04 TD
A.Node

ΔT= 74s   Alt. = 0°
Gam. = -1.0609   Mag. = 0.8911

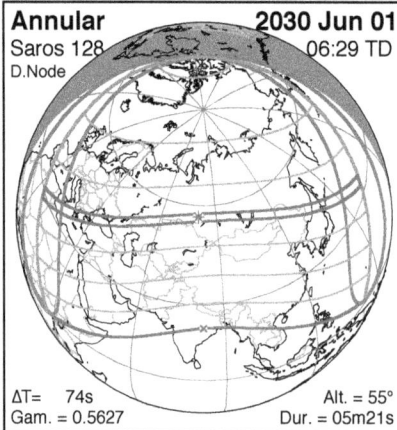

Annular   **2030 Jun 01**
Saros 128   06:29 TD
D.Node

ΔT= 74s   Alt. = 55°
Gam. = 0.5627   Dur. = 05m21s

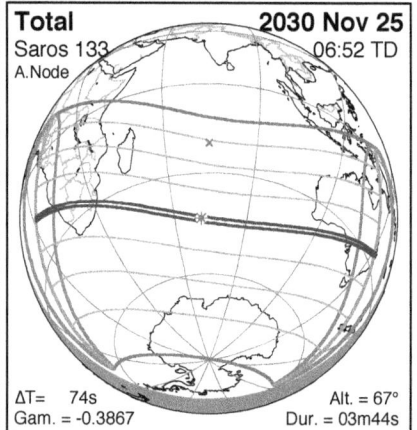

Total   **2030 Nov 25**
Saros 133   06:52 TD
A.Node

ΔT= 74s   Alt. = 67°
Gam. = -0.3867   Dur. = 03m44s

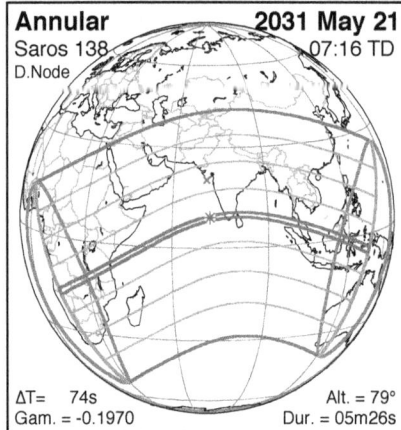

Annular   **2031 May 21**
Saros 138   07:16 TD
D.Node

ΔT= 74s   Alt. = 79°
Gam. = -0.1970   Dur. = 05m26s

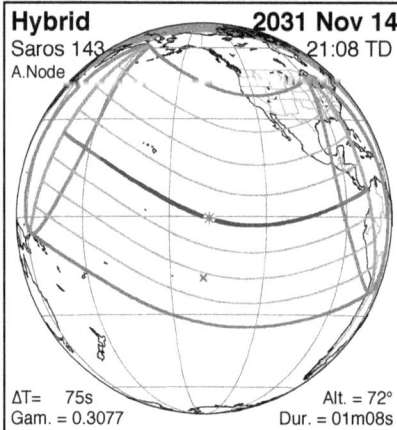

Hybrid   **2031 Nov 14**
Saros 143   21:08 TD
A.Node

ΔT= 75s   Alt. = 72°
Gam. = 0.3077   Dur. = 01m08s

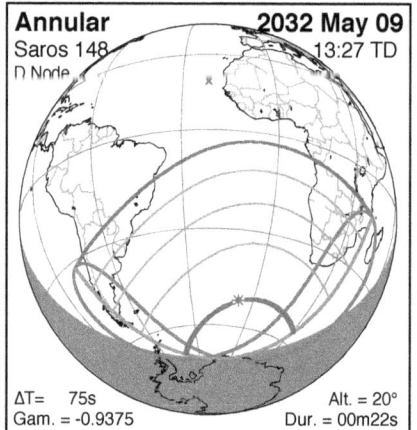

Annular   **2032 May 09**
Saros 148   13:27 TD
D.Node

ΔT= 75s   Alt. = 20°
Gam. = -0.9375   Dur. = 00m22s

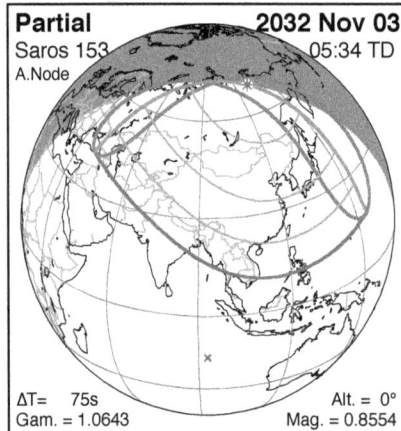

Partial   **2032 Nov 03**
Saros 153   05:34 TD
A.Node

ΔT= 75s   Alt. = 0°
Gam. = 1.0643   Mag. = 0.8554

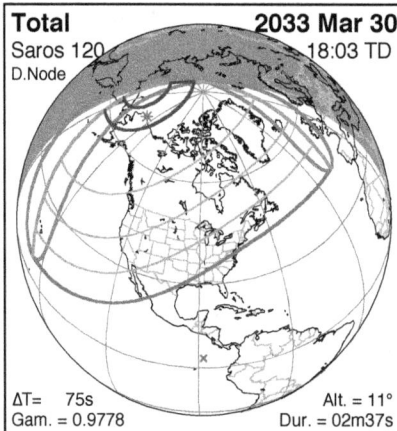

Total   **2033 Mar 30**
Saros 120   18:03 TD
D.Node

ΔT= 75s   Alt. = 11°
Gam. = 0.9778   Dur. = 02m37s

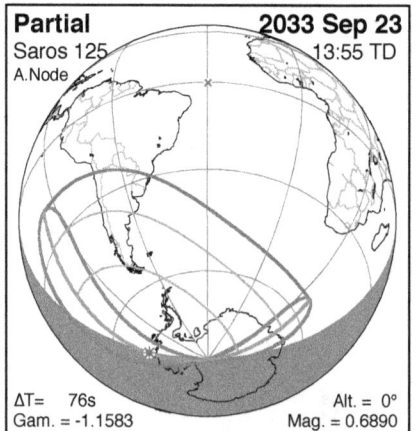

Partial   **2033 Sep 23**
Saros 125   13:55 TD
A.Node

ΔT= 76s   Alt. = 0°
Gam. = -1.1583   Mag. = 0.6890

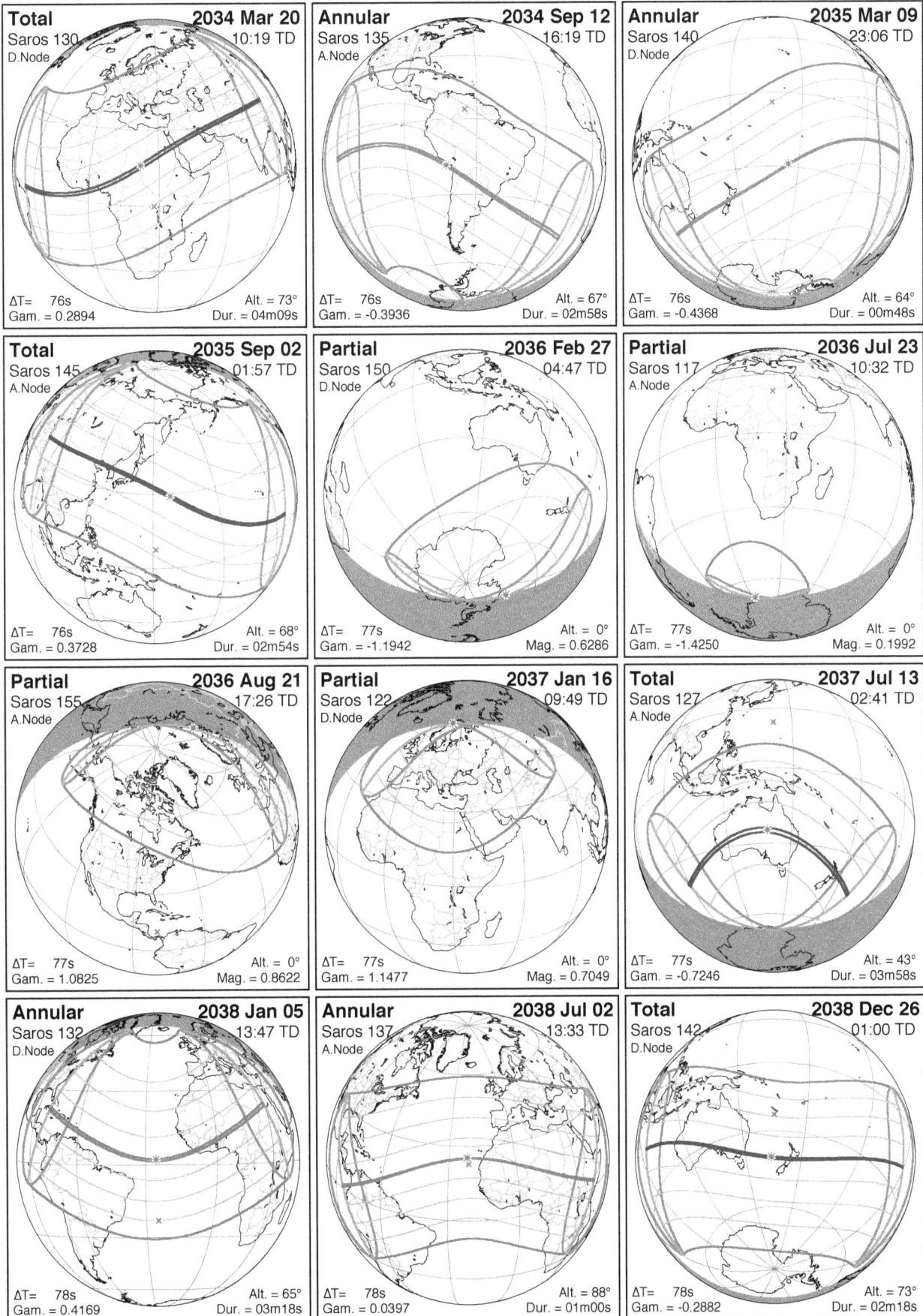

Total	2034 Mar 20
Saros 130	10:19 TD
D.Node	
ΔT = 76s	Alt. = 73°
Gam. = 0.2894	Dur. = 04m09s

Annular	2034 Sep 12
Saros 135	16:19 TD
A.Node	
ΔT = 76s	Alt. = 67°
Gam. = -0.3936	Dur. = 02m58s

Annular	2035 Mar 09
Saros 140	23:06 TD
D.Node	
ΔT = 76s	Alt. = 64°
Gam. = -0.4368	Dur. = 00m48s

Total	2035 Sep 02
Saros 145	01:57 TD
A.Node	
ΔT = 76s	Alt. = 68°
Gam. = 0.3728	Dur. = 02m54s

Partial	2036 Feb 27
Saros 150	04:47 TD
D.Node	
ΔT = 77s	Alt. = 0°
Gam. = -1.1942	Mag. = 0.6286

Partial	2036 Jul 23
Saros 117	10:32 TD
A.Node	
ΔT = 77s	Alt. = 0°
Gam. = -1.4250	Mag. = 0.1992

Partial	2036 Aug 21
Saros 155	17:26 TD
A.Node	
ΔT = 77s	Alt. = 0°
Gam. = 1.0825	Mag. = 0.8622

Partial	2037 Jan 16
Saros 122	09:49 TD
D.Node	
ΔT = 77s	Alt. = 0°
Gam. = 1.1477	Mag. = 0.7049

Total	2037 Jul 13
Saros 127	02:41 TD
A.Node	
ΔT = 77s	Alt. = 43°
Gam. = -0.7246	Dur. = 03m58s

Annular	2038 Jan 05
Saros 132	13:47 TD
D.Node	
ΔT = 78s	Alt. = 65°
Gam. = 0.4169	Dur. = 03m18s

Annular	2038 Jul 02
Saros 137	13:33 TD
A.Node	
ΔT = 78s	Alt. = 88°
Gam. = 0.0397	Dur. = 01m00s

Total	2038 Dec 26
Saros 142	01:00 TD
D.Node	
ΔT = 78s	Alt. = 73°
Gam. = -0.2882	Dur. = 02m18s

41

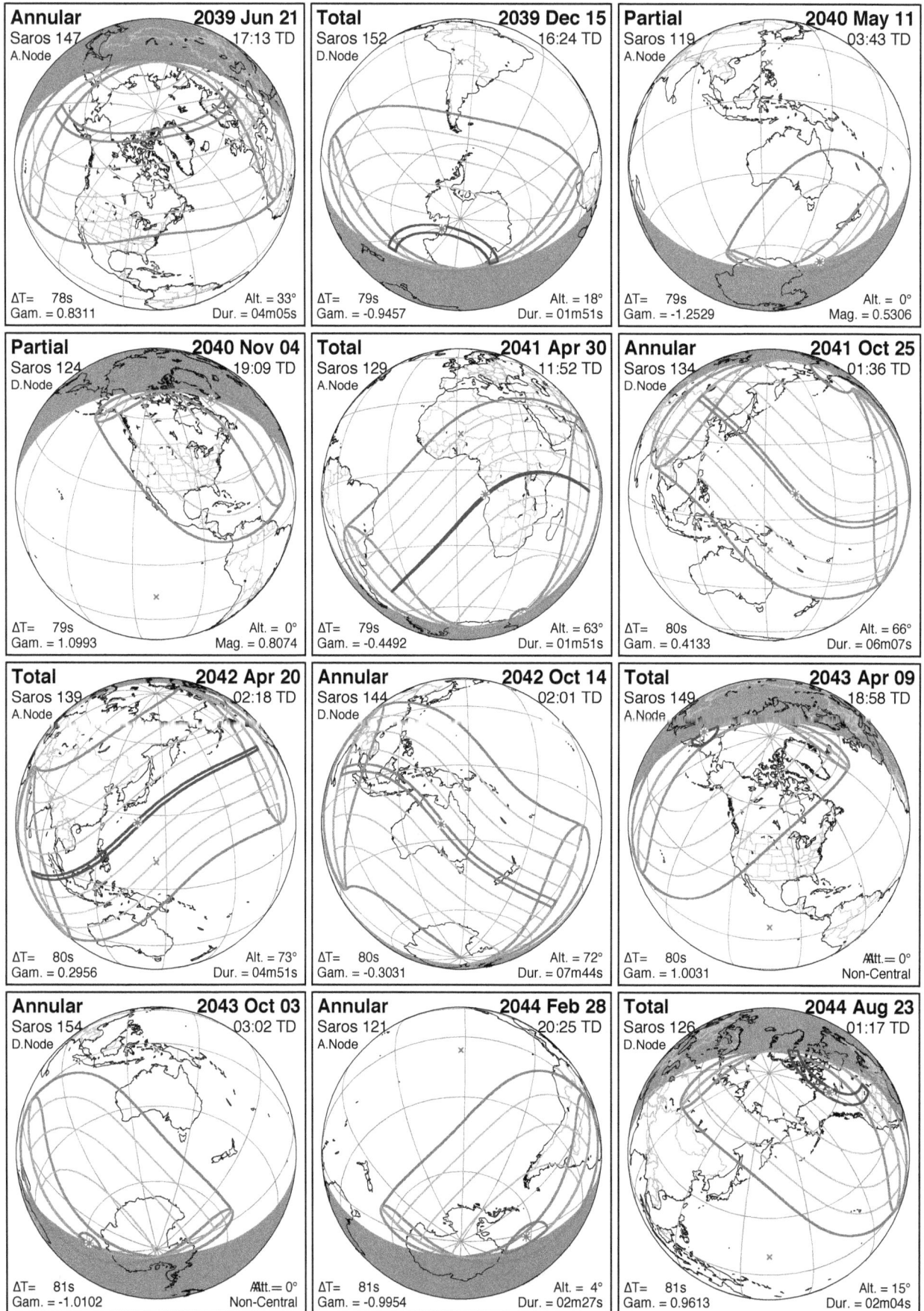

Annular **2039 Jun 21**	**Total** **2039 Dec 15**
Saros 147 17:13 TD	Saros 152 16:24 TD
A.Node	D.Node
ΔT= 78s Alt. = 33°	ΔT= 79s Alt. = 18°
Gam. = 0.8311 Dur. = 04m05s	Gam. = -0.9457 Dur. = 01m51s

Partial **2040 May 11**
Saros 119 03:43 TD
A.Node
ΔT= 79s Alt. = 0°
Gam. = -1.2529 Mag. = 0.5306

Partial **2040 Nov 04**	**Total** **2041 Apr 30**	**Annular** **2041 Oct 25**
Saros 124 19:09 TD	Saros 129 11:52 TD	Saros 134 01:36 TD
D.Node	A.Node	D.Node
ΔT= 79s Alt. = 0°	ΔT= 79s Alt. = 63°	ΔT= 80s Alt. = 66°
Gam. = 1.0993 Mag. = 0.8074	Gam. = -0.4492 Dur. = 01m51s	Gam. = 0.4133 Dur. = 06m07s

Total **2042 Apr 20**	**Annular** **2042 Oct 14**	**Total** **2043 Apr 09**
Saros 139 02:18 TD	Saros 144 02:01 TD	Saros 149 18:58 TD
A.Node	D.Node	A.Node
ΔT= 80s Alt. = 73°	ΔT= 80s Alt. = 72°	ΔT= 80s Alt.= 0°
Gam. = 0.2956 Dur. = 04m51s	Gam. = -0.3031 Dur. = 07m44s	Gam. = 1.0031 Non-Central

Annular **2043 Oct 03**	**Annular** **2044 Feb 28**	**Total** **2044 Aug 23**
Saros 154 03:02 TD	Saros 121 20:25 TD	Saros 126 01:17 TD
D.Node	A.Node	D.Node
ΔT= 81s Alt.= 0°	ΔT= 81s Alt. = 4°	ΔT= 81s Alt. = 15°
Gam. = -1.0102 Non-Central	Gam. = -0.9954 Dur. = 02m27s	Gam. = 0.9613 Dur. = 02m04s

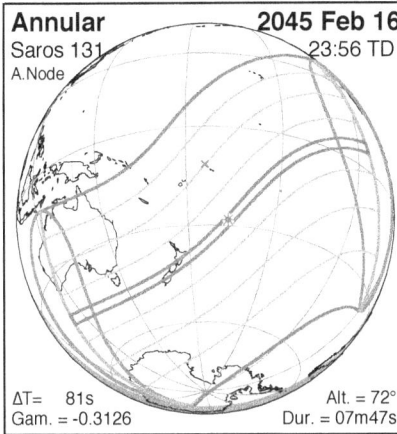

Annular   **2045 Feb 16**
Saros 131   23:56 TD
A.Node

ΔT= 81s   Alt. = 72°
Gam. = -0.3126   Dur. = 07m47s

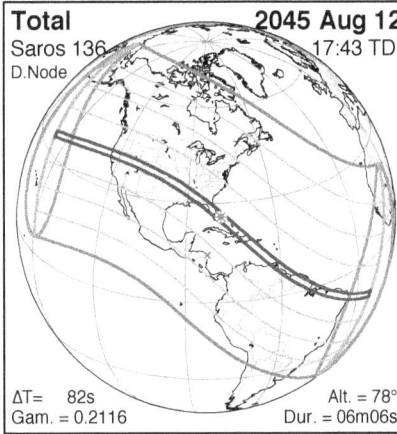

Total   **2045 Aug 12**
Saros 136   17:43 TD
D.Node

ΔT= 82s   Alt. = 78°
Gam. = 0.2116   Dur. = 06m06s

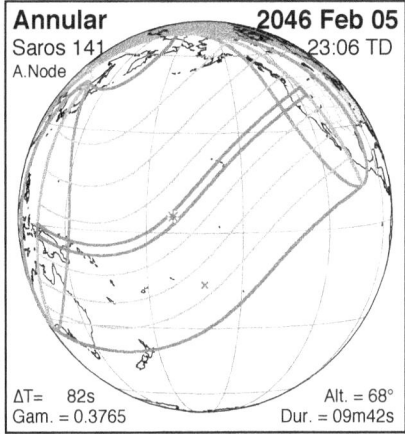

Annular   **2046 Feb 05**
Saros 141   23:06 TD
A.Node

ΔT= 82s   Alt. = 68°
Gam. = 0.3765   Dur. = 09m42s

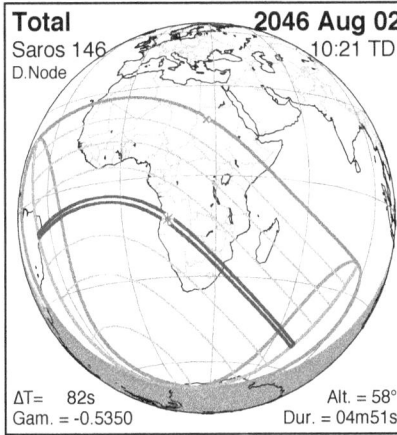

Total   **2046 Aug 02**
Saros 146   10:21 TD
D.Node

ΔT= 82s   Alt. = 58°
Gam. = -0.5350   Dur. = 04m51s

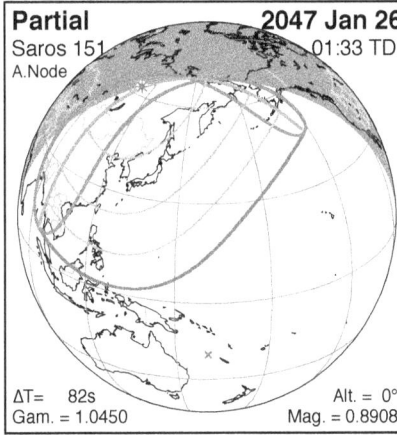

Partial   **2047 Jan 26**
Saros 151   01:33 TD
A.Node

ΔT= 82s   Alt. = 0°
Gam. = 1.0450   Mag. = 0.8908

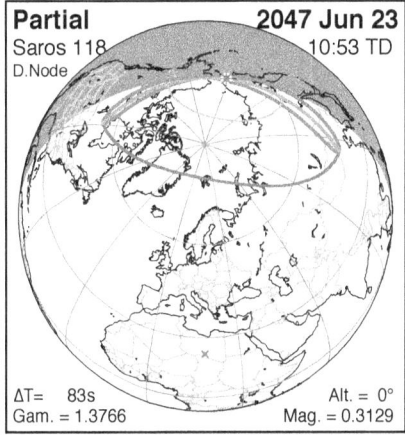

Partial   **2047 Jun 23**
Saros 118   10:53 TD
D.Node

ΔT= 83s   Alt. = 0°
Gam. = 1.3766   Mag. = 0.3129

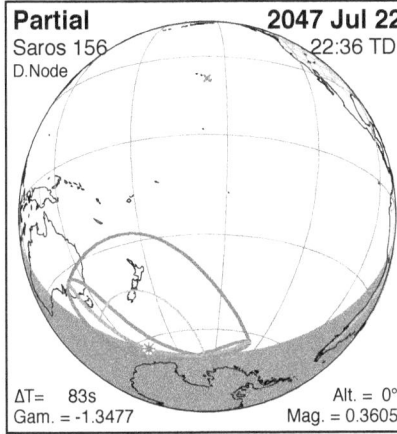

Partial   **2047 Jul 22**
Saros 156   22:36 TD
D.Node

ΔT= 83s   Alt. = 0°
Gam. = -1.3477   Mag. = 0.3605

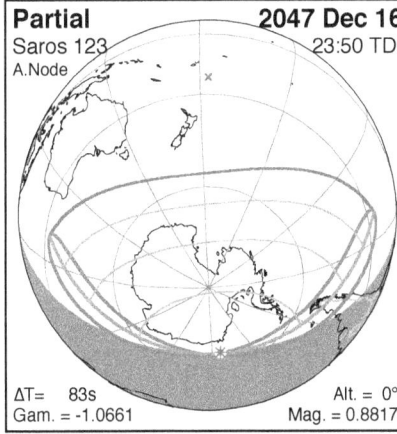

Partial   **2047 Dec 16**
Saros 123   23:50 TD
A.Node

ΔT= 83s   Alt. = 0°
Gam. = -1.0661   Mag. = 0.8817

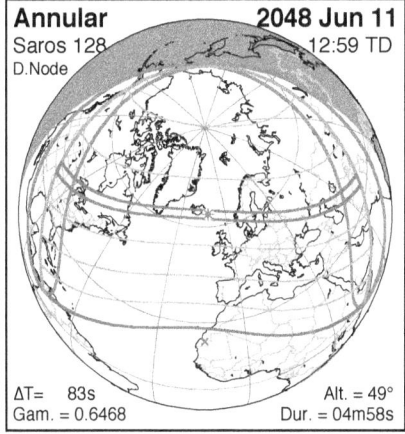

Annular   **2048 Jun 11**
Saros 128   12:59 TD
D.Node

ΔT= 83s   Alt. = 49°
Gam. = 0.6468   Dur. = 04m58s

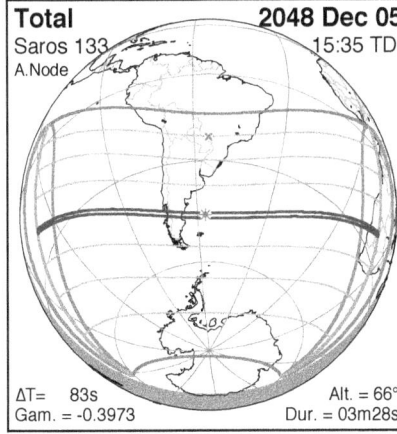

Total   **2048 Dec 05**
Saros 133   15:35 TD
A.Node

ΔT= 83s   Alt. = 66°
Gam. = -0.3973   Dur. = 03m28s

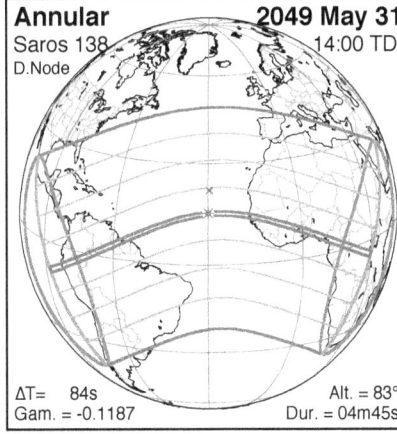

Annular   **2049 May 31**
Saros 138   14:00 TD
D.Node

ΔT= 84s   Alt. = 83°
Gam. = -0.1187   Dur. = 04m45s

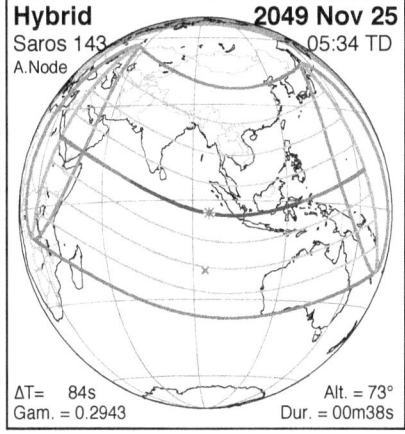

Hybrid   **2049 Nov 25**
Saros 143   05:34 TD
A.Node

ΔT= 84s   Alt. = 73°
Gam. = 0.2943   Dur. = 00m38s

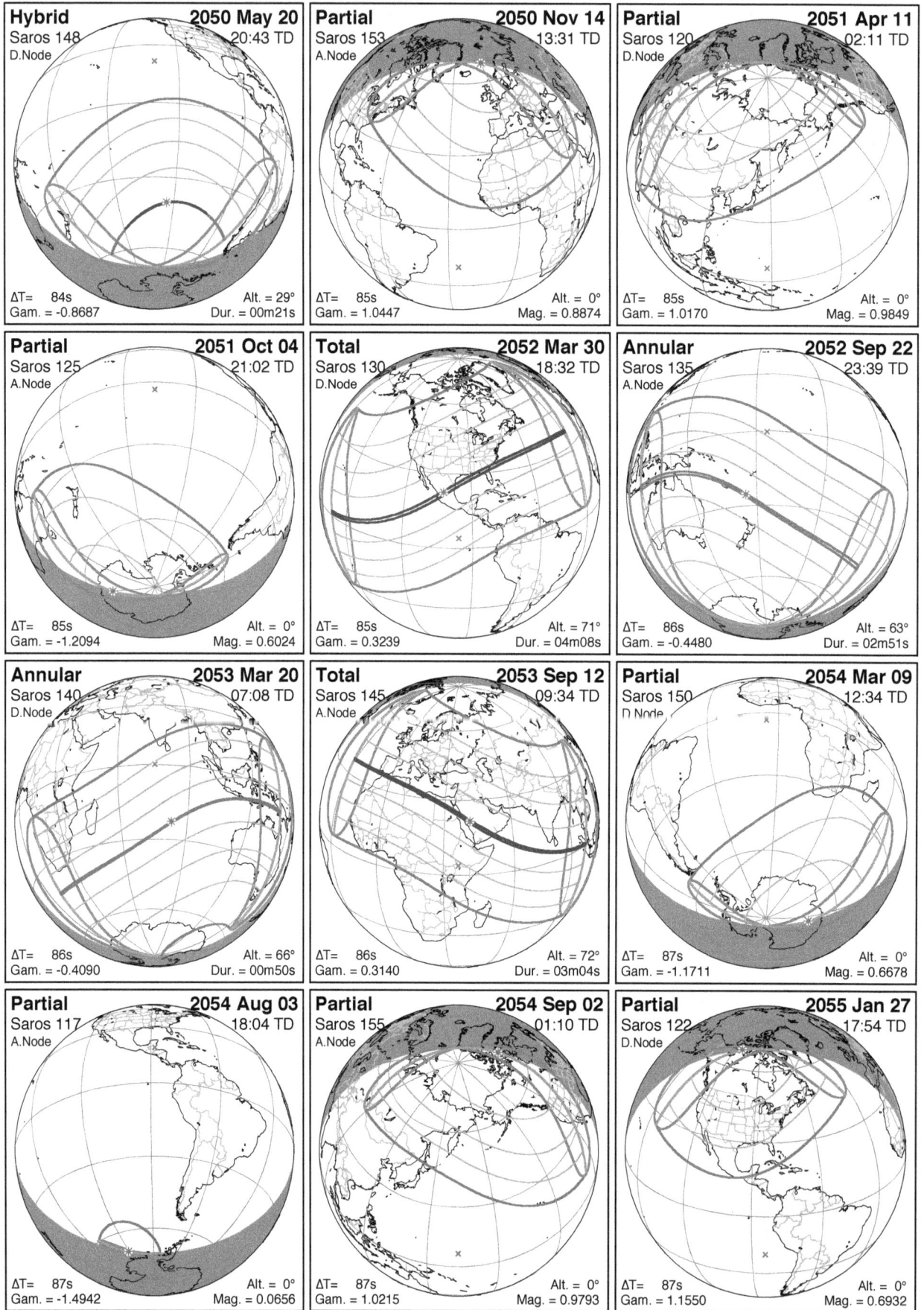

Hybrid **2050 May 20**	**Partial** **2050 Nov 14**	**Partial** **2051 Apr 11**
Saros 148 20:43 TD	Saros 153 13:31 TD	Saros 120 02:11 TD
D.Node	A.Node	D.Node
ΔT= 84s Alt. = 29°	ΔT= 85s Alt. = 0°	ΔT= 85s Alt. = 0°
Gam. = -0.8687 Dur. = 00m21s	Gam. = 1.0447 Mag. = 0.8874	Gam. = 1.0170 Mag. = 0.9849
Partial **2051 Oct 04**	**Total** **2052 Mar 30**	**Annular** **2052 Sep 22**
Saros 125 21:02 TD	Saros 130 18:32 TD	Saros 135 23:39 TD
A.Node	D.Node	A.Node
ΔT= 85s Alt. = 0°	ΔT= 85s Alt. = 71°	ΔT= 86s Alt. = 63°
Gam. = -1.2094 Mag. = 0.6024	Gam. = 0.3239 Dur. = 04m08s	Gam. = -0.4480 Dur. = 02m51s
Annular **2053 Mar 20**	**Total** **2053 Sep 12**	**Partial** **2054 Mar 09**
Saros 140 07:08 TD	Saros 145 09:34 TD	Saros 150 12:34 TD
D.Node	A.Node	D.Node
ΔT= 86s Alt. = 66°	ΔT= 86s Alt. = 72°	ΔT= 87s Alt. = 0°
Gam. = -0.4090 Dur. = 00m50s	Gam. = 0.3140 Dur. = 03m04s	Gam. = -1.1711 Mag. = 0.6678
Partial **2054 Aug 03**	**Partial** **2054 Sep 02**	**Partial** **2055 Jan 27**
Saros 117 18:04 TD	Saros 155 01:10 TD	Saros 122 17:54 TD
A.Node	A.Node	D.Node
ΔT= 87s Alt. = 0°	ΔT= 87s Alt. = 0°	ΔT= 87s Alt. = 0°
Gam. = -1.4942 Mag. = 0.0656	Gam. = 1.0215 Mag. = 0.9793	Gam. = 1.1550 Mag. = 0.6932

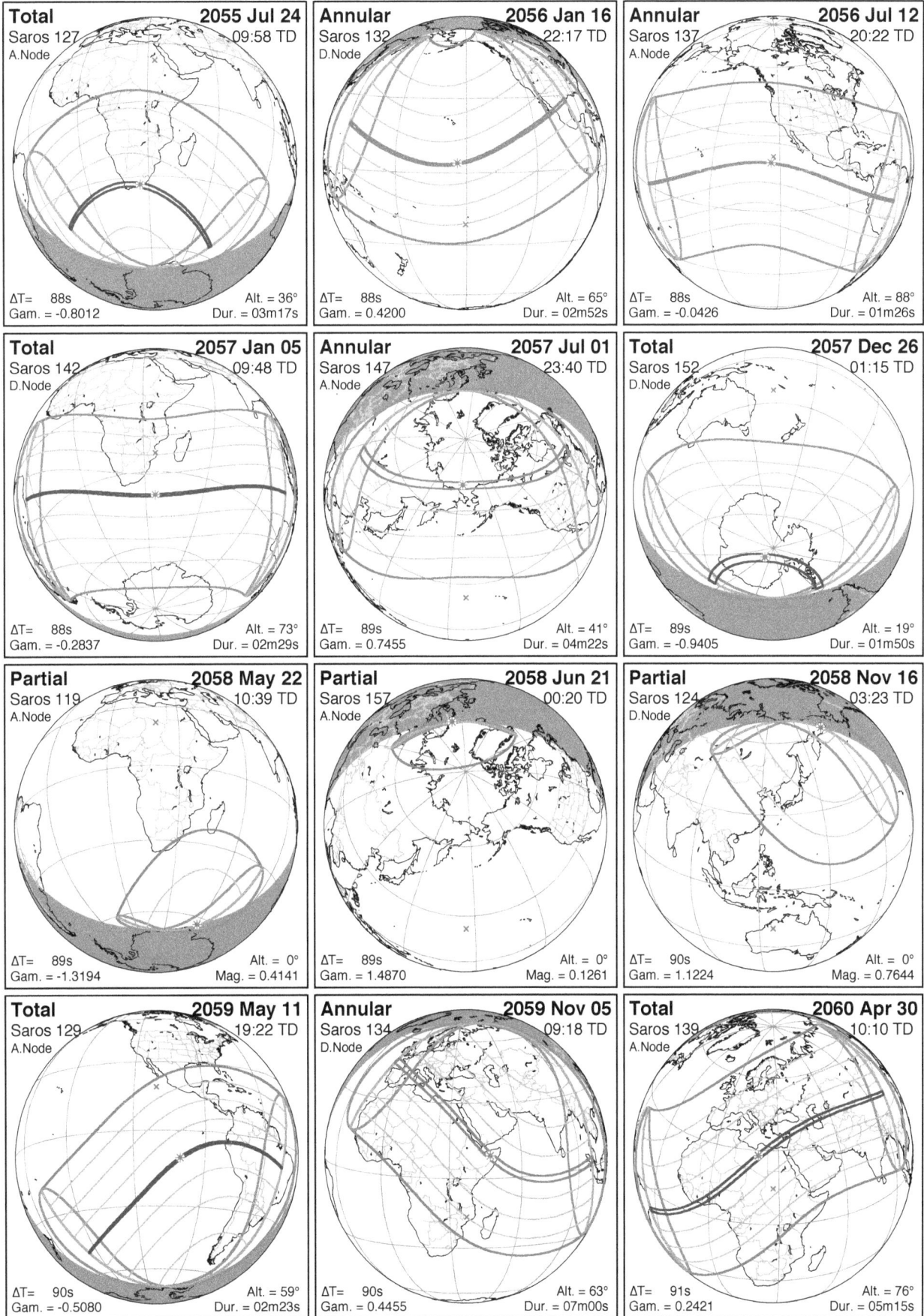

Total	2055 Jul 24
Saros 127	09:58 TD
A.Node	
ΔT= 88s	Alt. = 36°
Gam. = -0.8012	Dur. = 03m17s

Annular	2056 Jan 16
Saros 132	22:17 TD
D.Node	
ΔT= 88s	Alt. = 65°
Gam. = 0.4200	Dur. = 02m52s

Annular	2056 Jul 12
Saros 137	20:22 TD
A.Node	
ΔT= 88s	Alt. = 88°
Gam. = -0.0426	Dur. = 01m26s

Total	2057 Jan 05
Saros 142	09:48 TD
D.Node	
ΔT= 88s	Alt. = 73°
Gam. = -0.2837	Dur. = 02m29s

Annular	2057 Jul 01
Saros 147	23:40 TD
A.Node	
ΔT= 89s	Alt. = 41°
Gam. = 0.7455	Dur. = 04m22s

Total	2057 Dec 26
Saros 152	01:15 TD
D.Node	
ΔT= 89s	Alt. = 19°
Gam. = -0.9405	Dur. = 01m50s

Partial	2058 May 22
Saros 119	10:39 TD
A.Node	
ΔT= 89s	Alt. = 0°
Gam. = -1.3194	Mag. = 0.4141

Partial	2058 Jun 21
Saros 157	00:20 TD
A.Node	
ΔT= 89s	Alt. = 0°
Gam. = 1.4870	Mag. = 0.1261

Partial	2058 Nov 16
Saros 124	03:23 TD
D.Node	
ΔT= 90s	Alt. = 0°
Gam. = 1.1224	Mag. = 0.7644

Total	2059 May 11
Saros 129	19:22 TD
A.Node	
ΔT= 90s	Alt. = 59°
Gam. = -0.5080	Dur. = 02m23s

Annular	2059 Nov 05
Saros 134	09:18 TD
D.Node	
ΔT= 90s	Alt. = 63°
Gam. = 0.4455	Dur. = 07m00s

Total	2060 Apr 30
Saros 139	10:10 TD
A.Node	
ΔT= 91s	Alt. = 76°
Gam. = 0.2421	Dur. = 05m15s

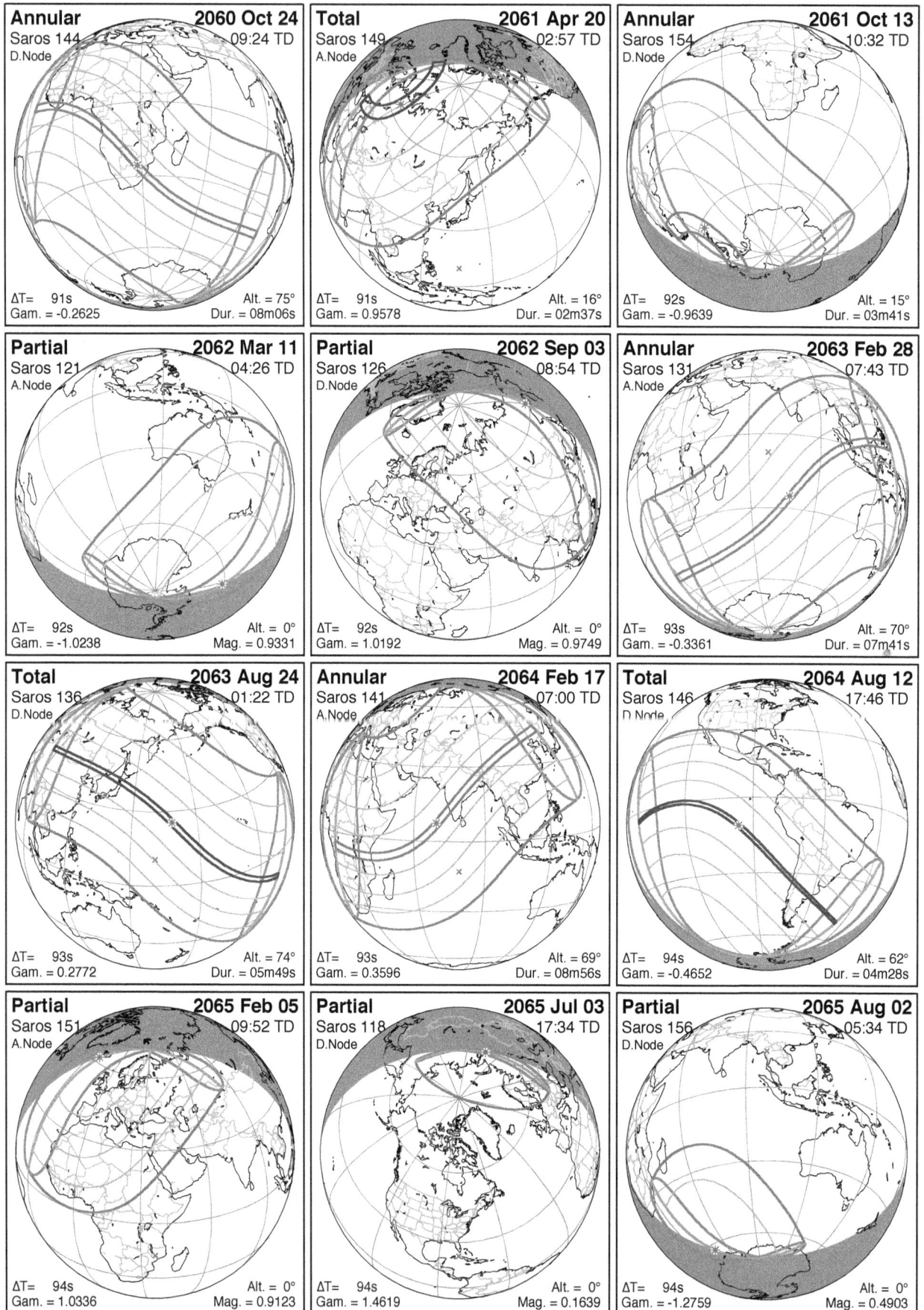

Annular **2060 Oct 24** Saros 144 09:24 TD D.Node ΔT= 91s Alt. = 75° Gam. = -0.2625 Dur. = 08m06s	**Total** **2061 Apr 20** Saros 149 02:57 TD A.Node ΔT= 91s Alt. = 16° Gam. = 0.9578 Dur. = 02m37s	**Annular** **2061 Oct 13** Saros 154 10:32 TD D.Node ΔT= 92s Alt. = 15° Gam. = -0.9639 Dur. = 03m41s
Partial **2062 Mar 11** Saros 121 04:26 TD A.Node ΔT= 92s Alt. = 0° Gam. = -1.0238 Mag. = 0.9331	**Partial** **2062 Sep 03** Saros 126 08:54 TD D.Node ΔT= 92s Alt. = 0° Gam. = 1.0192 Mag. = 0.9749	**Annular** **2063 Feb 28** Saros 131 07:43 TD A.Node ΔT= 93s Alt. = 70° Gam. = -0.3361 Dur. = 07m41s
Total **2063 Aug 24** Saros 136 01:22 TD D.Node ΔT= 93s Alt. = 74° Gam. = 0.2772 Dur. = 05m49s	**Annular** **2064 Feb 17** Saros 141 07:00 TD A.Node ΔT= 93s Alt. = 69° Gam. = 0.3596 Dur. = 08m56s	**Total** **2064 Aug 12** Saros 146 17:46 TD D.Node ΔT= 94s Alt. = 62° Gam. = -0.4652 Dur. = 04m28s
Partial **2065 Feb 05** Saros 151 09:52 TD A.Node ΔT= 94s Alt. = 0° Gam. = 1.0336 Mag. = 0.9123	**Partial** **2065 Jul 03** Saros 118 17:34 TD D.Node ΔT= 94s Alt. = 0° Gam. = 1.4619 Mag. = 0.1639	**Partial** **2065 Aug 02** Saros 156 05:34 TD D.Node ΔT= 94s Alt. = 0° Gam. = -1.2759 Mag. = 0.4903

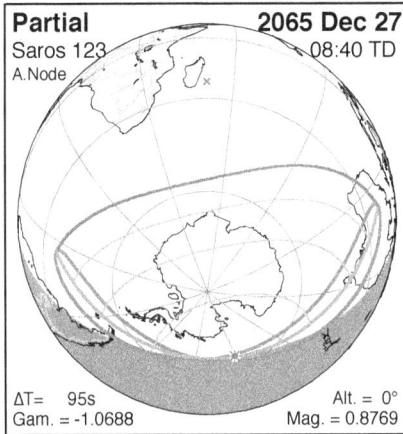

Partial   **2065 Dec 27**
Saros 123   08:40 TD
A.Node

ΔT= 95s   Alt. = 0°
Gam. = -1.0688   Mag. = 0.8769

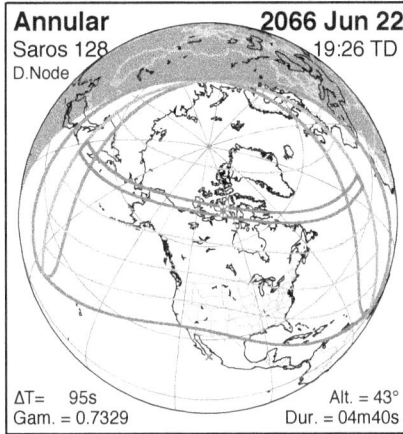

Annular   **2066 Jun 22**
Saros 128   19:26 TD
D.Node

ΔT= 95s   Alt. = 43°
Gam. = 0.7329   Dur. = 04m40s

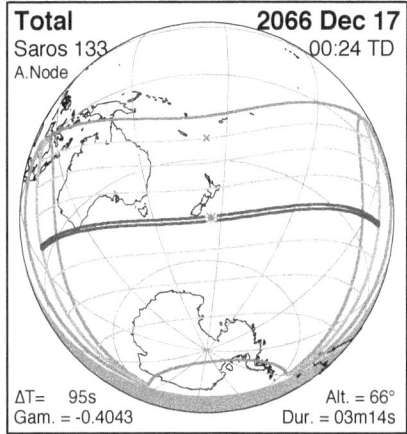

Total   **2066 Dec 17**
Saros 133   00:24 TD
A.Node

ΔT= 95s   Alt. = 66°
Gam. = -0.4043   Dur. = 03m14s

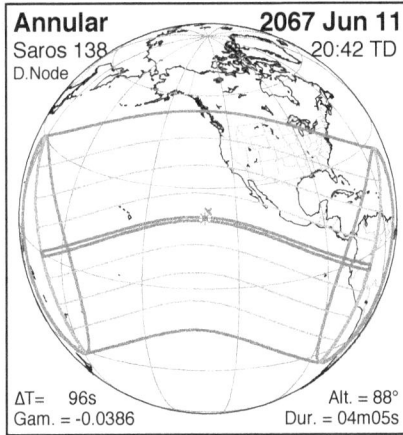

Annular   **2067 Jun 11**
Saros 138   20:42 TD
D.Node

ΔT= 96s   Alt. = 88°
Gam. = -0.0386   Dur. = 04m05s

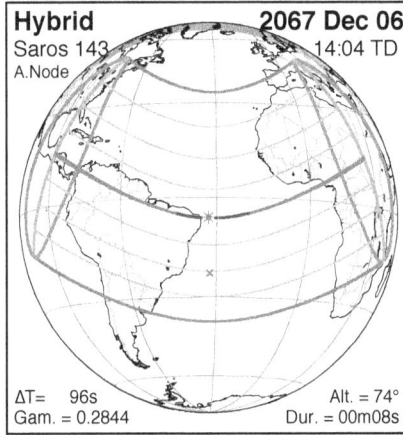

Hybrid   **2067 Dec 06**
Saros 143   14:04 TD
A.Node

ΔT= 96s   Alt. = 74°
Gam. = 0.2844   Dur. = 00m08s

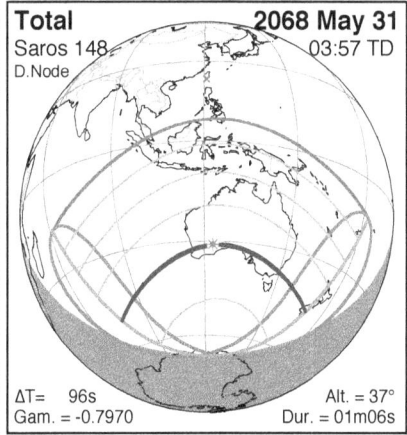

Total   **2068 May 31**
Saros 148   03:57 TD
D.Node

ΔT= 96s   Alt. = 37°
Gam. = -0.7970   Dur. = 01m06s

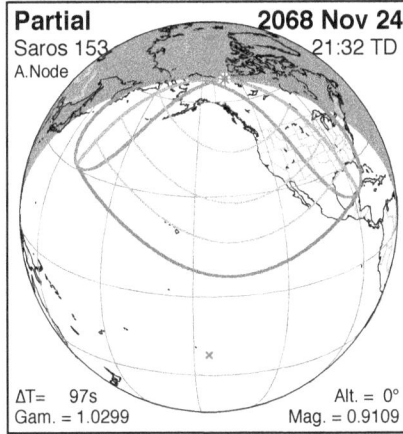

Partial   **2068 Nov 24**
Saros 153   21:32 TD
A.Node

ΔT= 97s   Alt. = 0°
Gam. = 1.0299   Mag. = 0.9109

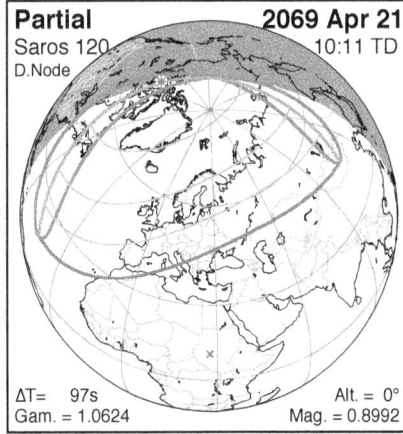

Partial   **2069 Apr 21**
Saros 120   10:11 TD
D.Node

ΔT= 97s   Alt. = 0°
Gam. = 1.0624   Mag. = 0.8992

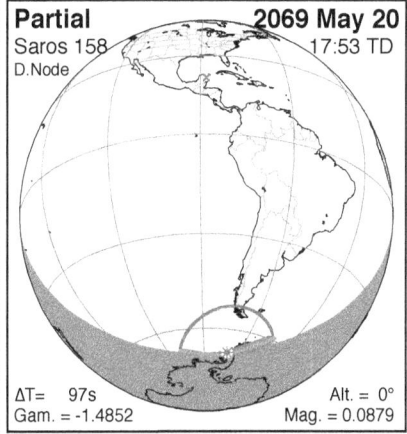

Partial   **2069 May 20**
Saros 158   17:53 TD
D.Node

ΔT= 97s   Alt. = 0°
Gam. = -1.4852   Mag. = 0.0879

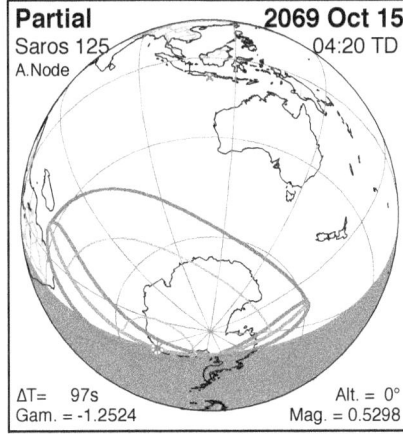

Partial   **2069 Oct 15**
Saros 125   04:20 TD
A.Node

ΔT= 97s   Alt. = 0°
Gam. = -1.2524   Mag. = 0.5298

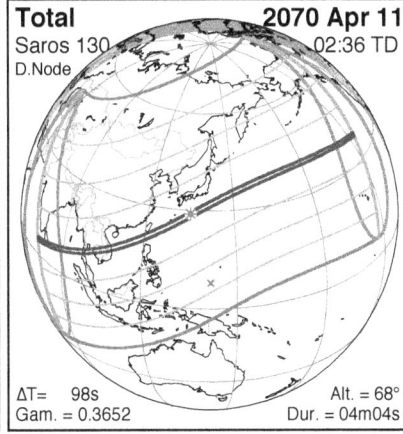

Total   **2070 Apr 11**
Saros 130   02:36 TD
D.Node

ΔT= 98s   Alt. = 68°
Gam. = 0.3652   Dur. = 04m04s

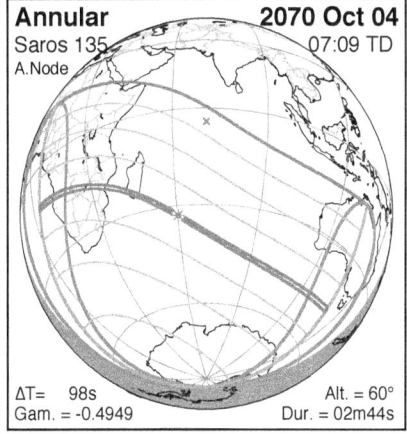

Annular   **2070 Oct 04**
Saros 135   07:09 TD
A.Node

ΔT= 98s   Alt. = 60°
Gam. = -0.4949   Dur. = 02m44s

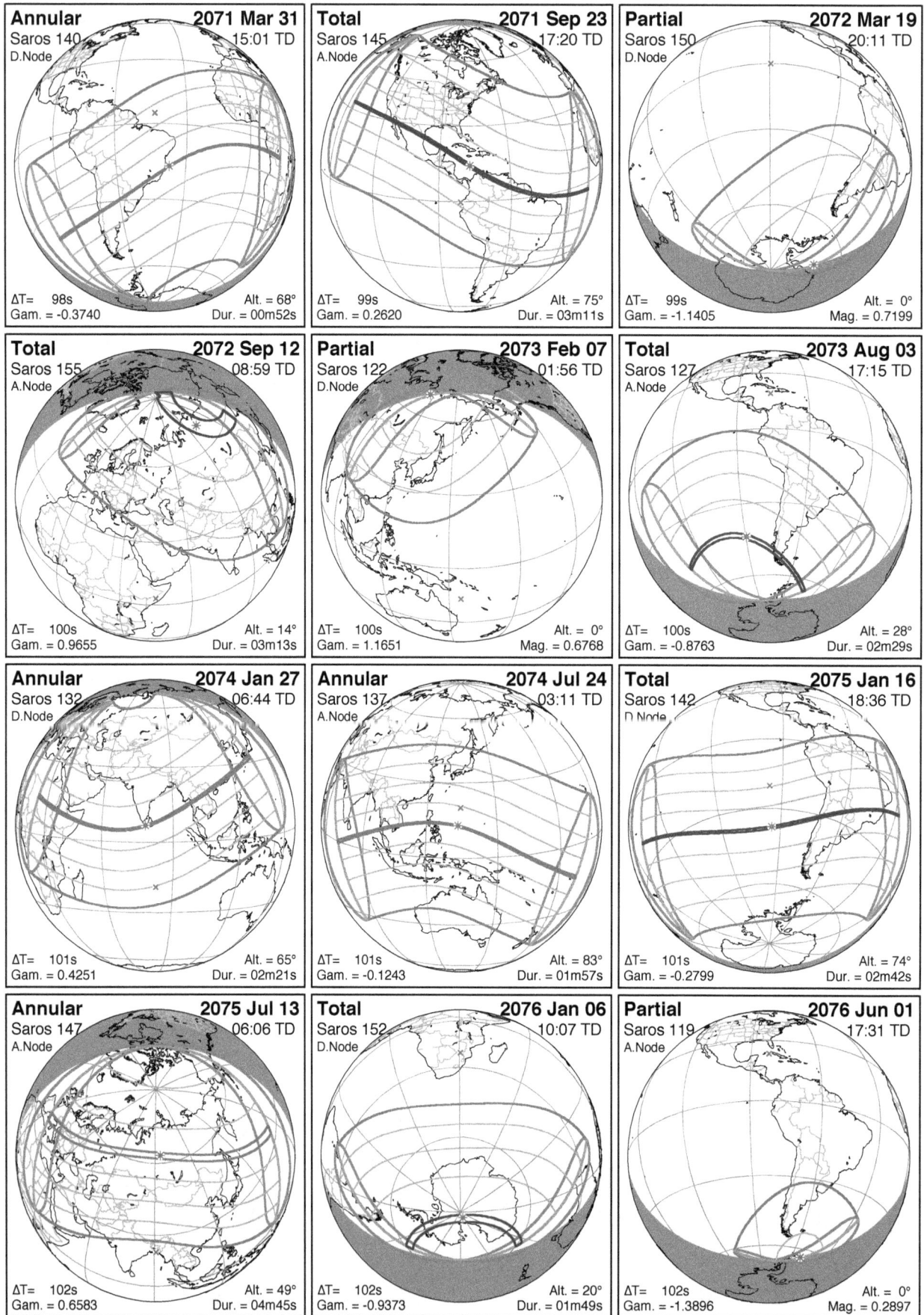

Annular	2071 Mar 31
Saros 140	15:01 TD
D.Node	
ΔT= 98s	Alt. = 68°
Gam. = -0.3740	Dur. = 00m52s

Total	2071 Sep 23
Saros 145	17:20 TD
A.Node	
ΔT= 99s	Alt. = 75°
Gam. = 0.2620	Dur. = 03m11s

Partial	2072 Mar 19
Saros 150	20:11 TD
D.Node	
ΔT= 99s	Alt. = 0°
Gam. = -1.1405	Mag. = 0.7199

Total	2072 Sep 12
Saros 155	08:59 TD
A.Node	
ΔT= 100s	Alt. = 14°
Gam. = 0.9655	Dur. = 03m13s

Partial	2073 Feb 07
Saros 122	01:56 TD
D.Node	
ΔT= 100s	Alt. = 0°
Gam. = 1.1651	Mag. = 0.6768

Total	2073 Aug 03
Saros 127	17:15 TD
A.Node	
ΔT= 100s	Alt. = 28°
Gam. = -0.8763	Dur. = 02m29s

Annular	2074 Jan 27
Saros 132	06:44 TD
D.Node	
ΔT= 101s	Alt. = 65°
Gam. = 0.4251	Dur. = 02m21s

Annular	2074 Jul 24
Saros 137	03:11 TD
A.Node	
ΔT= 101s	Alt. = 83°
Gam. = -0.1243	Dur. = 01m57s

Total	2075 Jan 16
Saros 142	18:36 TD
D.Node	
ΔT= 101s	Alt. = 74°
Gam. = -0.2799	Dur. = 02m42s

Annular	2075 Jul 13
Saros 147	06:06 TD
A.Node	
ΔT= 102s	Alt. = 49°
Gam. = 0.6583	Dur. = 04m45s

Total	2076 Jan 06
Saros 152	10:07 TD
D.Node	
ΔT= 102s	Alt. = 20°
Gam. = -0.9373	Dur. = 01m49s

Partial	2076 Jun 01
Saros 119	17:31 TD
A.Node	
ΔT= 102s	Alt. = 0°
Gam. = -1.3896	Mag. = 0.2897

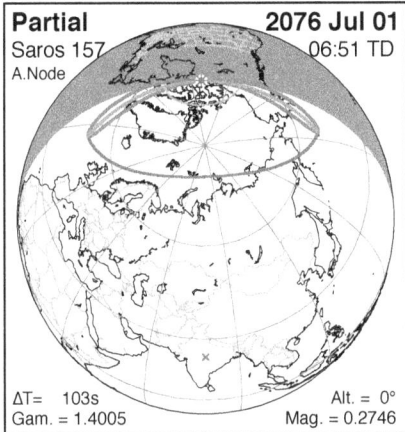

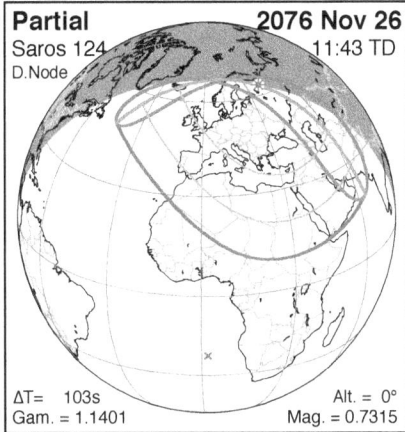

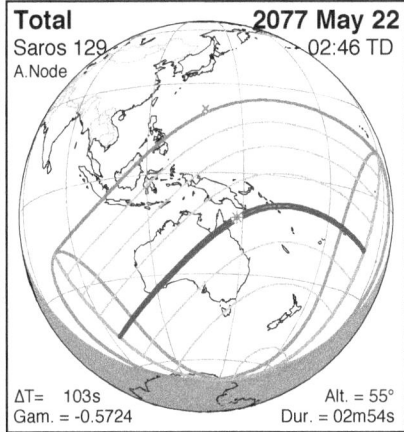

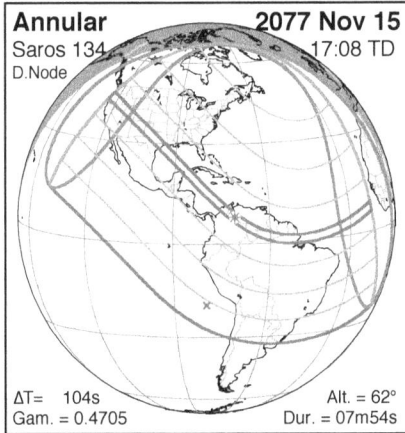

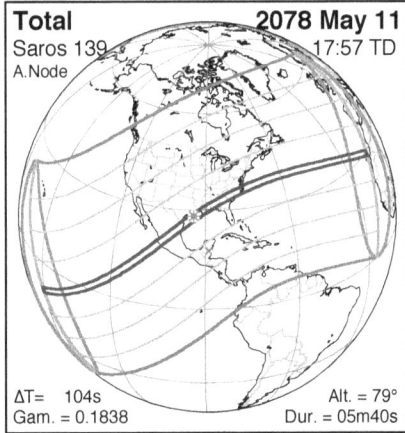

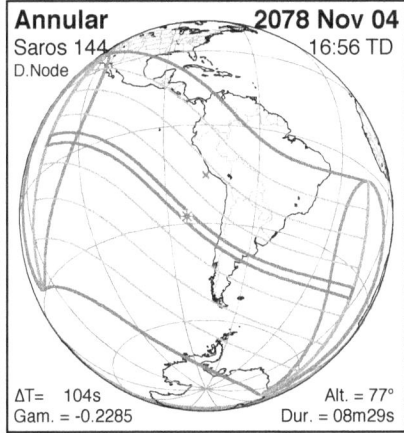

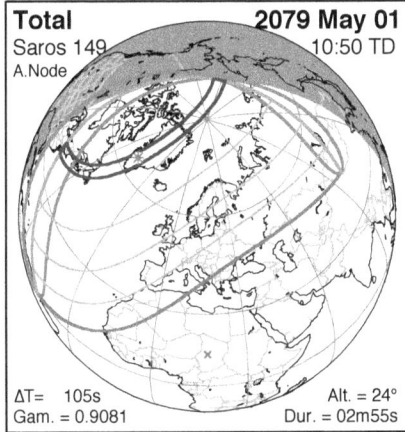

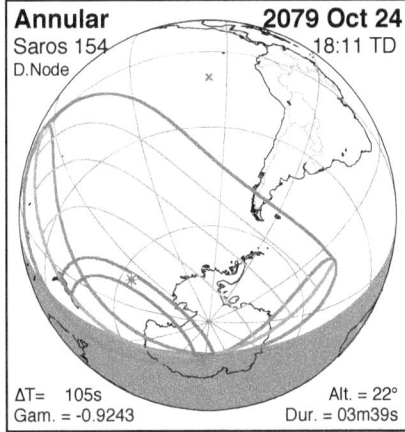

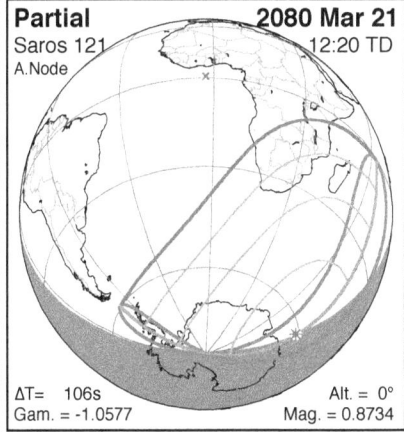

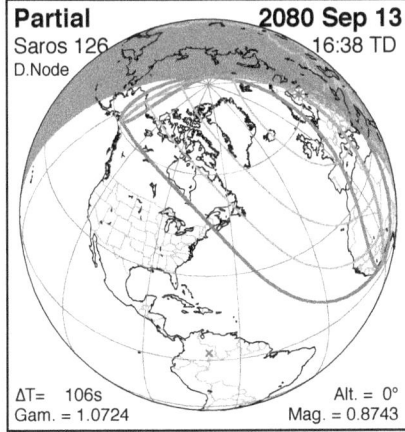

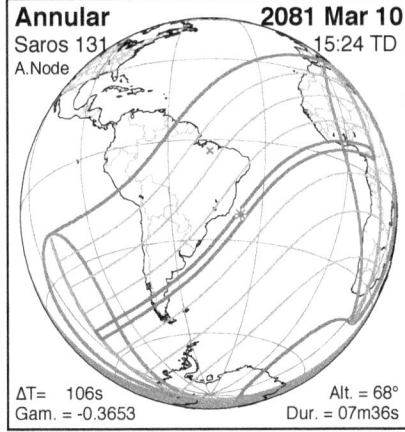

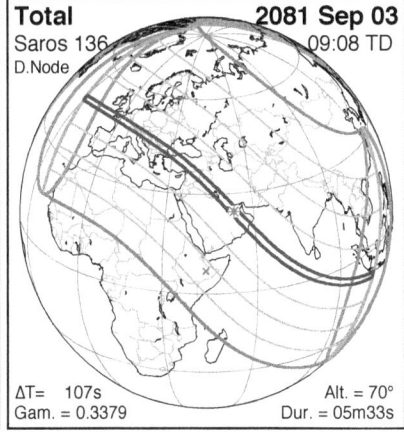

Partial	2076 Jul 01
Saros 157	06:51 TD
A.Node	
ΔT= 103s	Alt. = 0°
Gam. = 1.4005	Mag. = 0.2746

Partial	2076 Nov 26
Saros 124	11:43 TD
D.Node	
ΔT= 103s	Alt. = 0°
Gam. = 1.1401	Mag. = 0.7315

Total	2077 May 22
Saros 129	02:46 TD
A.Node	
ΔT= 103s	Alt. = 55°
Gam. = -0.5724	Dur. = 02m54s

Annular	2077 Nov 15
Saros 134	17:08 TD
D.Node	
ΔT= 104s	Alt. = 62°
Gam. = 0.4705	Dur. = 07m54s

Total	2078 May 11
Saros 139	17:57 TD
A.Node	
ΔT= 104s	Alt. = 79°
Gam. = 0.1838	Dur. = 05m40s

Annular	2078 Nov 04
Saros 144	16:56 TD
D.Node	
ΔT= 104s	Alt. = 77°
Gam. = -0.2285	Dur. = 08m29s

Total	2079 May 01
Saros 149	10:50 TD
A.Node	
ΔT= 105s	Alt. = 24°
Gam. = 0.9081	Dur. = 02m55s

Annular	2079 Oct 24
Saros 154	18:11 TD
D.Node	
ΔT= 105s	Alt. = 22°
Gam. = -0.9243	Dur. = 03m39s

Partial	2080 Mar 21
Saros 121	12:20 TD
A.Node	
ΔT= 106s	Alt. = 0°
Gam. = -1.0577	Mag. = 0.8734

Partial	2080 Sep 13
Saros 126	16:38 TD
D.Node	
ΔT= 106s	Alt. = 0°
Gam. = 1.0724	Mag. = 0.8743

Annular	2081 Mar 10
Saros 131	15:24 TD
A.Node	
ΔT= 106s	Alt. = 68°
Gam. = -0.3653	Dur. = 07m36s

Total	2081 Sep 03
Saros 136	09:08 TD
D.Node	
ΔT= 107s	Alt. = 70°
Gam. = 0.3379	Dur. = 05m33s

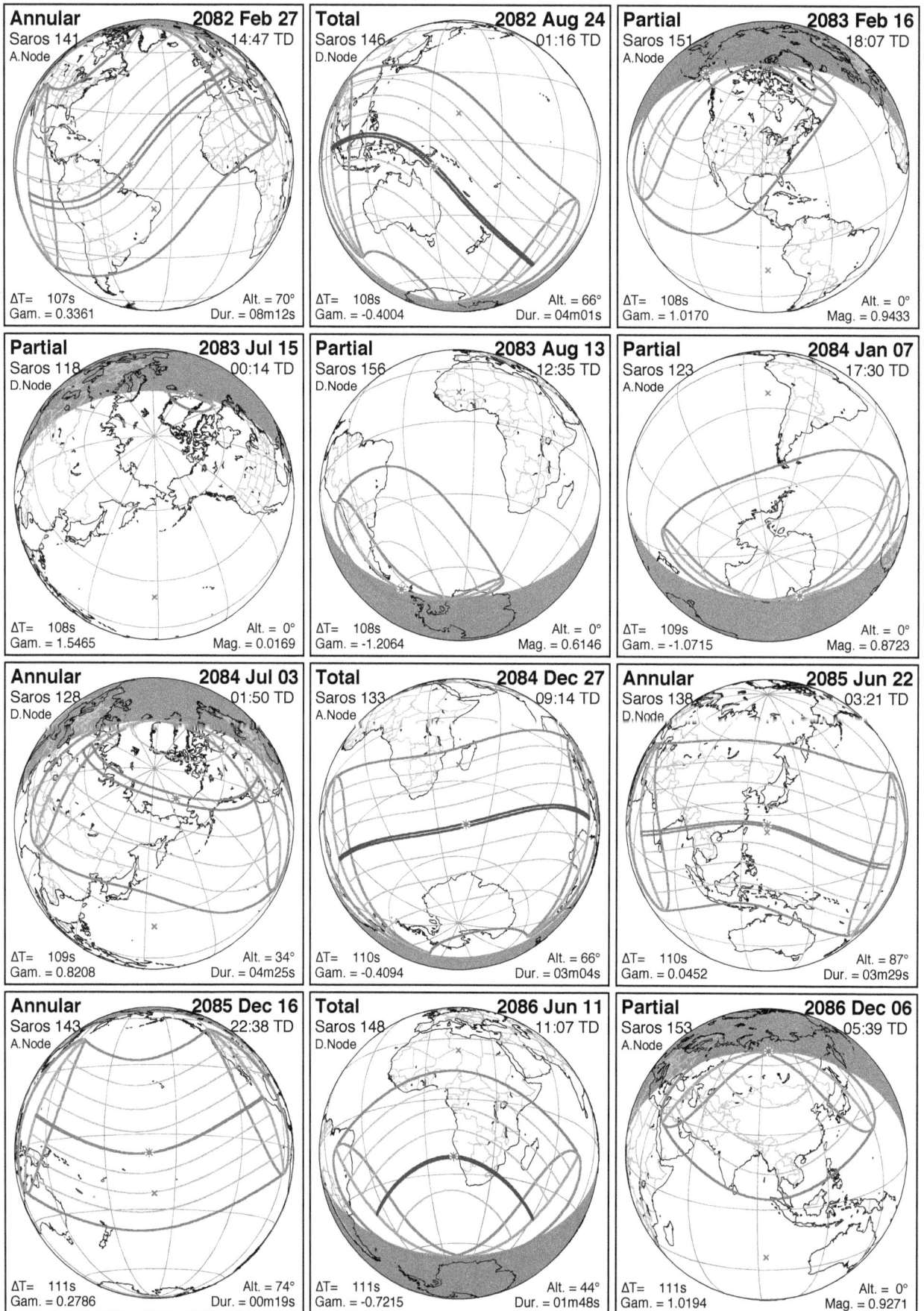

Annular 2082 Feb 27	**Total** 2082 Aug 24	**Partial** 2083 Feb 16
Saros 141 14:47 TD	Saros 146 01:16 TD	Saros 151 18:07 TD
A.Node	D.Node	A.Node
ΔT= 107s Alt. = 70°	ΔT= 108s Alt. = 66°	ΔT= 108s Alt. = 0°
Gam. = 0.3361 Dur. = 08m12s	Gam. = -0.4004 Dur. = 04m01s	Gam. = 1.0170 Mag. = 0.9433

Partial 2083 Jul 15	**Partial** 2083 Aug 13	**Partial** 2084 Jan 07
Saros 118 00:14 TD	Saros 156 12:35 TD	Saros 123 17:30 TD
D.Node	D.Node	A.Node
ΔT= 108s Alt. = 0°	ΔT= 108s Alt. = 0°	ΔT= 109s Alt. = 0°
Gam. = 1.5465 Mag. = 0.0169	Gam. = -1.2064 Mag. = 0.6146	Gam. = -1.0715 Mag. = 0.8723

Annular 2084 Jul 03	**Total** 2084 Dec 27	**Annular** 2085 Jun 22
Saros 128 01:50 TD	Saros 133 09:14 TD	Saros 138 03:21 TD
D.Node	A.Node	D.Node
ΔT= 109s Alt. = 34°	ΔT= 110s Alt. = 66°	ΔT= 110s Alt. = 87°
Gam. = 0.8208 Dur. = 04m25s	Gam. = -0.4094 Dur. = 03m04s	Gam. = 0.0452 Dur. = 03m29s

Annular 2085 Dec 16	**Total** 2086 Jun 11	**Partial** 2086 Dec 06
Saros 143 22:38 TD	Saros 148 11:07 TD	Saros 153 05:39 TD
A.Node	D.Node	A.Node
ΔT= 111s Alt. = 74°	ΔT= 111s Alt. = 44°	ΔT= 111s Alt. = 0°
Gam. = 0.2786 Dur. = 00m19s	Gam. = -0.7215 Dur. = 01m48s	Gam. = 1.0194 Mag. = 0.9271

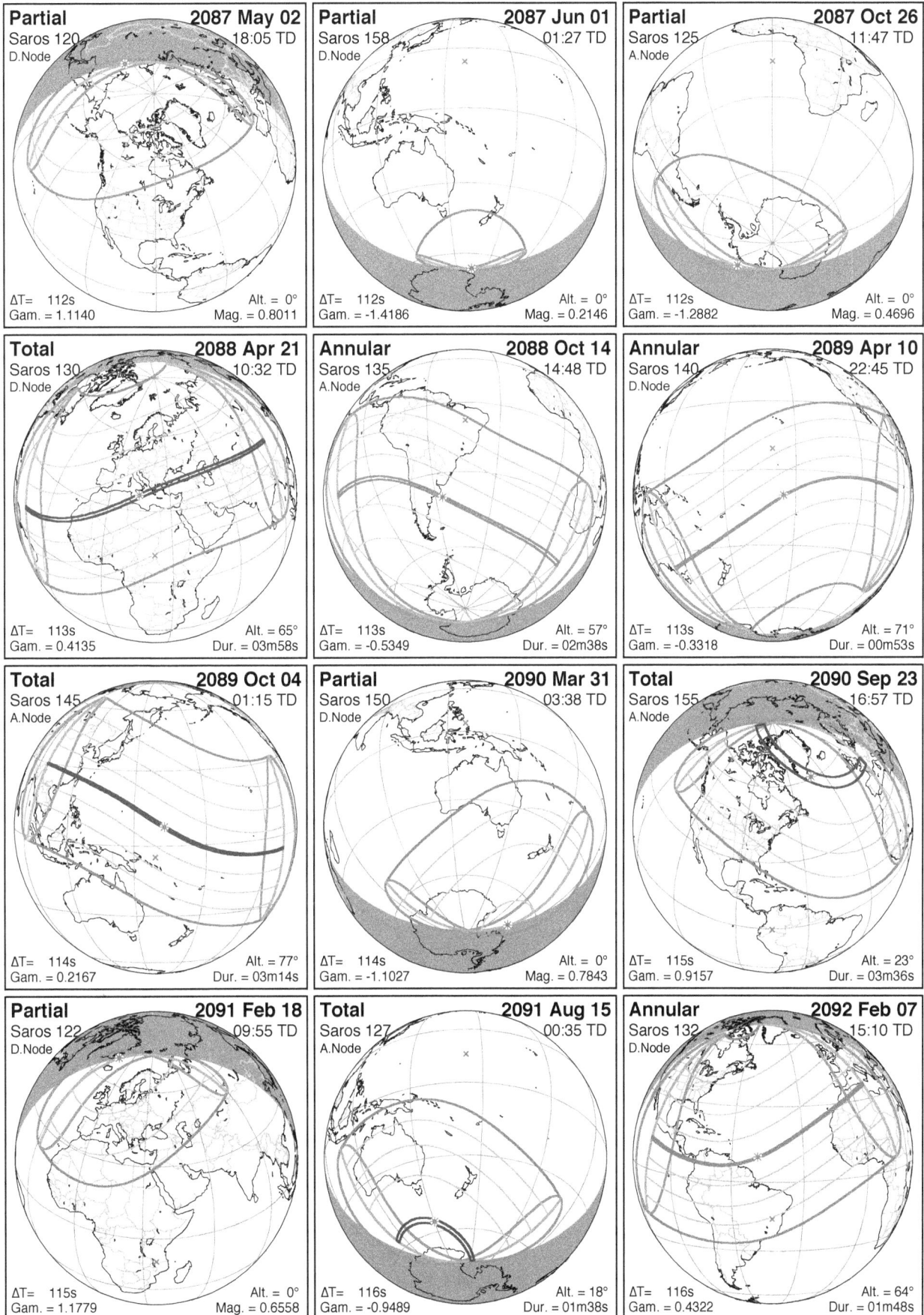

Partial	**2087 May 02**
Saros 120	18:05 TD
D.Node	
ΔT= 112s	Alt. = 0°
Gam. = 1.1140	Mag. = 0.8011

Partial	**2087 Jun 01**
Saros 158	01:27 TD
D.Node	
ΔT= 112s	Alt. = 0°
Gam. = -1.4186	Mag. = 0.2146

Partial	**2087 Oct 26**
Saros 125	11:47 TD
A.Node	
ΔT= 112s	Alt. = 0°
Gam. = -1.2882	Mag. = 0.4696

Total	**2088 Apr 21**
Saros 130	10:32 TD
D.Node	
ΔT= 113s	Alt. = 65°
Gam. = 0.4135	Dur. = 03m58s

Annular	**2088 Oct 14**
Saros 135	14:48 TD
A.Node	
ΔT= 113s	Alt. = 57°
Gam. = -0.5349	Dur. = 02m38s

Annular	**2089 Apr 10**
Saros 140	22:45 TD
D.Node	
ΔT= 113s	Alt. = 71°
Gam. = -0.3318	Dur. = 00m53s

Total	**2089 Oct 04**
Saros 145	01:15 TD
A.Node	
ΔT= 114s	Alt. = 77°
Gam. = 0.2167	Dur. = 03m14s

Partial	**2090 Mar 31**
Saros 150	03:38 TD
D.Node	
ΔT= 114s	Alt. = 0°
Gam. = -1.1027	Mag. = 0.7843

Total	**2090 Sep 23**
Saros 155	16:57 TD
A.Node	
ΔT= 115s	Alt. = 23°
Gam. = 0.9157	Dur. = 03m36s

Partial	**2091 Feb 18**
Saros 122	09:55 TD
D.Node	
ΔT= 115s	Alt. = 0°
Gam. = 1.1779	Mag. = 0.6558

Total	**2091 Aug 15**
Saros 127	00:35 TD
A.Node	
ΔT= 116s	Alt. = 18°
Gam. = -0.9489	Dur. = 01m38s

Annular	**2092 Feb 07**
Saros 132	15:10 TD
D.Node	
ΔT= 116s	Alt. = 64°
Gam. = 0.4322	Dur. = 01m48s

51

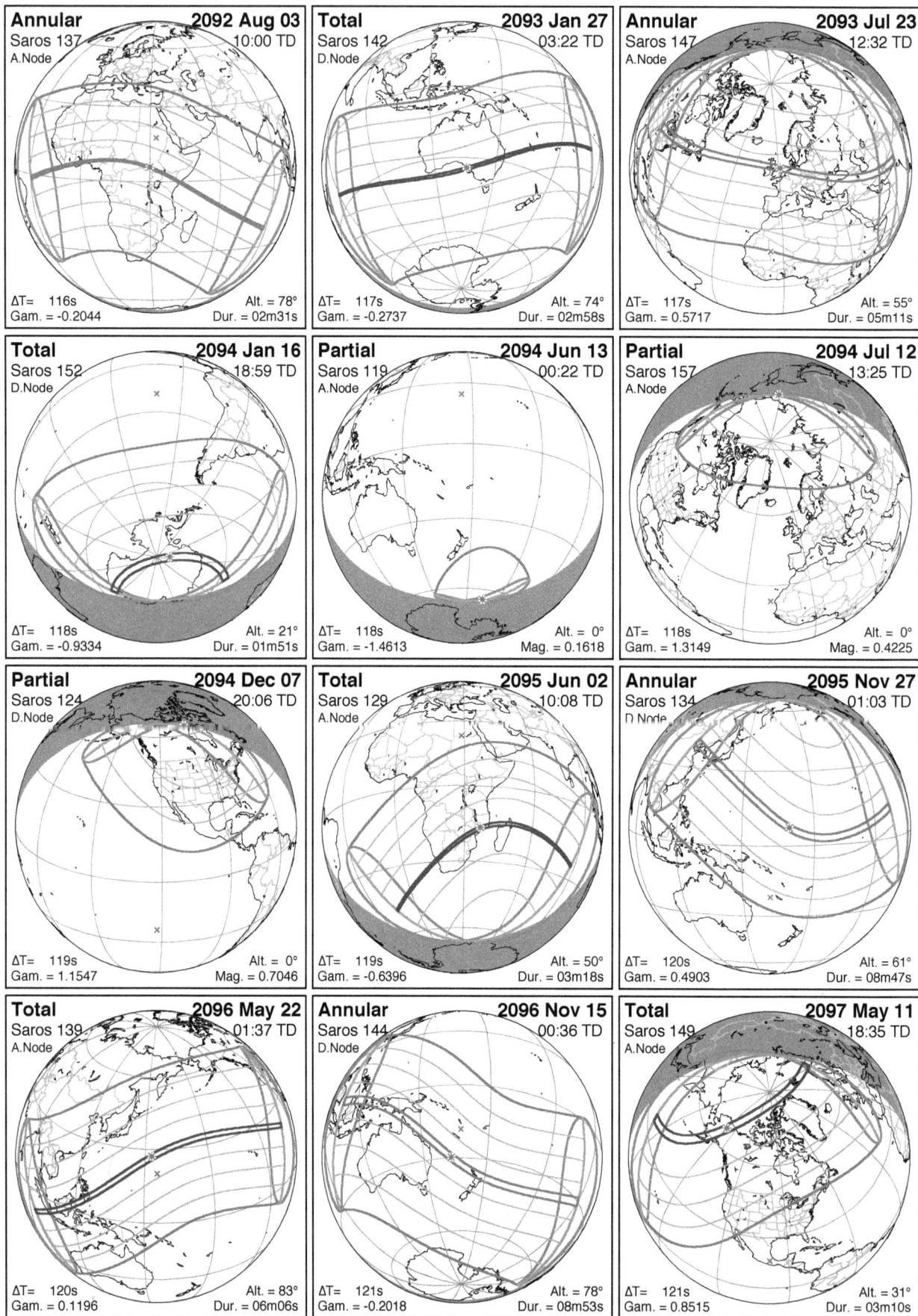

Annular	2092 Aug 03
Saros 137	10:00 TD
A.Node	
ΔT= 116s	Alt. = 78°
Gam. = -0.2044	Dur. = 02m31s

Total	2093 Jan 27
Saros 142	03:22 TD
D.Node	
ΔT= 117s	Alt. = 74°
Gam. = -0.2737	Dur. = 02m58s

Annular	2093 Jul 23
Saros 147	12:32 TD
A.Node	
ΔT= 117s	Alt. = 55°
Gam. = 0.5717	Dur. = 05m11s

Total	2094 Jan 16
Saros 152	18:59 TD
D.Node	
ΔT= 118s	Alt. = 21°
Gam. = -0.9334	Dur. = 01m51s

Partial	2094 Jun 13
Saros 119	00:22 TD
A.Node	
ΔT= 118s	Alt. = 0°
Gam. = -1.4613	Mag. = 0.1618

Partial	2094 Jul 12
Saros 157	13:25 TD
A.Node	
ΔT= 118s	Alt. = 0°
Gam. = 1.3149	Mag. = 0.4225

Partial	2094 Dec 07
Saros 124	20:06 TD
D.Node	
ΔT= 119s	Alt. = 0°
Gam. = 1.1547	Mag. = 0.7046

Total	2095 Jun 02
Saros 129	10:08 TD
A.Node	
ΔT= 119s	Alt. = 50°
Gam. = -0.6396	Dur. = 03m18s

Annular	2095 Nov 27
Saros 134	01:03 TD
D.Node	
ΔT= 120s	Alt. = 61°
Gam. = 0.4903	Dur. = 08m47s

Total	2096 May 22
Saros 139	01:37 TD
A.Node	
ΔT= 120s	Alt. = 83°
Gam. = 0.1196	Dur. = 06m06s

Annular	2096 Nov 15
Saros 144	00:36 TD
D.Node	
ΔT= 121s	Alt. = 78°
Gam. = -0.2018	Dur. = 08m53s

Total	2097 May 11
Saros 149	18:35 TD
A.Node	
ΔT= 121s	Alt. = 31°
Gam. = 0.8515	Dur. = 03m10s

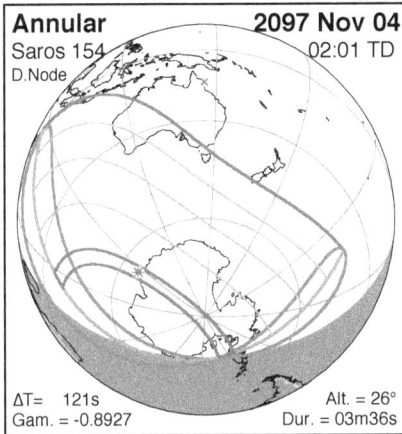

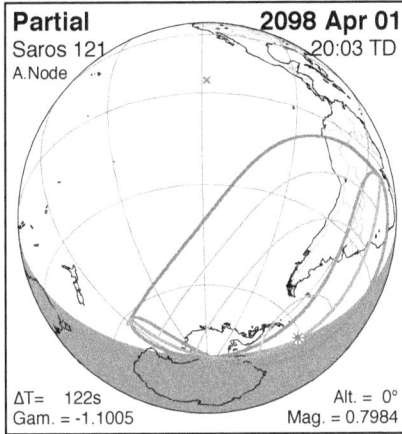

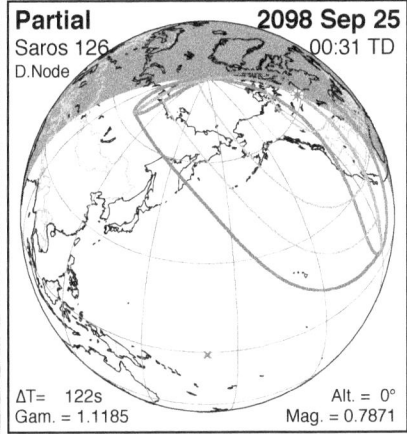

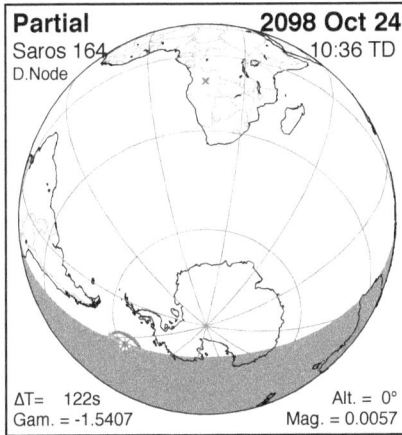

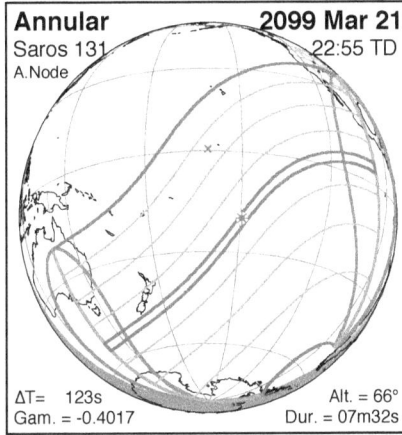

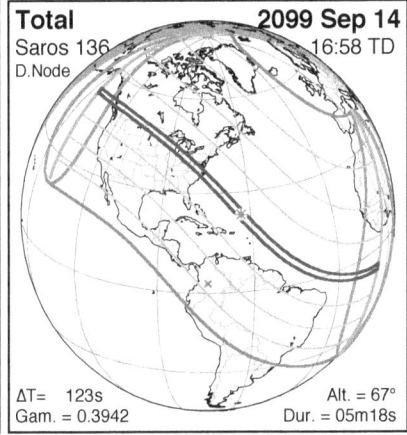

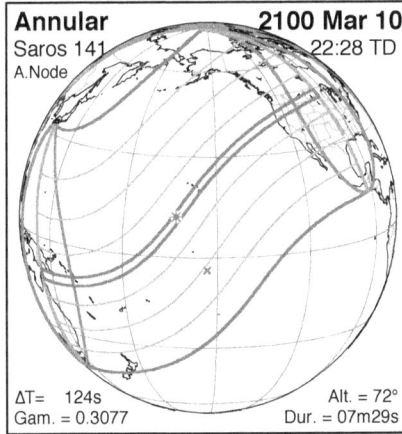

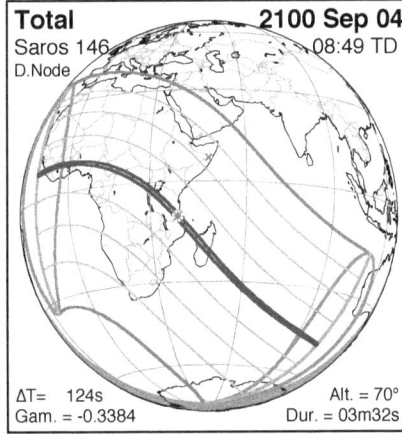

Annular	2097 Nov 04
Saros 154	02:01 TD
D.Node	
ΔT= 121s	Alt. = 26°
Gam. = -0.8927	Dur. = 03m36s

Partial	2098 Apr 01
Saros 121	20:03 TD
A.Node	
ΔT= 122s	Alt. = 0°
Gam. = -1.1005	Mag. = 0.7984

Partial	2098 Sep 25
Saros 126	00:31 TD
D.Node	
ΔT= 122s	Alt. = 0°
Gam. = 1.1185	Mag. = 0.7871

Partial	2098 Oct 24
Saros 164	10:36 TD
D.Node	
ΔT= 122s	Alt. = 0°
Gam. = -1.5407	Mag. = 0.0057

Annular	2099 Mar 21
Saros 131	22:55 TD
A.Node	
ΔT= 123s	Alt. = 66°
Gam. = -0.4017	Dur. = 07m32s

Total	2099 Sep 14
Saros 136	16:58 TD
D.Node	
ΔT= 123s	Alt. = 67°
Gam. = 0.3942	Dur. = 05m18s

Annular	2100 Mar 10
Saros 141	22:28 TD
A.Node	
ΔT= 124s	Alt. = 72°
Gam. = 0.3077	Dur. = 07m29s

Total	2100 Sep 04
Saros 146	08:49 TD
D.Node	
ΔT= 124s	Alt. = 70°
Gam. = -0.3384	Dur. = 03m32s

Appendix C

Large Global Solar Eclipse Maps: 2017 to 2066

Key to Large Global Solar Eclipse Maps

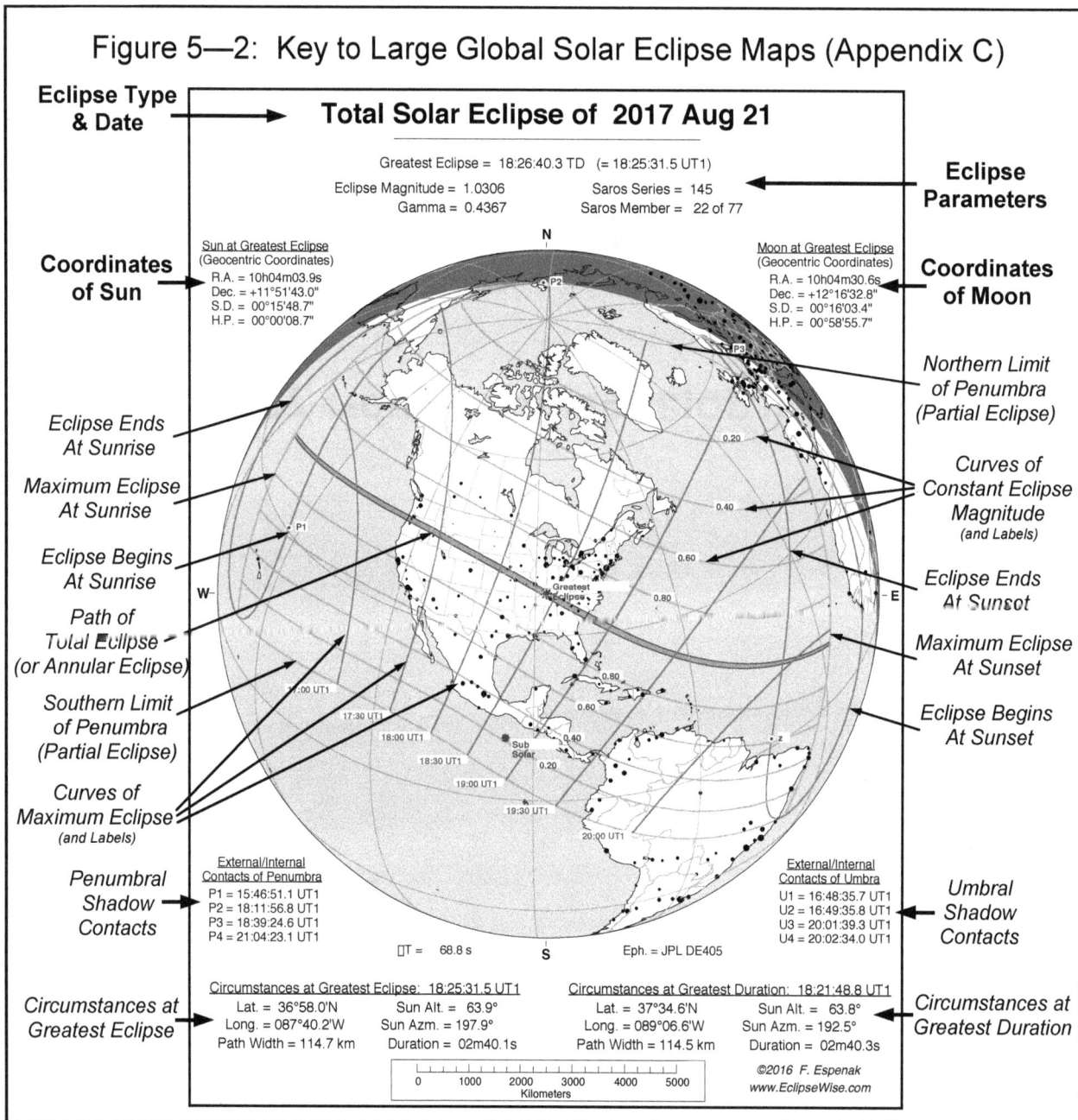

Figure 5—2: Key to Large Global Solar Eclipse Maps (Appendix C)

Eclipse Type & Date →

Total Solar Eclipse of 2017 Aug 21

Greatest Eclipse = 18:26:40.3 TD (= 18:25:31.5 UT1)
Eclipse Magnitude = 1.0306   Saros Series = 145
Gamma = 0.4367   Saros Member = 22 of 77

Eclipse Parameters

Coordinates of Sun

Sun at Greatest Eclipse
(Geocentric Coordinates)
R.A. = 10h04m03.9s
Dec. = +11°51'43.0"
S.D. = 00°15'48.7"
H.P. = 00°00'08.7"

Moon at Greatest Eclipse
(Geocentric Coordinates)
R.A. = 10h04m30.6s
Dec. = +12°16'32.8"
S.D. = 00°16'03.4"
H.P. = 00°58'55.7"

Coordinates of Moon

Eclipse Ends At Sunrise
Maximum Eclipse At Sunrise
Eclipse Begins At Sunrise
Path of Total Eclipse (or Annular Eclipse)
Southern Limit of Penumbra (Partial Eclipse)
Curves of Maximum Eclipse (and Labels)

Northern Limit of Penumbra (Partial Eclipse)
Curves of Constant Eclipse Magnitude (and Labels)
Eclipse Ends At Sunset
Maximum Eclipse At Sunset
Eclipse Begins At Sunset

Penumbral Shadow Contacts →

External/Internal Contacts of Penumbra
P1 = 15:46:51.1 UT1
P2 = 18:11:56.8 UT1
P3 = 18:39:24.6 UT1
P4 = 21:04:23.1 UT1

ΔT = 68.8 s   Eph. = JPL DE405

External/Internal Contacts of Umbra
U1 = 16:48:35.7 UT1
U2 = 16:49:35.8 UT1
U3 = 20:01:39.3 UT1
U4 = 20:02:34.0 UT1

Umbral Shadow Contacts

Circumstances at Greatest Eclipse

Circumstances at Greatest Eclipse: 18:25:31.5 UT1
Lat. = 36°58.0'N   Sun Alt. = 63.9°
Long. = 087°40.2'W   Sun Azm. = 197.9°
Path Width = 114.7 km   Duration = 02m40.1s

Circumstances at Greatest Duration: 18:21:48.8 UT1
Lat. = 37°34.6'N   Sun Alt. = 63.8°
Long. = 089°06.6'W   Sun Azm. = 192.5°
Path Width = 114.5 km   Duration = 02m40.3s

Circumstances at Greatest Duration

©2016 F. Espenak
www.EclipseWise.com

Annular Solar Eclipse of 2017 Feb 26

Greatest Eclipse = 14:54:32.8 TD (= 14:53:24.3 UT1)

Eclipse Magnitude = 0.9922   Saros Series = 140
Gamma = -0.4578   Saros Member = 29 of 71

Sun at Greatest Eclipse
(Geocentric Coordinates)
R.A. = 22h39m23.1s
Dec. = -08°29'38.8"
S.D. = 00°16'09.0"
H.P. = 00°00'08.9"

Moon at Greatest Eclipse
(Geocentric Coordinates)
R.A. = 22h39m53.2s
Dec. = -08°55'03.6"
S.D. = 00°15'47.8"
H.P. = 00°57'58.6"

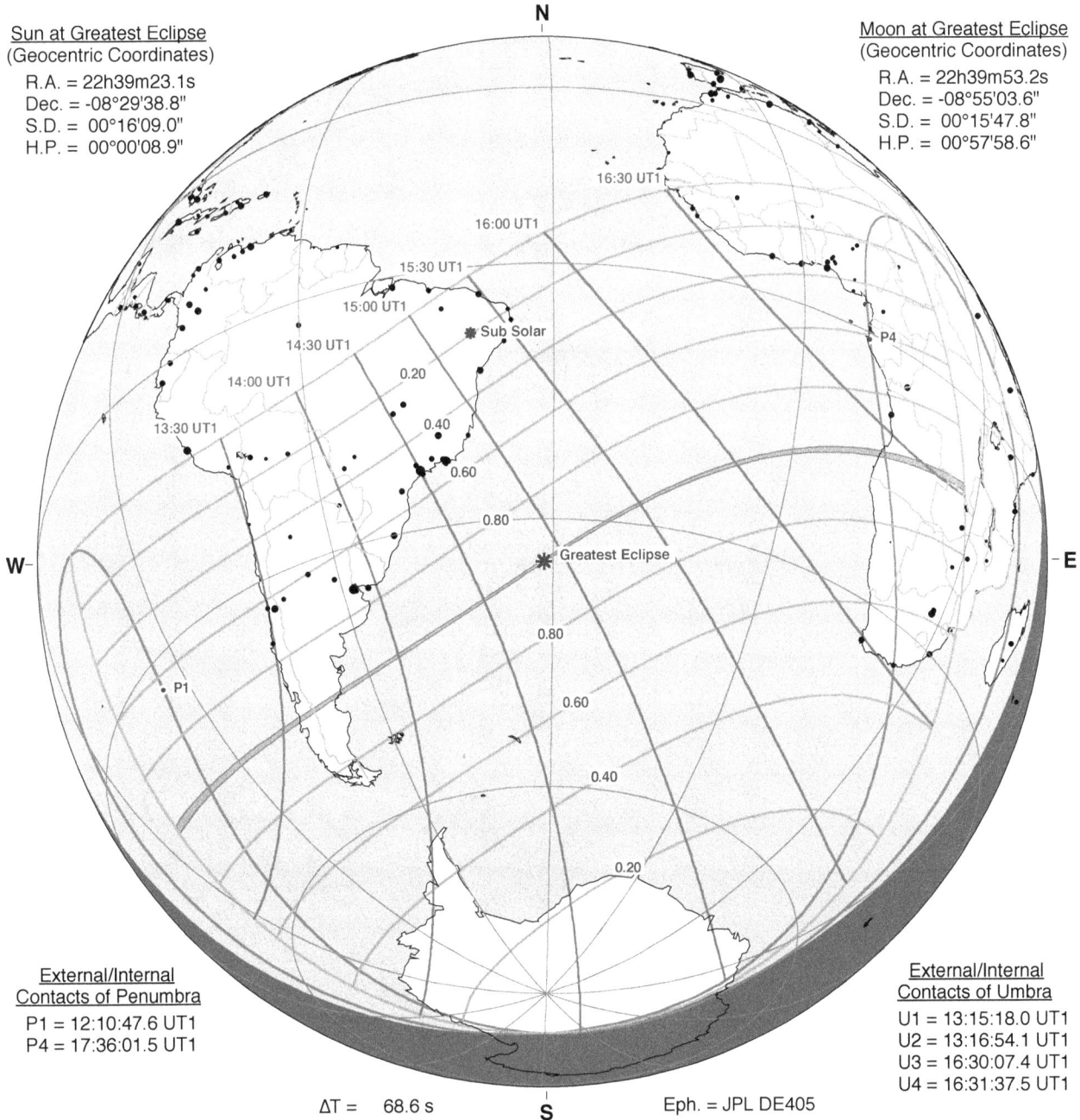

N

16:30 UT1
16:00 UT1
15:30 UT1
15:00 UT1
14:30 UT1   Sub Solar
14:00 UT1   0.20
13:30 UT1   0.40
0.60
0.80
W   Greatest Eclipse   E
0.80
P1   0.60
0.40
0.20
P4

External/Internal
Contacts of Penumbra
P1 = 12:10:47.6 UT1
P4 = 17:36:01.5 UT1

External/Internal
Contacts of Umbra
U1 = 13:15:18.0 UT1
U2 = 13:16:54.1 UT1
U3 = 16:30:07.4 UT1
U4 = 16:31:37.5 UT1

ΔT = 68.6 s   S   Eph. = JPL DE405

Circumstances at Greatest Eclipse: 14:53:24.3 UT1

Lat. = 34°40.8'S	Sun Alt. = 62.6°
Long. = 031°11.4'W	Sun Azm. = 340.5°
Path Width = 30.6 km	Duration = 00m44.0s

Circumstances at Greatest Duration: 13:16:06.0 UT1

Lat. = 43°07.5'S	Sun Alt. = 0.0°
Long. = 113°52.9'W	Sun Azm. = 101.7°
Path Width = 96.3 km	Duration = 01m22.4s

©2016 F. Espenak
www.EclipseWise.com

0 1000 2000 3000 4000 5000
Kilometers

Total Solar Eclipse of 2017 Aug 21

Greatest Eclipse = 18:26:40.3 TD   (= 18:25:31.5 UT1)

Eclipse Magnitude = 1.0306   Saros Series = 145

Gamma = 0.4367   Saros Member = 22 of 77

Sun at Greatest Eclipse
(Geocentric Coordinates)
R.A. = 10h04m03.9s
Dec. = +11°51'43.0"
S.D. = 00°15'48.7"
H.P. = 00°00'08.7"

Moon at Greatest Eclipse
(Geocentric Coordinates)
R.A. = 10h04m30.6s
Dec. = +12°16'32.8"
S.D. = 00°16'03.4"
H.P. = 00°58'55.7"

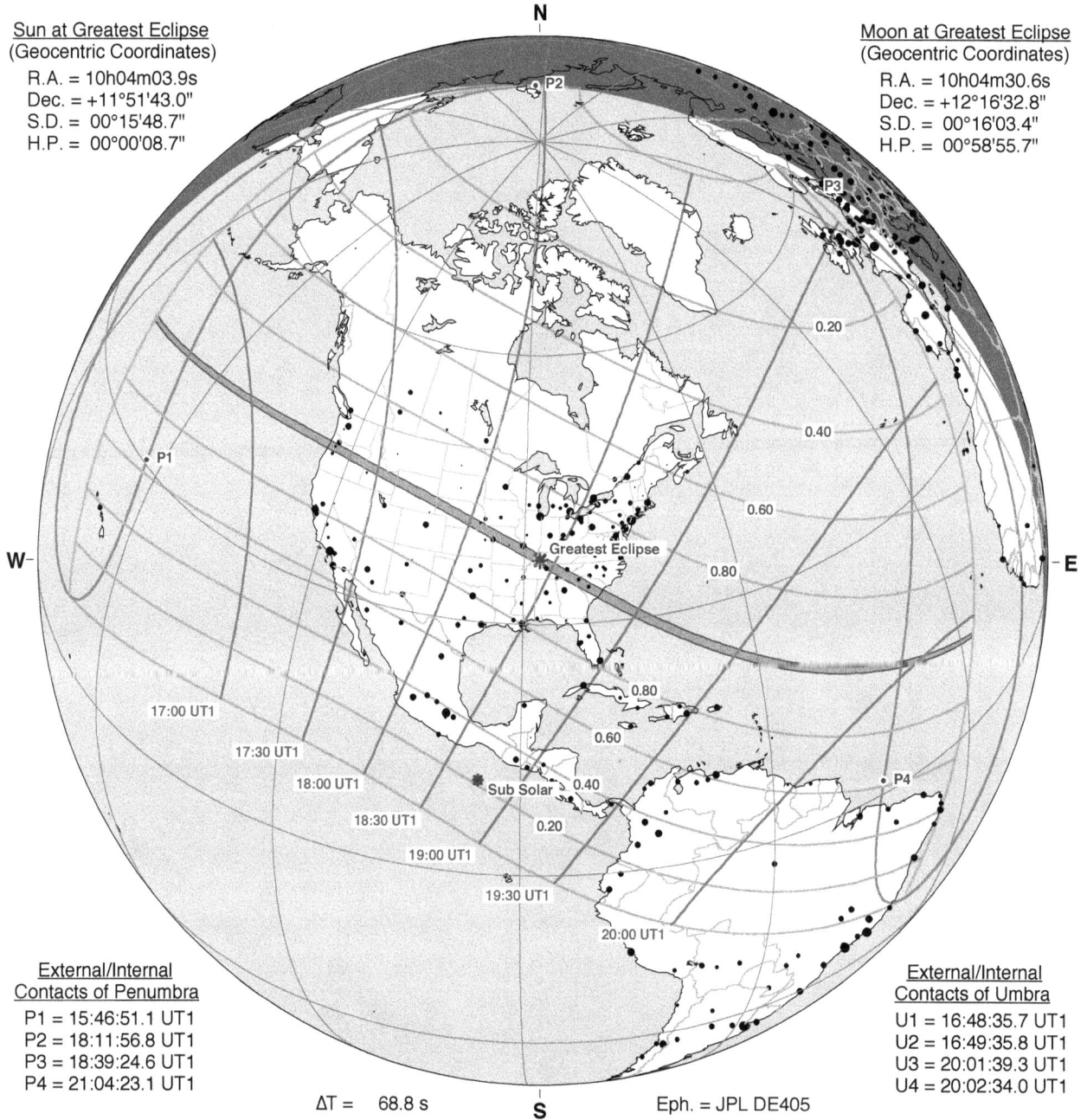

External/Internal
Contacts of Penumbra
P1 = 15:46:51.1 UT1
P2 = 18:11:56.8 UT1
P3 = 18:39:24.6 UT1
P4 = 21:04:23.1 UT1

External/Internal
Contacts of Umbra
U1 = 16:48:35.7 UT1
U2 = 16:49:35.8 UT1
U3 = 20:01:39.3 UT1
U4 = 20:02:34.0 UT1

ΔT = 68.8 s

Eph. = JPL DE405

Circumstances at Greatest Eclipse: 18:25:31.5 UT1

Lat. = 36°58.0'N	Sun Alt. = 63.9°
Long. = 087°40.2'W	Sun Azm. = 197.9°
Path Width = 114.7 km	Duration = 02m40.1s

Circumstances at Greatest Duration: 18:21:48.8 UT1

Lat. = 37°34.6'N	Sun Alt. = 63.8°
Long. = 089°06.6'W	Sun Azm. = 192.5°
Path Width = 114.5 km	Duration = 02m40.3s

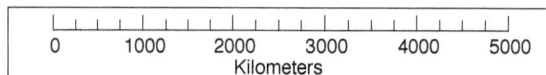

0   1000   2000   3000   4000   5000
Kilometers

©2016 F. Espenak
www.EclipseWise.com

Partial Solar Eclipse of 2018 Feb 15

Greatest Eclipse = 20:52:33.3 TD   (= 20:51:24.3 UT1)

Eclipse Magnitude = 0.5991   Saros Series = 150
Gamma = -1.2116   Saros Member = 17 of 71

Sun at Greatest Eclipse
(Geocentric Coordinates)
R.A. = 21h57m18.8s
Dec. = -12°28'07.3"
S.D. = 00°16'11.4"
H.P. = 00°00'08.9"

Moon at Greatest Eclipse
(Geocentric Coordinates)
R.A. = 21h58m26.9s
Dec. = -13°32'29.9"
S.D. = 00°14'59.4"
H.P. = 00°55'00.9"

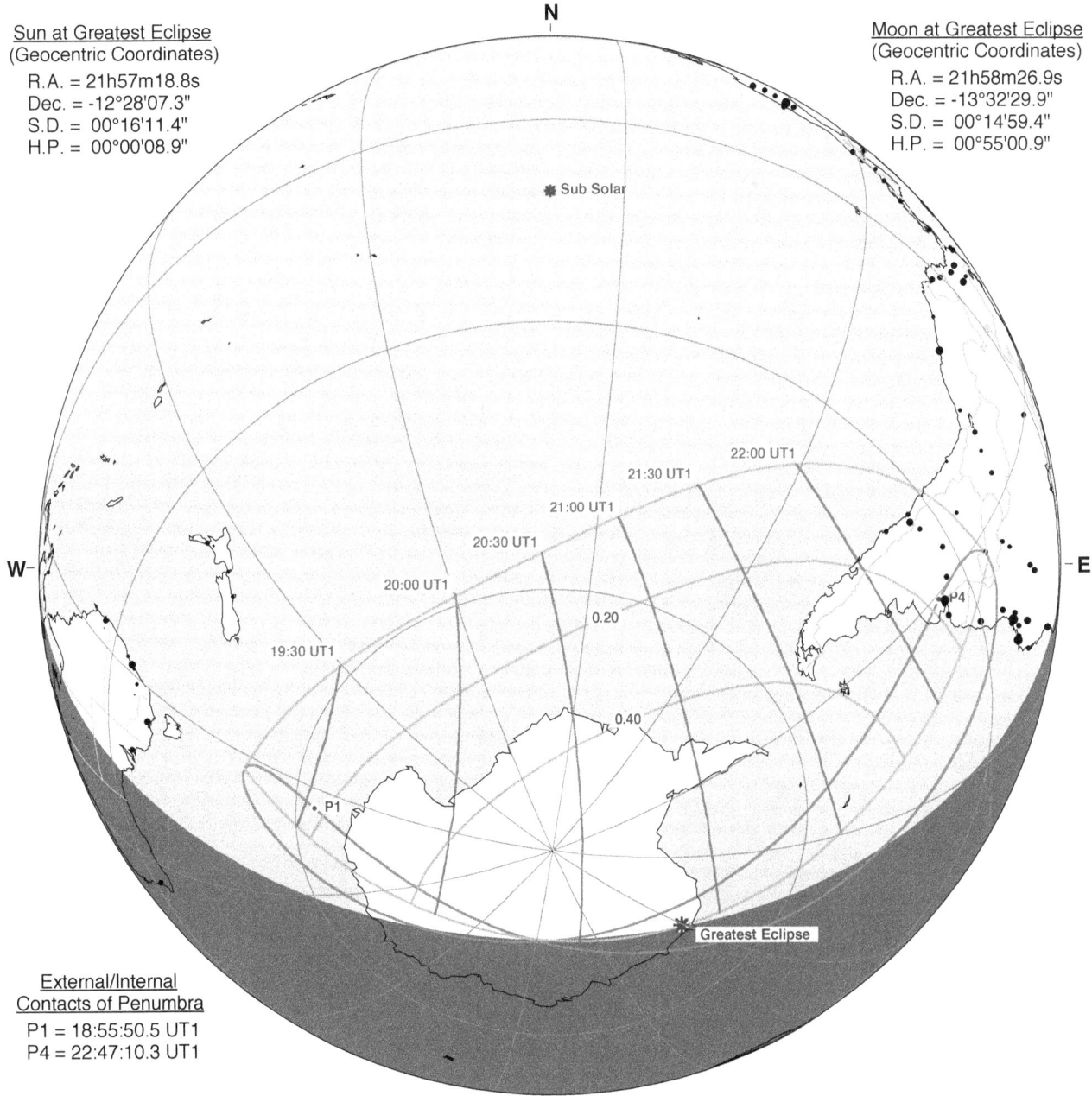

N

Sub Solar

22:00 UT1
21:30 UT1
21:00 UT1
20:30 UT1
20:00 UT1
0.20
19:30 UT1
0.40
P1
P4

W

E

Greatest Eclipse

External/Internal Contacts of Penumbra
P1 = 18:55:50.5 UT1
P4 = 22:47:10.3 UT1

ΔT = 69.0 s   S   Eph. = JPL DE405

Circumstances at Greatest Eclipse: 20:51:24.3 UT1
Lat. = 71°01.6'S   Sun Alt. =   0.0°
Long. = 000°38.5'E   Sun Azm. = 228.4°

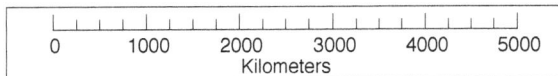

0   1000   2000   3000   4000   5000
Kilometers

©2016 F. Espenak
www.EclipseWise.com

Partial Solar Eclipse of 2018 Jul 13

Greatest Eclipse = 03:02:16.1 TD (= 03:01:07.0 UT1)

Eclipse Magnitude = 0.3365   Saros Series = 117

Gamma = -1.3542   Saros Member = 69 of 71

Sun at Greatest Eclipse
(Geocentric Coordinates)
R.A. = 07h29m31.1s
Dec. = +21°50'30.6"
S.D. = 00°15'44.0"
H.P. = 00°00'08.7"

Moon at Greatest Eclipse
(Geocentric Coordinates)
R.A. = 07h29m10.9s
Dec. = +20°27'46.1"
S.D. = 00°16'42.8"
H.P. = 01°01'20.4"

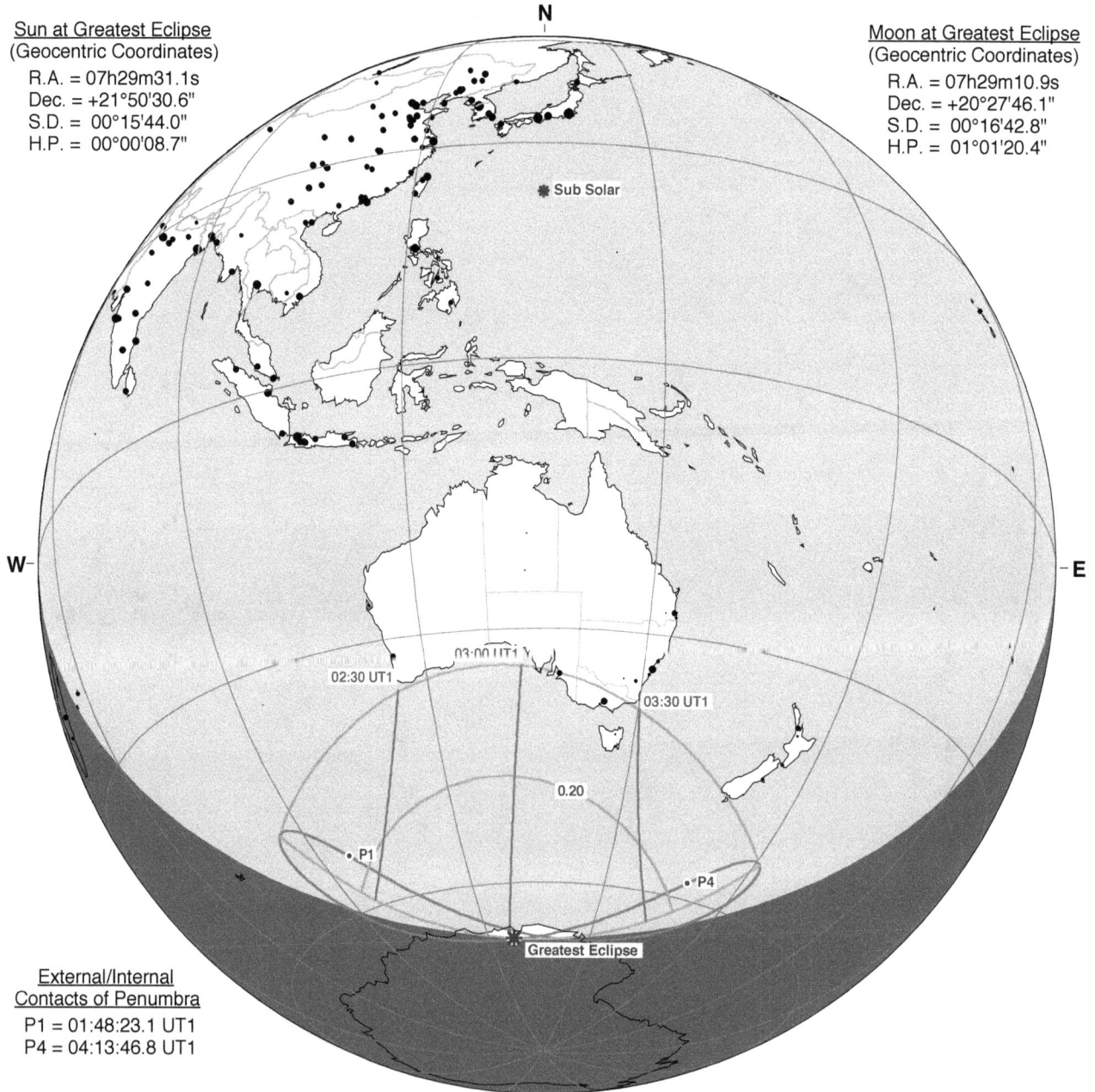

N

Sub Solar

W

E

03:00 UT1

02:30 UT1

03:30 UT1

0.20

P1

P4

Greatest Eclipse

**External/Internal
Contacts of Penumbra**
P1 = 01:48:23.1 UT1
P4 = 04:13:46.8 UT1

ΔT = 69.2 s

S

Eph. = JPL DE405

Circumstances at Greatest Eclipse: 03:01:07.0 UT1
Lat. = 67°55.4'S   Sun Alt. = 0.0°
Long. = 127°25.8'E   Sun Azm. = 8.1°

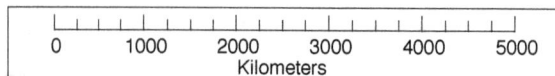

0   1000   2000   3000   4000   5000
Kilometers

©2016 F. Espenak
www.EclipseWise.com

Partial Solar Eclipse of 2018 Aug 11

Greatest Eclipse = 09:47:28.0 TD   (= 09:46:18.8 UT1)

Eclipse Magnitude = 0.7368   Saros Series = 155
Gamma = 1.1476   Saros Member = 6 of 71

Sun at Greatest Eclipse
(Geocentric Coordinates)
R.A. = 09h24m28.1s
Dec. = +15°13'19.1"
S.D. = 00°15'46.8"
H.P. = 00°00'08.7"

Moon at Greatest Eclipse
(Geocentric Coordinates)
R.A. = 09h25m31.3s
Dec. = +16°21'40.4"
S.D. = 00°16'40.0"
H.P. = 01°01'10.1"

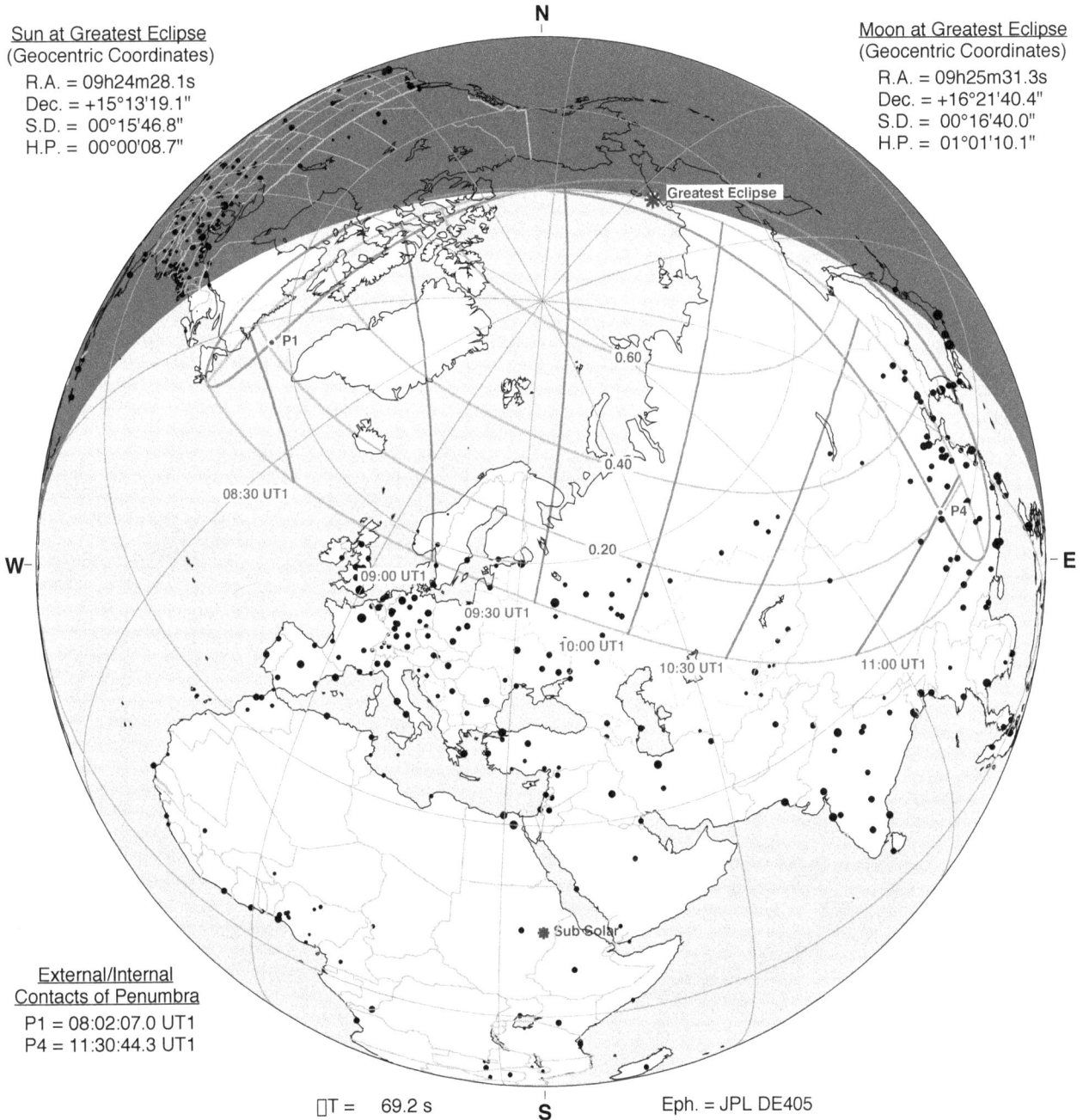

N

Greatest Eclipse

0.60
0.40
0.20

08:30 UT1
09:00 UT1
09:30 UT1
10:00 UT1
10:30 UT1
11:00 UT1

P1
P4

W   E

Sub-Solar

External/Internal
Contacts of Penumbra
P1 = 08:02:07.0 UT1
P4 = 11:30:44.3 UT1

ΔT = 69.2 s

S

Eph. = JPL DE405

Circumstances at Greatest Eclipse: 09:46:18.8 UT1
Lat. = 70°23.3'N   Sun Alt. = 0.0°
Long. = 174°28.1'E   Sun Azm. = 321.5°

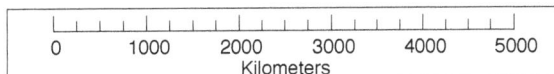

0   1000   2000   3000   4000   5000
Kilometers

©2016 F. Espenak
www.EclipseWise.com

Partial Solar Eclipse of 2019 Jan 06

Greatest Eclipse = 01:42:37.7 TD (= 01:41:28.3 UT1)

Eclipse Magnitude = 0.7146   Saros Series = 122
Gamma = 1.1417   Saros Member = 58 of 70

Sun at Greatest Eclipse
(Geocentric Coordinates)
R.A. = 19h06m57.4s
Dec. = -22°32'36.6"
S.D. = 00°16'15.9"
H.P. = 00°00'08.9"

Moon at Greatest Eclipse
(Geocentric Coordinates)
R.A. = 19h06m53.0s
Dec. = -21°30'36.4"
S.D. = 00°14'50.4"
H.P. = 00°54'27.6"

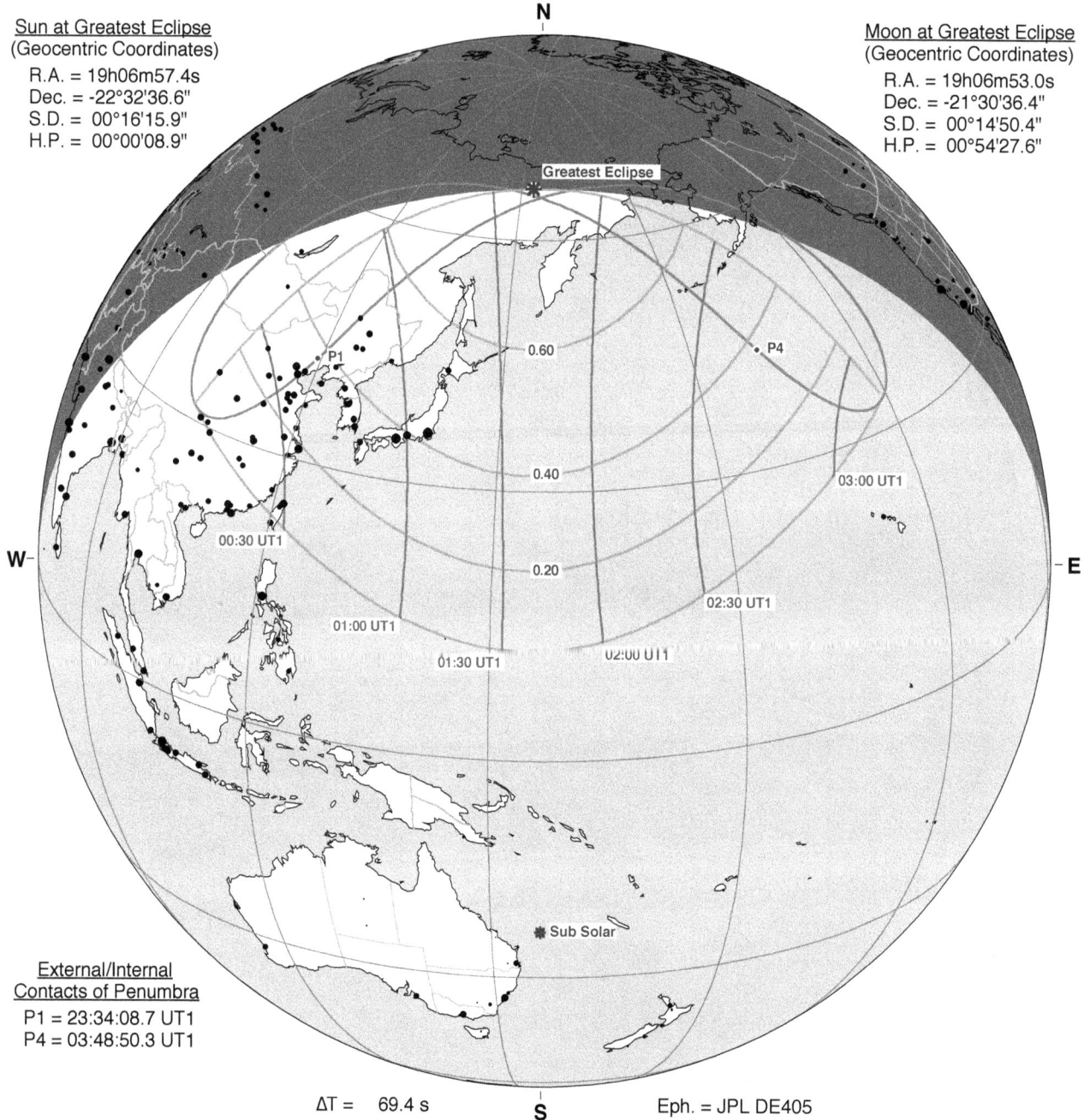

N

Greatest Eclipse

0.60

P1

P4

0.40

03:00 UT1

00:30 UT1

0.20

W

E

02:30 UT1

01:00 UT1

02:00 UT1

01:30 UT1

Sub Solar

External/Internal
Contacts of Penumbra
P1 = 23:34:08.7 UT1
P4 = 03:48:50.3 UT1

ΔT = 69.4 s   S   Eph. = JPL DE405

Circumstances at Greatest Eclipse: 01:41:28.3 UT1
Lat. = 67°26.1'N   Sun Alt. = 0.0°
Long. = 153°33.7'E   Sun Azm. = 177.7°

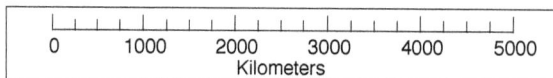

0 1000 2000 3000 4000 5000
Kilometers

©2016 F. Espenak
www.EclipseWise.com

Total Solar Eclipse of 2019 Jul 02

Greatest Eclipse = 19:24:07.5 TD (= 19:22:57.9 UT1)

Eclipse Magnitude = 1.0459   Saros Series = 127

Gamma = -0.6466   Saros Member = 58 of 82

Sun at Greatest Eclipse
(Geocentric Coordinates)

R.A. = 06h46m14.8s
Dec. = +23°00'36.4"
S.D. = 00°15'43.8"
H.P. = 00°00'08.6"

Moon at Greatest Eclipse
(Geocentric Coordinates)

R.A. = 06h46m17.9s
Dec. = +22°22'09.2"
S.D. = 00°16'14.9"
H.P. = 00°59'37.8"

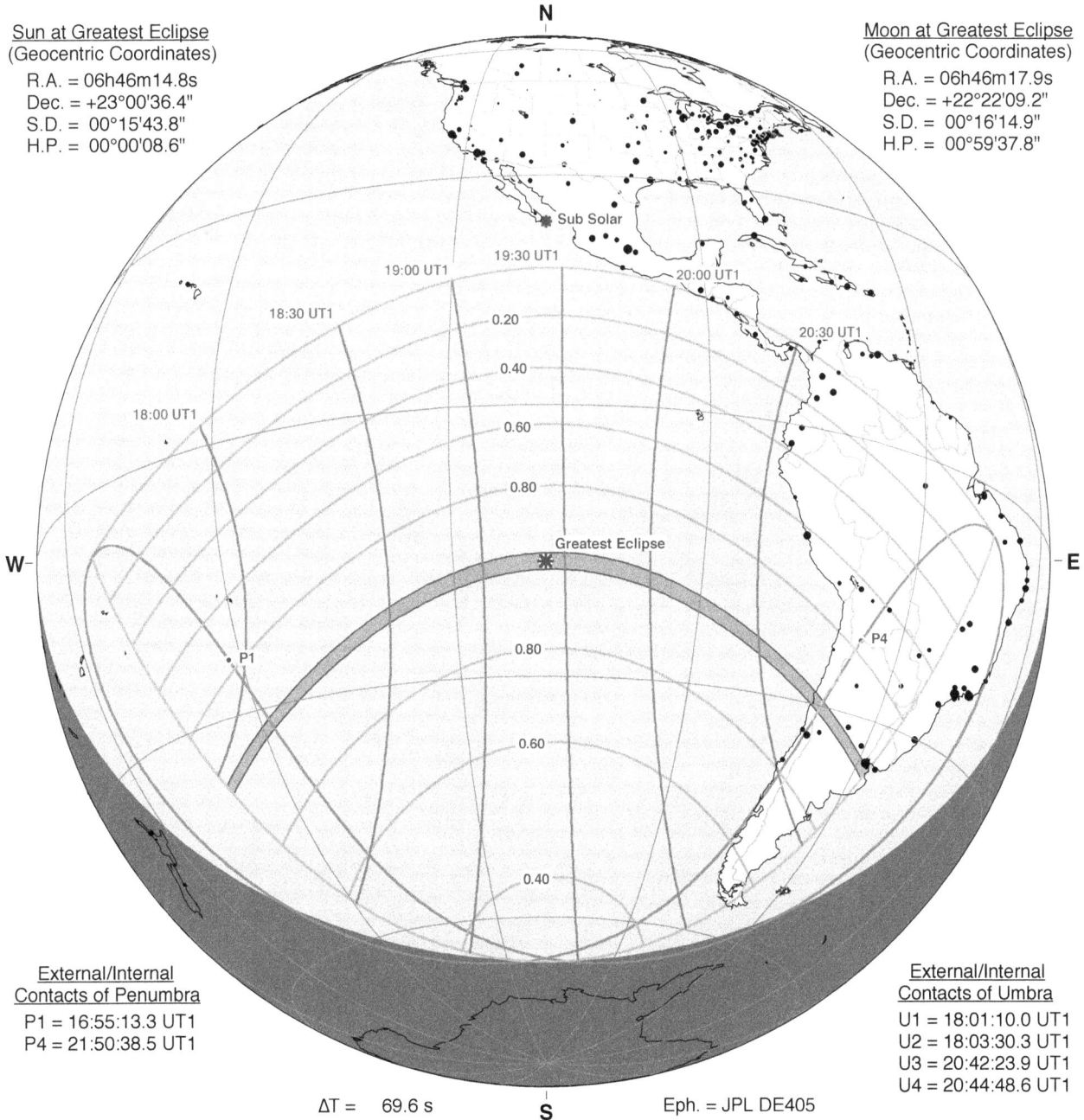

N

Sub Solar

19:00 UT1
19:30 UT1
20:00 UT1
18:30 UT1
20:30 UT1
0.20
18:00 UT1
0.40
0.60
0.80
Greatest Eclipse
W — E
0.80
P1
P4
0.60
0.40

S

External/Internal Contacts of Penumbra

P1 = 16:55:13.3 UT1
P4 = 21:50:38.5 UT1

External/Internal Contacts of Umbra

U1 = 18:01:10.0 UT1
U2 = 18:03:30.3 UT1
U3 = 20:42:23.9 UT1
U4 = 20:44:48.6 UT1

ΔT =   69.6 s   Eph. = JPL DE405

Circumstances at Greatest Eclipse: 19:22:57.9 UT1

Lat. = 17°23.3'S   Sun Alt. = 49.6°
Long. = 108°59.9'W   Sun Azm. = 359.0°
Path Width = 200.6 km   Duration = 04m32.8s

Circumstances at Greatest Duration: 19:24:09.3 UT1

Lat. = 17°22.8'S   Sun Alt. = 49.6°
Long. = 108°37.3'W   Sun Azm. = 358.0°
Path Width = 200.7 km   Duration = 04m32.8s

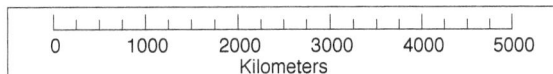

©2016 F. Espenak
www.EclipseWise.com

0   1000   2000   3000   4000   5000
Kilometers

Annular Solar Eclipse of 2019 Dec 26

Greatest Eclipse = 05:18:53.1 TD (= 05:17:43.3 UT1)

Eclipse Magnitude = 0.9701	Saros Series = 132
Gamma = 0.4135	Saros Member = 46 of 71

Sun at Greatest Eclipse
(Geocentric Coordinates)
R.A. = 18h17m56.7s
Dec. = -23°22'19.2"
S.D. = 00°16'15.7"
H.P. = 00°00'08.9"

Moon at Greatest Eclipse
(Geocentric Coordinates)
R.A. = 18h18m03.7s
Dec. = -22°58'50.4"
S.D. = 00°15'33.0"
H.P. = 00°57'04.0"

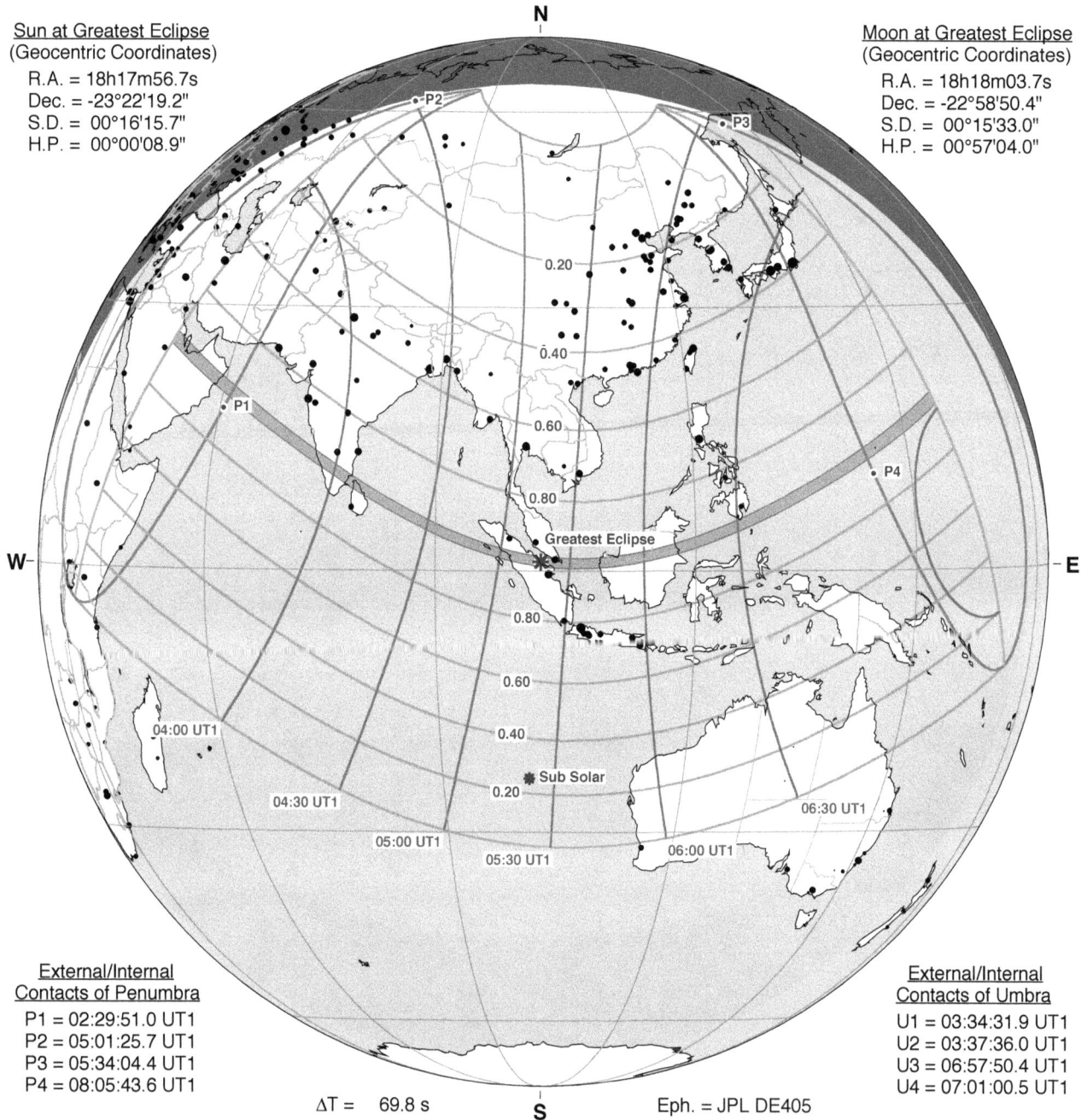

N

P2 P3

0.20

0.40

P1

0.60

0.80

P4

Greatest Eclipse

W — E

0.80

0.60

0.40

Sub Solar
0.20

04:00 UT1

04:30 UT1

05:00 UT1

05:30 UT1

06:00 UT1

06:30 UT1

External/Internal
Contacts of Penumbra
P1 = 02:29:51.0 UT1
P2 = 05:01:25.7 UT1
P3 = 05:34:04.4 UT1
P4 = 08:05:43.6 UT1

ΔT = 69.8 s

S

Eph. = JPL DE405

External/Internal
Contacts of Umbra
U1 = 03:34:31.9 UT1
U2 = 03:37:36.0 UT1
U3 = 06:57:50.4 UT1
U4 = 07:01:00.5 UT1

Circumstances at Greatest Eclipse: 05:17:43.3 UT1		Circumstances at Greatest Duration: 05:28:29.5 UT1	
Lat. = 01°00.5'N	Sun Alt. = 65.6°	Lat. = 00°46.7'N	Sun Alt. = 64.9°
Long. = 102°14.9'E	Sun Azm. = 183.6°	Long. = 105°05.1'E	Sun Azm. = 195.6°
Path Width = 117.9 km	Duration = 03m39.5s	Path Width = 118.6 km	Duration = 03m40.0s

0 1000 2000 3000 4000 5000
Kilometers

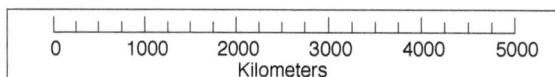

©2016 F. Espenak
www.EclipseWise.com

Annular Solar Eclipse of 2020 Jun 21

Greatest Eclipse = 06:41:15.4 TD   (= 06:40:05.4 UT1)

Eclipse Magnitude = 0.9940   Saros Series = 137
Gamma = 0.1209   Saros Member = 36 of 70

Sun at Greatest Eclipse
(Geocentric Coordinates)
R.A. = 06h01m33.0s
Dec. = +23°26'09.7"
S.D. = 00°15'44.2"
H.P. = 00°00'08.7"

Moon at Greatest Eclipse
(Geocentric Coordinates)
R.A. = 06h01m30.2s
Dec. = +23°32'56.7"
S.D. = 00°15'24.0"
H.P. = 00°56'31.1"

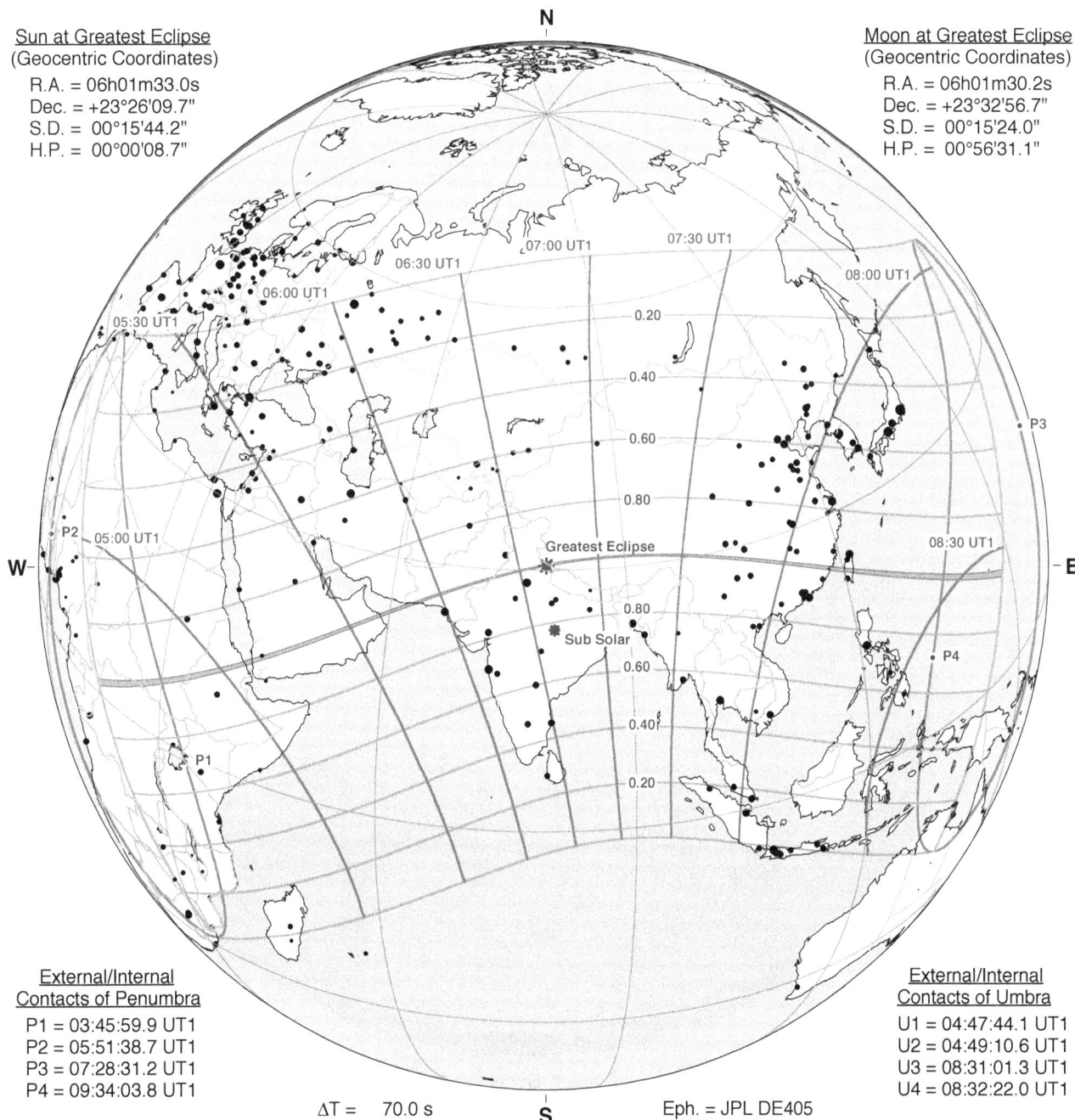

N

05:30 UT1
06:00 UT1
06:30 UT1
07:00 UT1
07:30 UT1
08:00 UT1
08:30 UT1
05:00 UT1
P2
P3
W
E
Greatest Eclipse
Sub Solar
P1
P4
0.20
0.40
0.60
0.80

S

External/Internal Contacts of Penumbra
P1 = 03:45:59.9 UT1
P2 = 05:51:38.7 UT1
P3 = 07:28:31.2 UT1
P4 = 09:34:03.8 UT1

External/Internal Contacts of Umbra
U1 = 04:47:44.1 UT1
U2 = 04:49:10.6 UT1
U3 = 08:31:01.3 UT1
U4 = 08:32:22.0 UT1

ΔT = 70.0 s   Eph. = JPL DE405

Circumstances at Greatest Eclipse: 06:40:05.4 UT1
Lat. = 30°31.2'N   Sun Alt. = 82.9°
Long. = 079°40.0'E   Sun Azm. = 174.2°
Path Width = 21.2 km   Duration = 00m38.2s

Circumstances at Greatest Duration: 04:48:27.4 UT1
Lat. = 01°16.1'N   Sun Alt. = 0.0°
Long. = 017°47.9'E   Sun Azm. = 66.6°
Path Width = 85.2 km   Duration = 01m22.4s

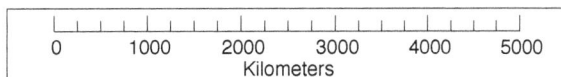

©2016 F. Espenak
www.EclipseWise.com

0 1000 2000 3000 4000 5000
Kilometers

65

Total Solar Eclipse of 2020 Dec 14

Greatest Eclipse = 16:14:39.4 TD   (= 16:13:29.1 UT1)

Eclipse Magnitude = 1.0254   Saros Series = 142
Gamma = -0.2939   Saros Member = 23 of 72

Sun at Greatest Eclipse
(Geocentric Coordinates)
R.A. = 17h30m05.9s
Dec. = -23°15'32.3"
S.D. = 00°16'14.9"
H.P. = 00°00'08.9"

Moon at Greatest Eclipse
(Geocentric Coordinates)
R.A. = 17h29m54.3s
Dec. = -23°32'58.8"
S.D. = 00°16'23.7"
H.P. = 01°00'10.4"

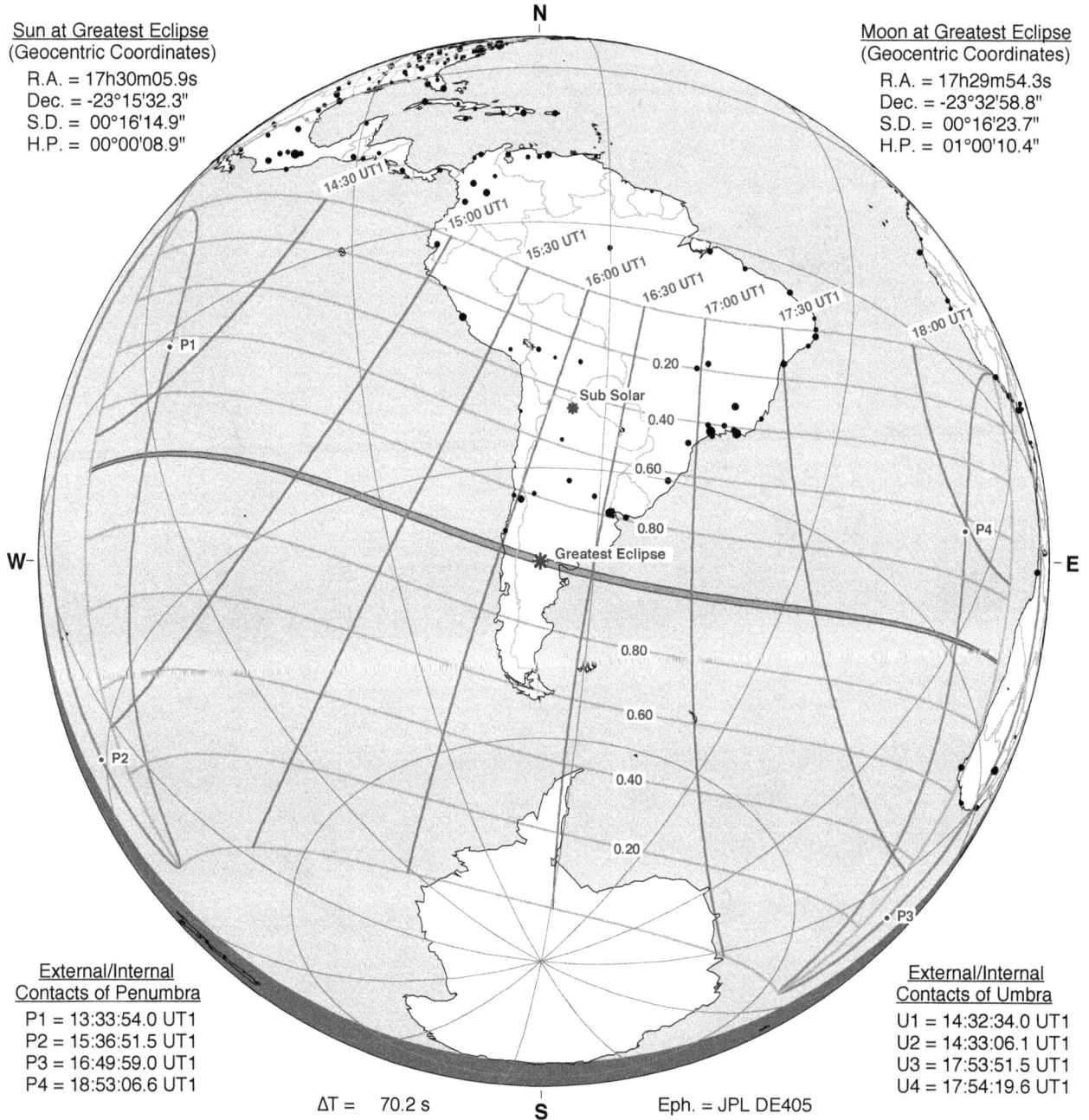

External/Internal
Contacts of Penumbra
P1 = 13:33:54.0 UT1
P2 = 15:36:51.5 UT1
P3 = 16:49:59.0 UT1
P4 = 18:53:06.6 UT1

External/Internal
Contacts of Umbra
U1 = 14:32:34.0 UT1
U2 = 14:33:06.1 UT1
U3 = 17:53:51.5 UT1
U4 = 17:54:19.6 UT1

ΔT =   70.2 s   Eph. = JPL DE405

Circumstances at Greatest Eclipse: 16:13:29.1 UT1
Lat. = 40°20.1'S   Sun Alt. =   72.7°
Long. = 067°57.4'W   Sun Azm. = 10.3°
Path Width = 90.2 km   Duration = 02m09.7s

Circumstances at Greatest Duration: 16:13:34.9 UT1
Lat. = 40°20.7'S   Sun Alt. =   72.7°
Long. = 067°54.8'W   Sun Azm. = 10.1°
Path Width = 90.2 km   Duration = 02m09.7s

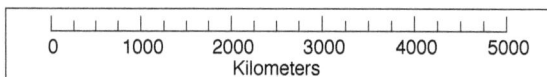

©2016 F. Espenak
www.EclipseWise.com

Annular Solar Eclipse of 2021 Jun 10

Greatest Eclipse = 10:43:06.7 TD   (= 10:41:56.3 UT1)

Eclipse Magnitude = 0.9435	Saros Series = 147
Gamma = 0.9152	Saros Member = 23 of 80

Sun at Greatest Eclipse
(Geocentric Coordinates)
R.A. = 05h15m31.4s
Dec. = +23°02'37.1"
S.D. = 00°15'45.2"
H.P. = 00°00'08.7"

Moon at Greatest Eclipse
(Geocentric Coordinates)
R.A. = 05h14m53.6s
Dec. = +23°51'21.6"
S.D. = 00°14'46.8"
H.P. = 00°54'14.5"

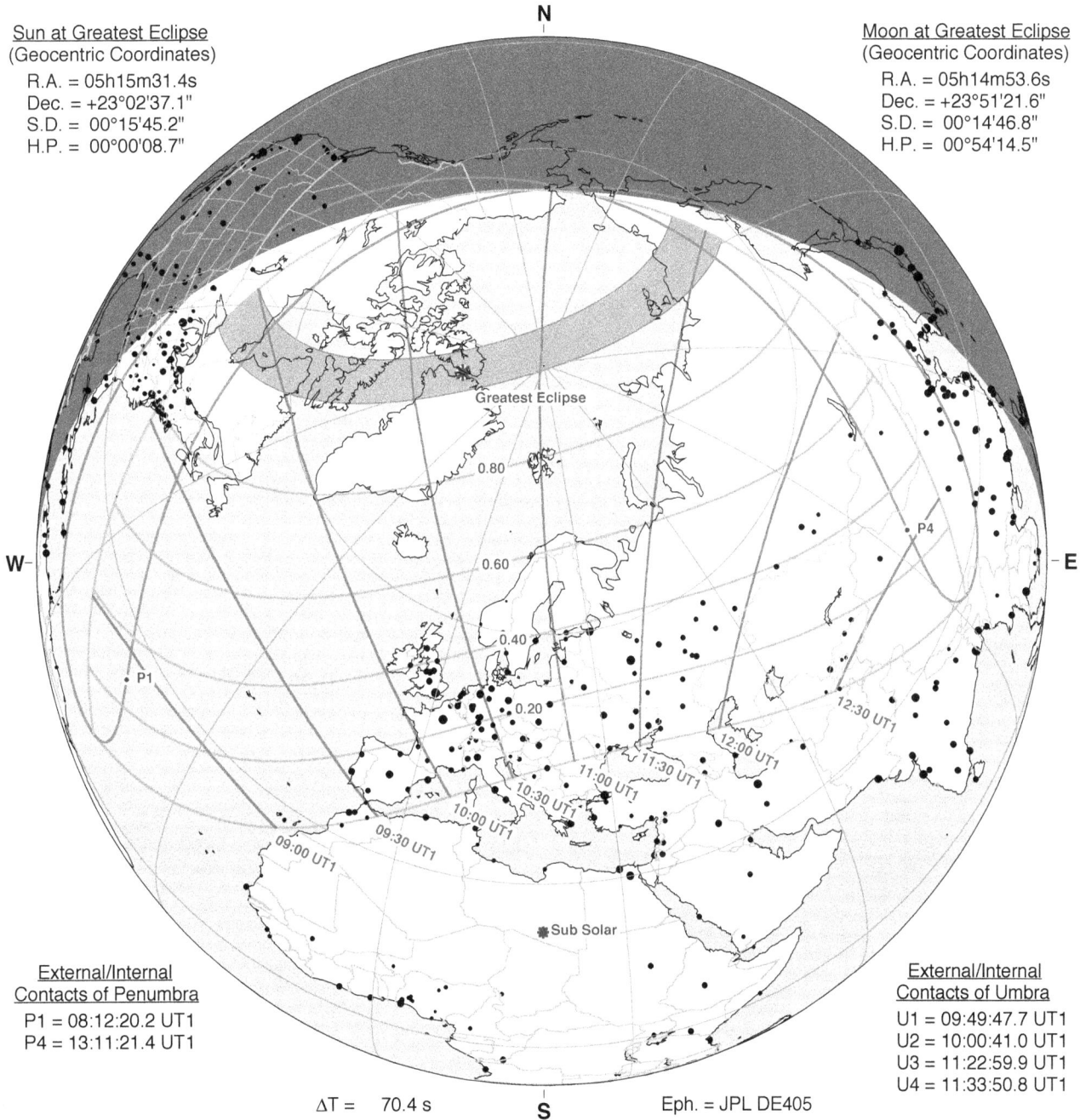

N

Greatest Eclipse

0.80

0.60

P4

W

E

0.40

0.20

12:30 UT1

12:00 UT1

11:30 UT1

11:00 UT1

10:30 UT1

P1

10:00 UT1

09:30 UT1

09:00 UT1

Sub Solar

**External/Internal
Contacts of Penumbra**
P1 = 08:12:20.2 UT1
P4 = 13:11:21.4 UT1

**External/Internal
Contacts of Umbra**
U1 = 09:49:47.7 UT1
U2 = 10:00:41.0 UT1
U3 = 11:22:59.9 UT1
U4 = 11:33:50.8 UT1

ΔT = 70.4 s

S

Eph. = JPL DE405

Circumstances at Greatest Eclipse: 10:41:56.3 UT1

Lat. = 80°48.9'N	Sun Alt. = 23.3°
Long. = 066°46.1'W	Sun Azm. = 89.9°
Path Width = 526.8 km	Duration = 03m51.2s

Circumstances at Greatest Duration: 10:41:57.4 UT1

Lat. = 80°49.4'N	Sun Alt. = 23.3°
Long. = 066°46.4'W	Sun Azm. = 89.9°
Path Width = 526.8 km	Duration = 03m51.2s

0   1000   2000   3000   4000   5000
Kilometers

©2016 F. Espenak
www.EclipseWise.com

67

Total Solar Eclipse of 2021 Dec 04

Greatest Eclipse = 07:34:37.9 TD   (= 07:33:27.3 UT1)

Eclipse Magnitude = 1.0367	Saros Series = 152
Gamma = -0.9526	Saros Member = 13 of 70

Sun at Greatest Eclipse
(Geocentric Coordinates)
R.A. = 16h43m32.4s
Dec. = -22°16'29.4"
S.D. = 00°16'13.6"
H.P. = 00°00'08.9"

Moon at Greatest Eclipse
(Geocentric Coordinates)
R.A. = 16h42m35.0s
Dec. = -23°13'22.3"
S.D. = 00°16'44.7"
H.P. = 01°01'27.3"

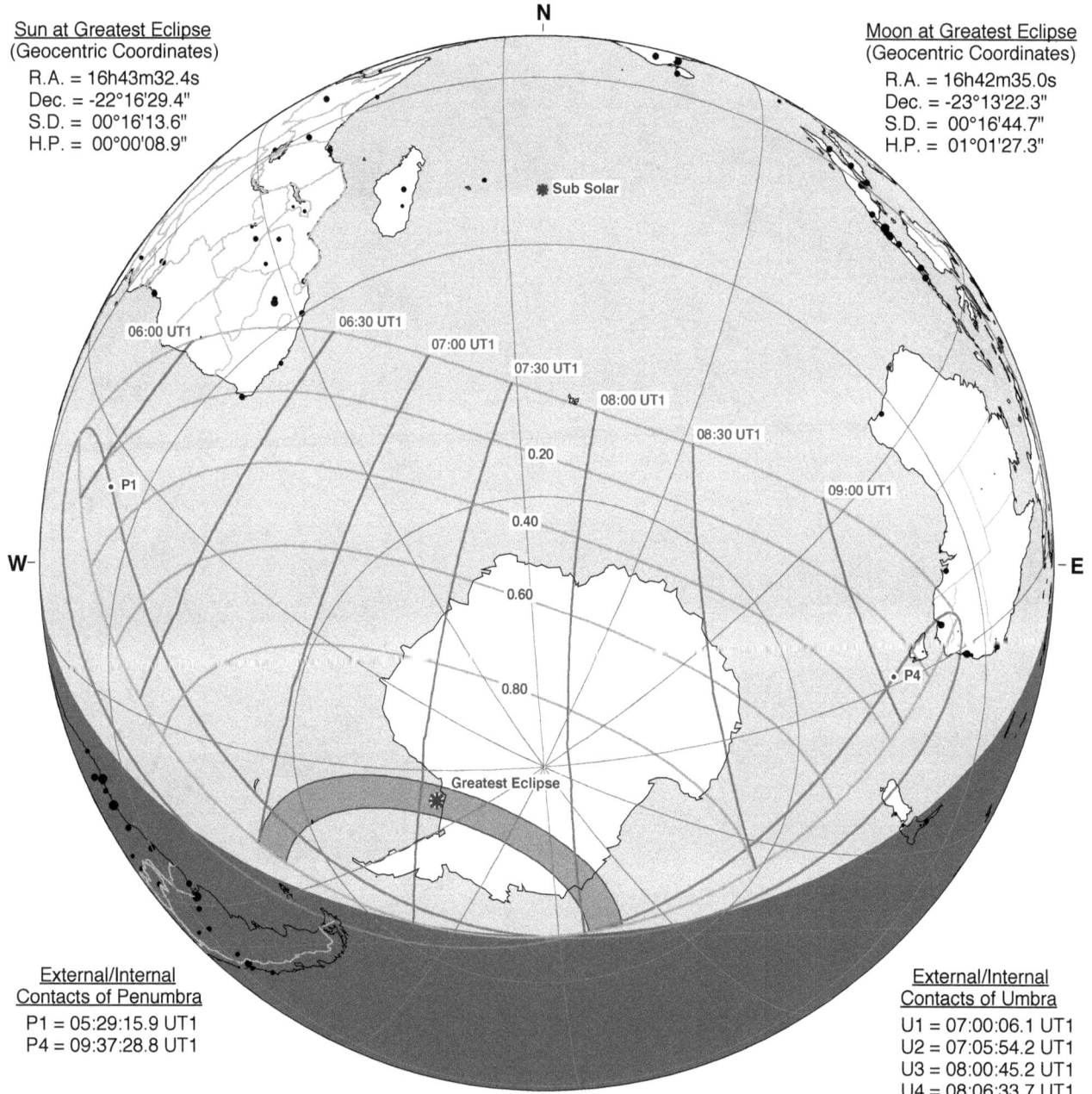

External/Internal
Contacts of Penumbra
P1 = 05:29:15.9 UT1
P4 = 09:37:28.8 UT1

External/Internal
Contacts of Umbra
U1 = 07:00:06.1 UT1
U2 = 07:05:54.2 UT1
U3 = 08:00:45.2 UT1
U4 = 08:06:33.7 UT1

ΔT = 70.6 s   Eph. = JPL DE405

Circumstances at Greatest Eclipse: 07:33:27.3 UT1

Lat. = 76°46.6'S	Sun Alt. = 17.2°
Long. = 046°13.7'W	Sun Azm. = 114.8°
Path Width = 418.8 km	Duration = 01m54.4s

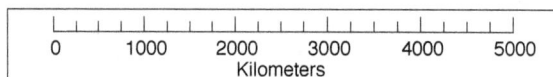

Circumstances at Greatest Duration: 07:33:29.6 UT1

Lat. = 76°47.7'S	Sun Alt. = 17.2°
Long. = 046°16.7'W	Sun Azm. = 114.8°
Path Width = 418.7 km	Duration = 01m54.4s

©2016 F. Espenak
www.EclipseWise.com

Partial Solar Eclipse of 2022 Apr 30

Greatest Eclipse = 20:42:36.5 TD (= 20:41:25.8 UT1)

Eclipse Magnitude = 0.6396   Saros Series = 119
Gamma = -1.1901   Saros Member = 66 of 71

Sun at Greatest Eclipse
(Geocentric Coordinates)
R.A. = 02h32m15.6s
Dec. = +14°57'53.5"
S.D. = 00°15'52.6"
H.P. = 00°00'08.7"

Moon at Greatest Eclipse
(Geocentric Coordinates)
R.A. = 02h34m04.8s
Dec. = +13°57'48.8"
S.D. = 00°15'04.0"
H.P. = 00°55'17.7"

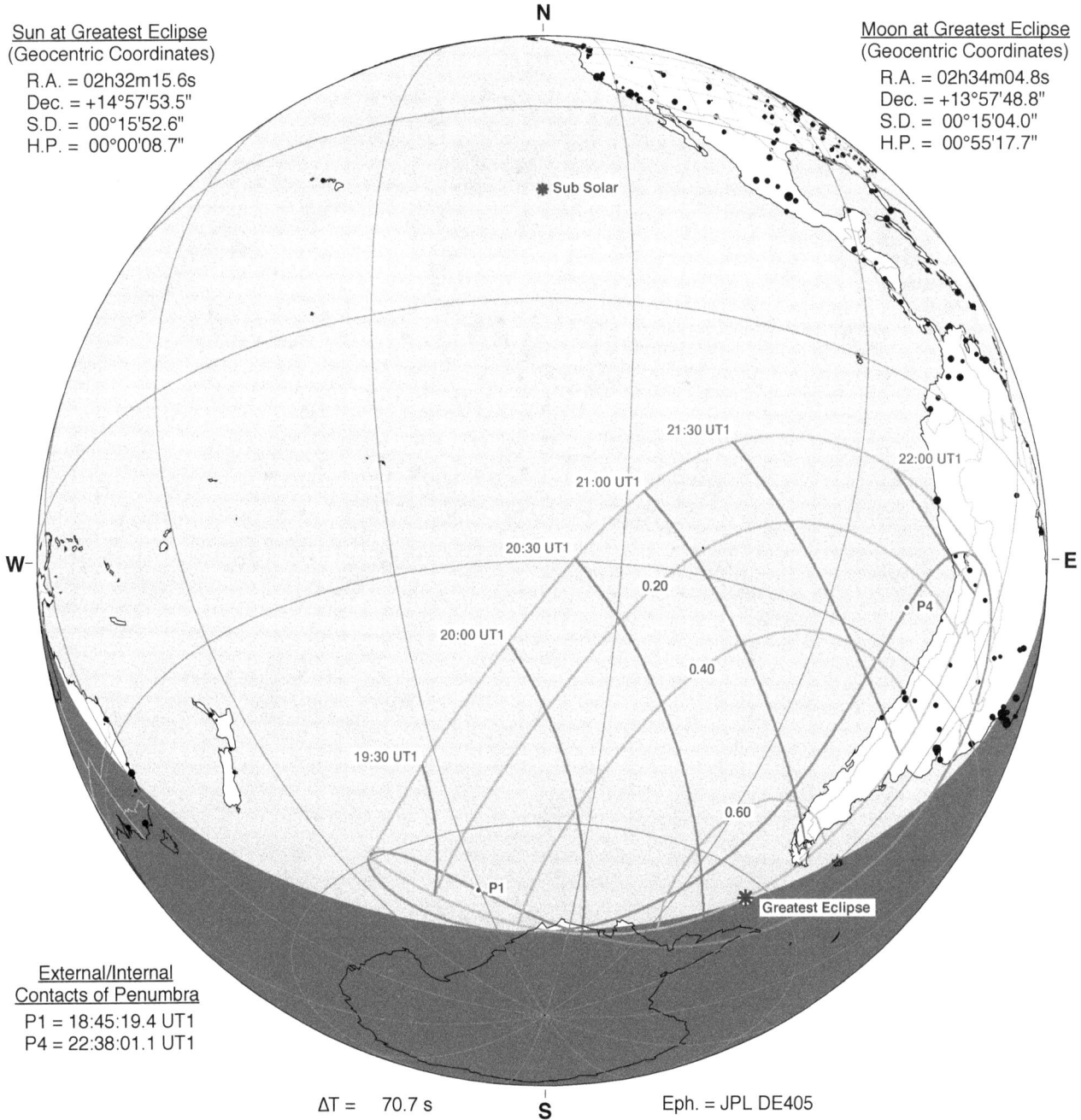

N

✳ Sub Solar

21:30 UT1
21:00 UT1
22:00 UT1
20:30 UT1
0.20
W
E
20:00 UT1
0.40
P4
19:30 UT1
0.60
P1
✳ Greatest Eclipse

External/Internal
Contacts of Penumbra
P1 = 18:45:19.4 UT1
P4 = 22:38:01.1 UT1

ΔT = 70.7 s   S   Eph. = JPL DE405

Circumstances at Greatest Eclipse: 20:41:25.8 UT1
Lat. = 62°07.0'S   Sun Alt. = 0.0°
Long. = 071°28.7'W   Sun Azm. = 303.5°

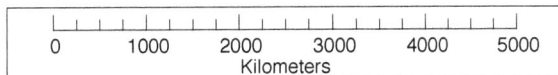

0 1000 2000 3000 4000 5000
Kilometers

©2016 F. Espenak
www.EclipseWise.com

Partial Solar Eclipse of 2022 Oct 25

Greatest Eclipse = 11:01:20.0 TD (= 11:00:09.1 UT1)

Eclipse Magnitude = 0.8619   Saros Series = 124
Gamma = 1.0701   Saros Member = 55 of 73

Sun at Greatest Eclipse
(Geocentric Coordinates)
R.A. = 13h59m20.5s
Dec. = -12°10'17.0"
S.D. = 00°16'05.0"
H.P. = 00°00'08.8"

Moon at Greatest Eclipse
(Geocentric Coordinates)
R.A. = 14h01m10.9s
Dec. = -11°14'16.0"
S.D. = 00°15'52.6"
H.P. = 00°58'16.0"

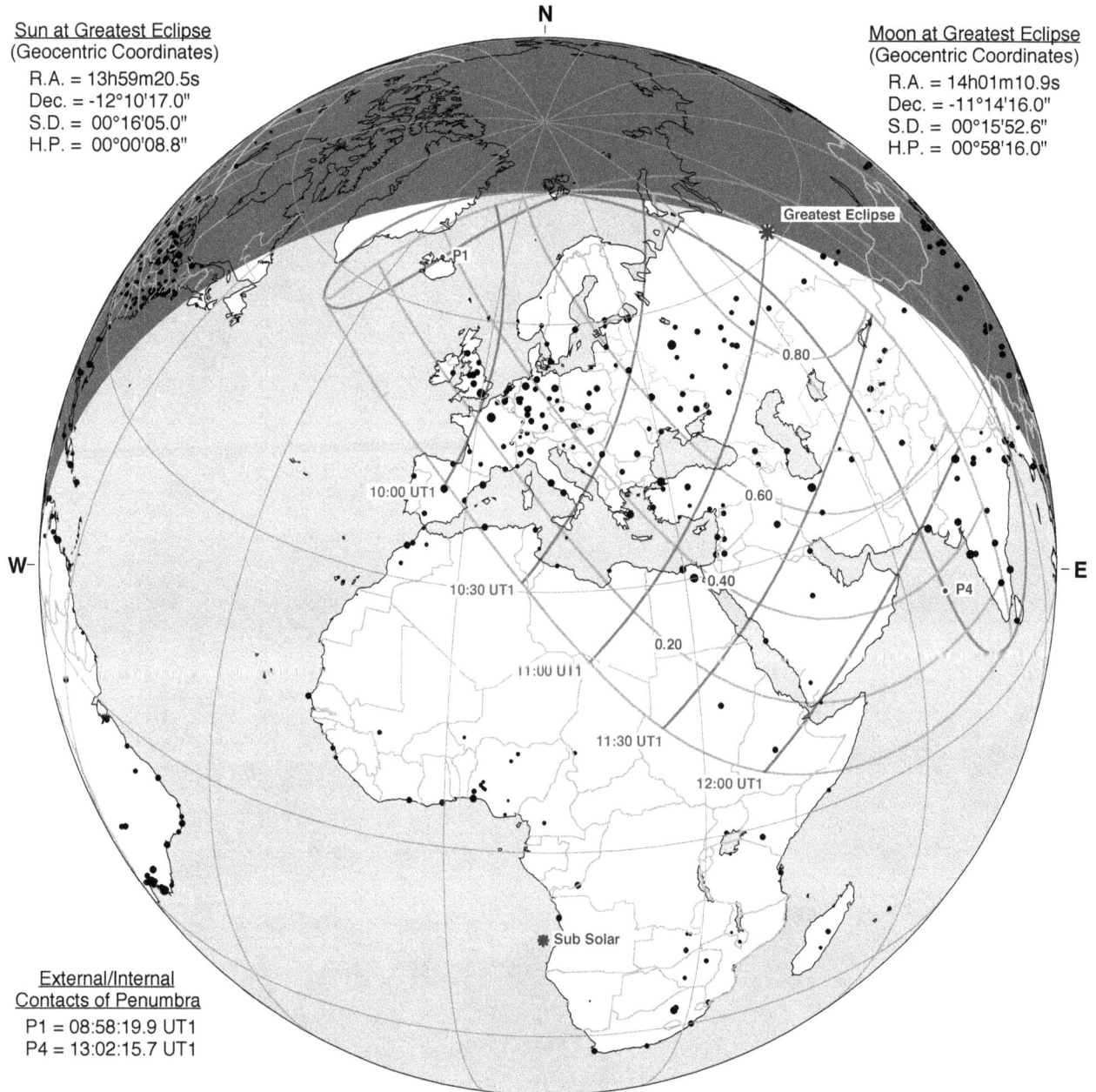

N

Greatest Eclipse

P1

0.80

0.60

10:00 UT1

0.40

W — — E

10:30 UT1

P4

0.20

11:00 UT1

11:30 UT1

12:00 UT1

Sub Solar

**External/Internal
Contacts of Penumbra**
P1 = 08:58:19.9 UT1
P4 = 13:02:15.7 UT1

ΔT = 70.9 s   S   Eph. = JPL DE405

Circumstances at Greatest Eclipse: 11:00:09.1 UT1
Lat. = 61°38.9'N   Sun Alt. = 0.0°
Long. = 077°20.8'E   Sun Azm. = 243.6°

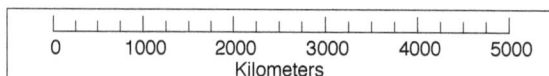

0   1000   2000   3000   4000   5000
Kilometers

Hybrid Solar Eclipse of 2023 Apr 20

Greatest Eclipse = 04:17:56.0 TD (= 04:16:44.9 UT1)

Eclipse Magnitude = 1.0132

Gamma = -0.3952

Saros Series = 129

Saros Member = 52 of 80

Sun at Greatest Eclipse
(Geocentric Coordinates)
R.A. = 01h51m01.7s
Dec. = +11°24'54.1"
S.D. = 00°15'55.4"
H.P. = 00°00'08.8"

Moon at Greatest Eclipse
(Geocentric Coordinates)
R.A. = 01h51m43.2s
Dec. = +11°04'16.7"
S.D. = 00°15'53.6"
H.P. = 00°58'19.9"

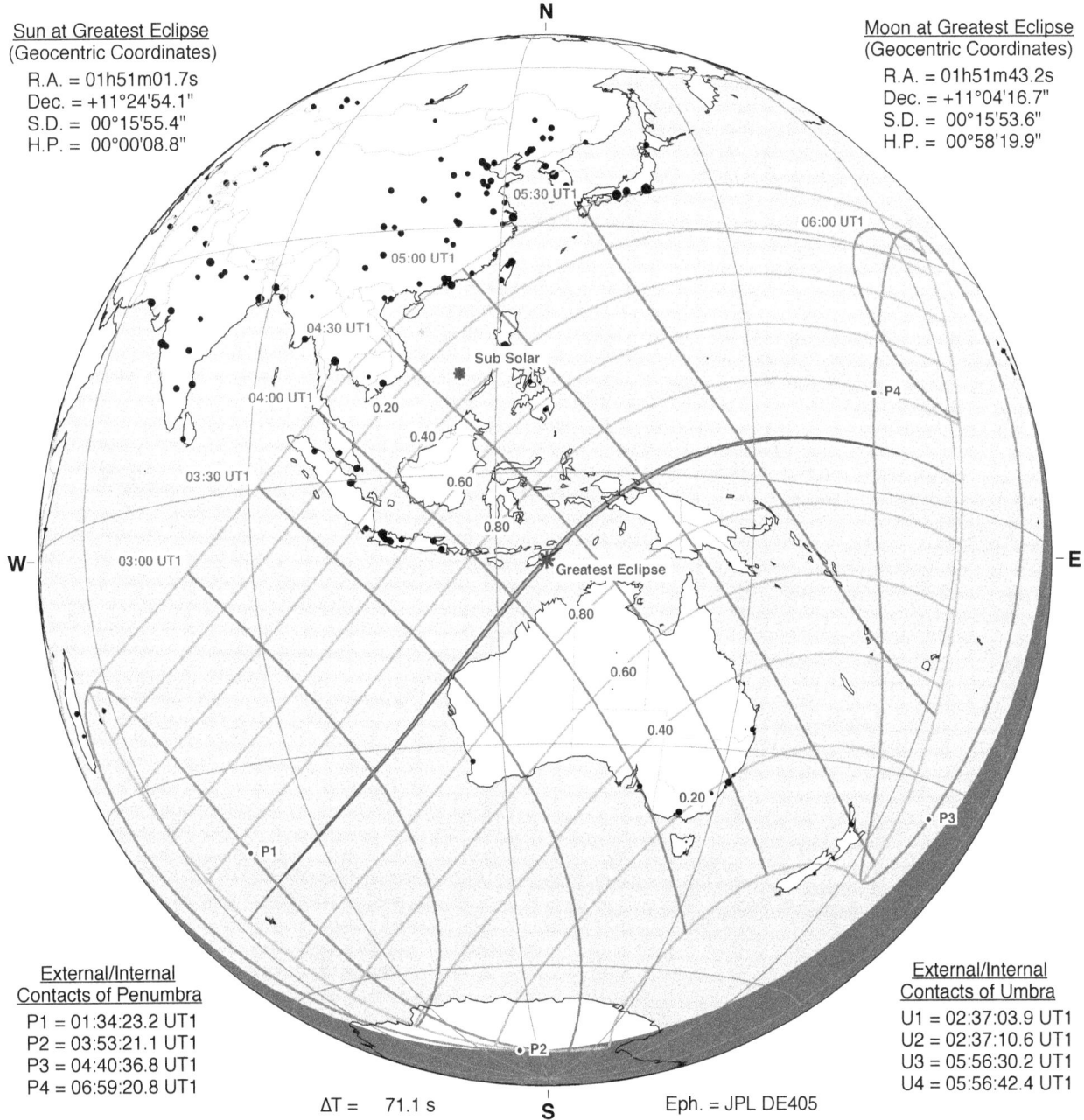

External/Internal
Contacts of Penumbra
P1 = 01:34:23.2 UT1
P2 = 03:53:21.1 UT1
P3 = 04:40:36.8 UT1
P4 = 06:59:20.8 UT1

External/Internal
Contacts of Umbra
U1 = 02:37:03.9 UT1
U2 = 02:37:10.6 UT1
U3 = 05:56:30.2 UT1
U4 = 05:56:42.4 UT1

ΔT = 71.1 s

Eph. = JPL DE405

Circumstances at Greatest Eclipse: 04:16:44.9 UT1

Lat. = 09°35.7'S	Sun Alt. = 66.7°
Long. = 125°46.8'E	Sun Azm. = 334.0°
Path Width = 49.0 km	Duration = 01m16.1s

Circumstances at Greatest Duration: 04:16:15.6 UT1

Lat. = 09°42.5'S	Sun Alt. = 66.7°
Long. = 125°40.1'E	Sun Azm. = 334.6°
Path Width = 48.9 km	Duration = 01m16.1s

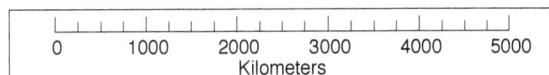

©2016 F. Espenak
www.EclipseWise.com

Annular Solar Eclipse of 2023 Oct 14

Greatest Eclipse = 18:00:40.6 TD   (= 17:59:29.3 UT1)

Eclipse Magnitude = 0.9520   Saros Series = 134
Gamma = 0.3753   Saros Member = 44 of 71

Sun at Greatest Eclipse
(Geocentric Coordinates)
R.A. = 13h18m05.4s
Dec. = -08°14'36.7"
S.D. = 00°16'02.0"
H.P. = 00°00'08.8"

Moon at Greatest Eclipse
(Geocentric Coordinates)
R.A. = 13h18m44.3s
Dec. = -07°56'18.9"
S.D. = 00°15'02.9"
H.P. = 00°55'13.8"

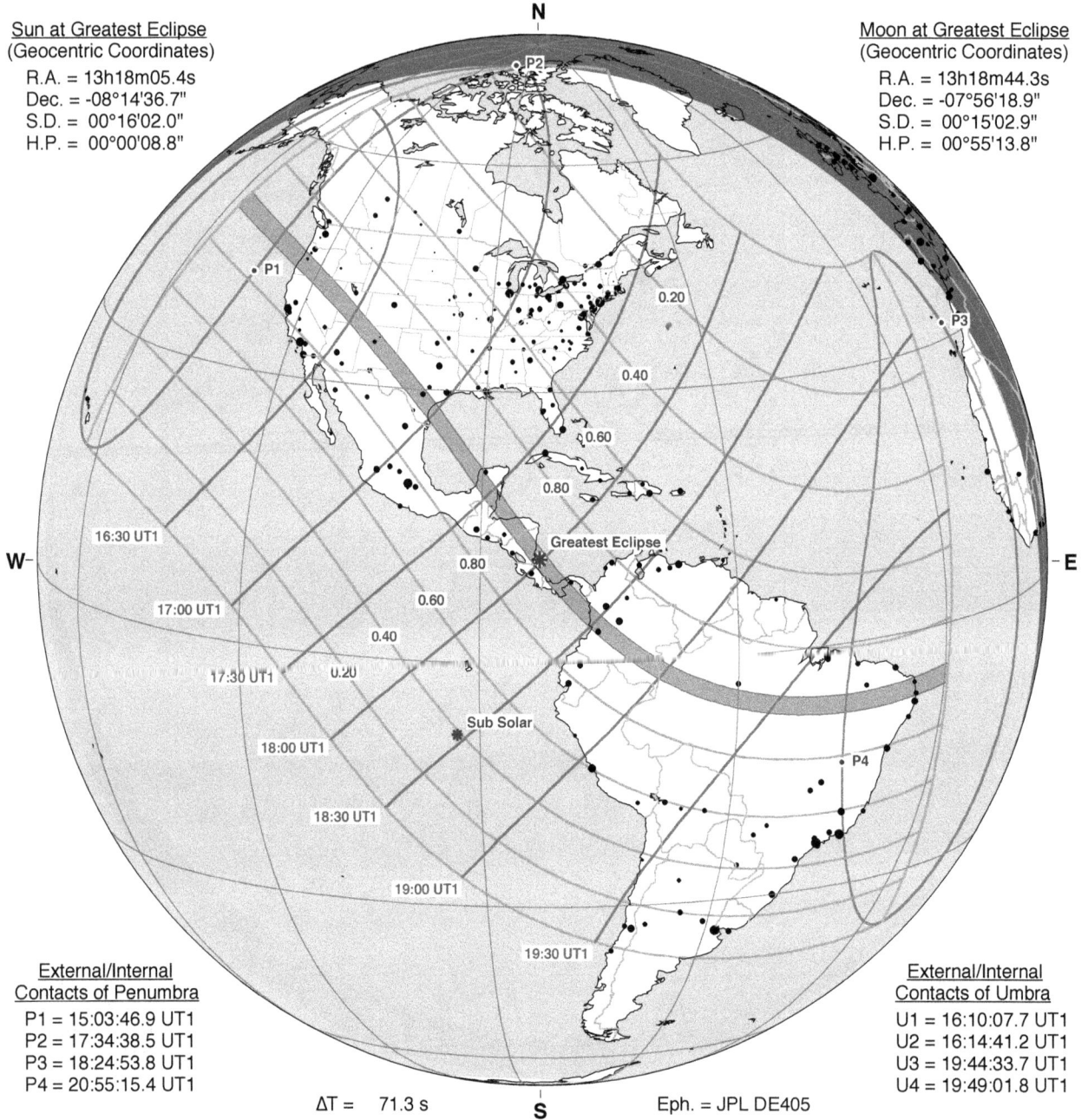

External/Internal
Contacts of Penumbra
P1 = 15:03:46.9 UT1
P2 = 17:34:38.5 UT1
P3 = 18:24:53.8 UT1
P4 = 20:55:15.4 UT1

External/Internal
Contacts of Umbra
U1 = 16:10:07.7 UT1
U2 = 16:14:41.2 UT1
U3 = 19:44:33.7 UT1
U4 = 19:49:01.8 UT1

ΔT = 71.3 s   Eph. = JPL DE405

Circumstances at Greatest Eclipse: 17:59:29.3 UT1
Lat. = 11°22.1'N   Sun Alt. = 67.9°
Long. = 083°06.1'W   Sun Azm. = 208.0°
Path Width = 187.4 km   Duration = 05m17.2s

Circumstances at Greatest Duration: 18:13:09.2 UT1
Lat. = 08°14.6'N   Sun Alt. = 66.8°
Long. = 080°24.1'W   Sun Azm. = 225.1°
Path Width = 191.1 km   Duration = 05m17.8s

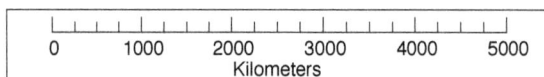

©2016 F. Espenak
www.EclipseWise.com

Total Solar Eclipse of 2024 Apr 08

Greatest Eclipse = 18:18:29.4 TD (= 18:17:17.9 UT1)

Eclipse Magnitude = 1.0566   Saros Series = 139

Gamma = 0.3431   Saros Member = 30 of 71

Sun at Greatest Eclipse
(Geocentric Coordinates)
R.A. = 01h11m36.9s
Dec. = +07°35'29.4"
S.D. = 00°15'58.2"
H.P. = 00°00'08.8"

Moon at Greatest Eclipse
(Geocentric Coordinates)
R.A. = 01h10m57.5s
Dec. = +07°53'55.5"
S.D. = 00°16'36.3"
H.P. = 01°00'56.6"

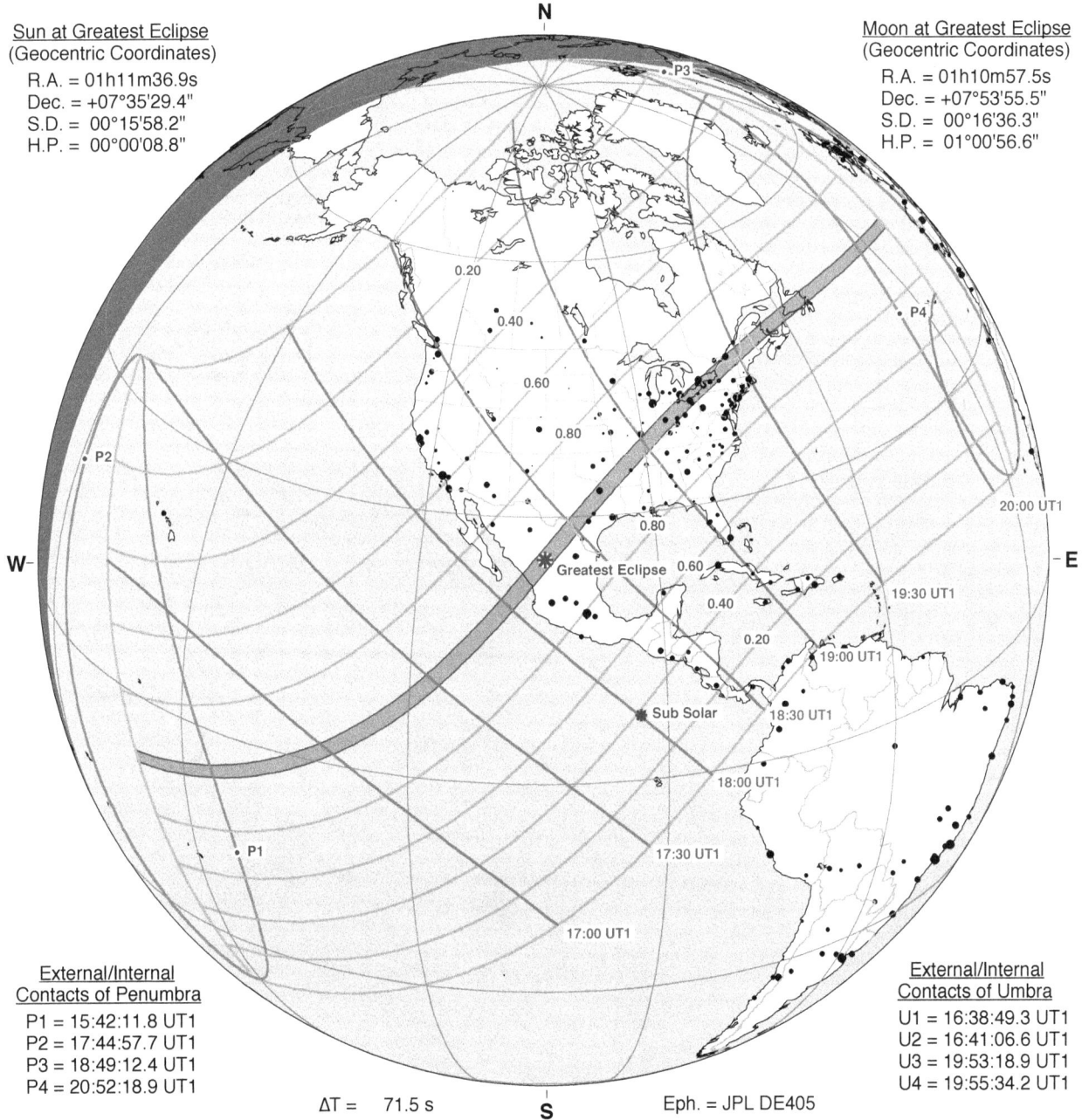

External/Internal
Contacts of Penumbra
P1 = 15:42:11.8 UT1
P2 = 17:44:57.7 UT1
P3 = 18:49:12.4 UT1
P4 = 20:52:18.9 UT1

External/Internal
Contacts of Umbra
U1 = 16:38:49.3 UT1
U2 = 16:41:06.6 UT1
U3 = 19:53:18.9 UT1
U4 = 19:55:34.2 UT1

ΔT = 71.5 s   Eph. = JPL DE405

Circumstances at Greatest Eclipse: 18:17:17.9 UT1

Lat. = 25°17.4'N	Sun Alt. = 69.8°
Long. = 104°08.3'W	Sun Azm. = 149.4°
Path Width = 197.5 km	Duration = 04m28.1s

Circumstances at Greatest Duration: 18:19:32.7 UT1

Lat. = 25°55.8'N	Sun Alt. = 69.7°
Long. = 103°31.1'W	Sun Azm. = 153.3°
Path Width = 197.0 km	Duration = 04m28.2s

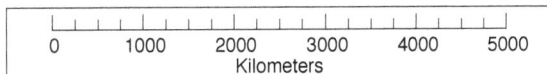

©2016 F. Espenak
www.EclipseWise.com

Annular Solar Eclipse of 2024 Oct 02

Greatest Eclipse = 18:46:13.3 TD   (= 18:45:01.6 UT1)

Eclipse Magnitude = 0.9326   Saros Series = 144
Gamma = -0.3509   Saros Member = 17 of 70

Sun at Greatest Eclipse
(Geocentric Coordinates)
R.A. = 12h36m58.9s
Dec. = -03°59'03.9"
S.D. = 00°15'58.9"
H.P. = 00°00'08.8"

Moon at Greatest Eclipse
(Geocentric Coordinates)
R.A. = 12h36m22.3s
Dec. = -04°15'35.4"
S.D. = 00°14'41.8"
H.P. = 00°53'56.4"

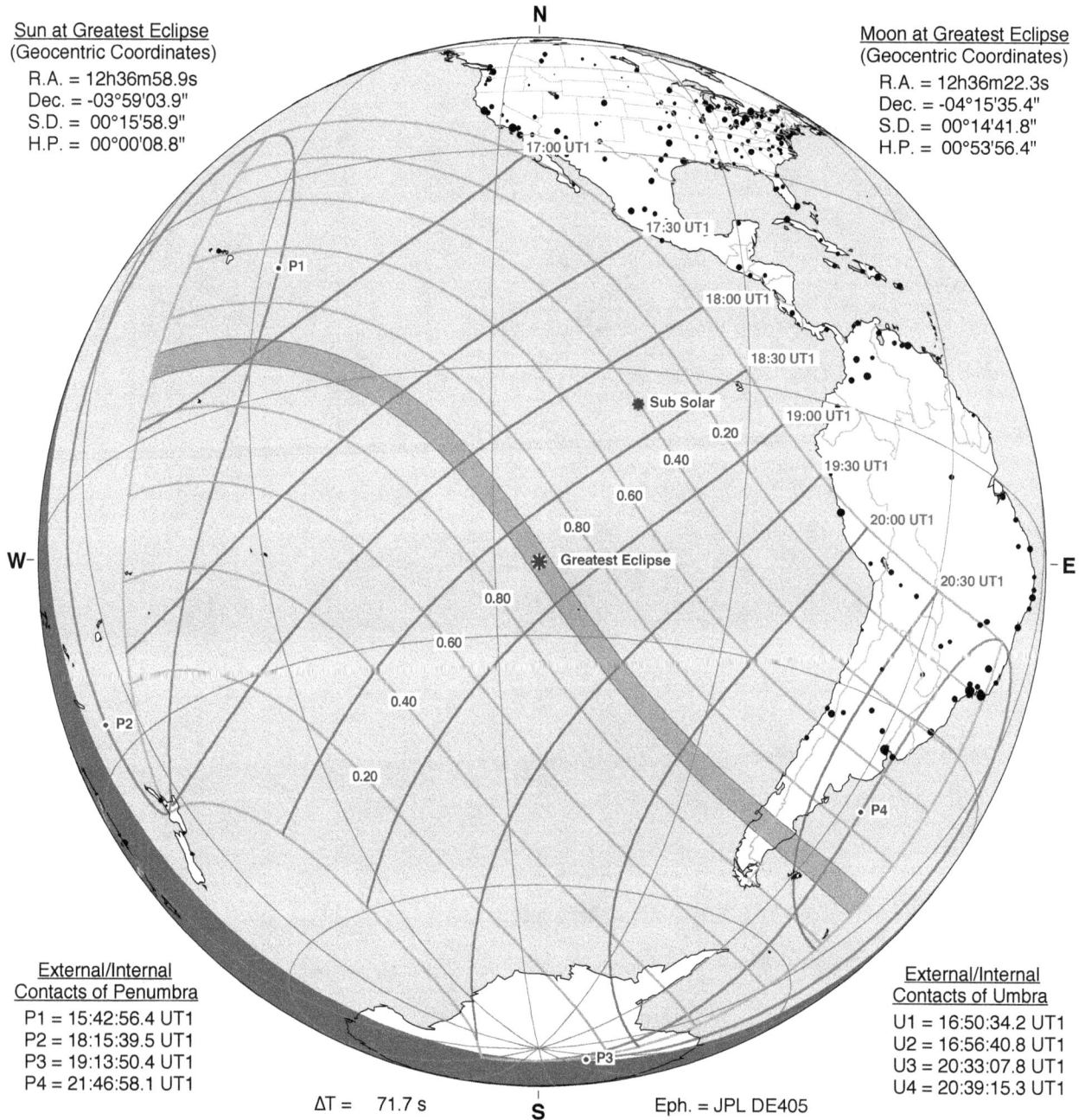

External/Internal Contacts of Penumbra
P1 = 15:42:56.4 UT1
P2 = 18:15:39.5 UT1
P3 = 19:13:50.4 UT1
P4 = 21:46:58.1 UT1

ΔT = 71.7 s

Eph. = JPL DE405

External/Internal Contacts of Umbra
U1 = 16:50:34.2 UT1
U2 = 16:56:40.8 UT1
U3 = 20:33:07.8 UT1
U4 = 20:39:15.3 UT1

Circumstances at Greatest Eclipse: 18:45:01.6 UT1
Lat. = 21°57.2'S   Sun Alt. = 69.3°
Long. = 114°30.5'W   Sun Azm. = 31.1°
Path Width = 266.5 km   Duration = 07m25.1s

Circumstances at Greatest Duration: 18:53:00.1 UT1
Lat. = 23°57.8'S   Sun Alt. = 68.9°
Long. = 112°56.3'W   Sun Azm. = 19.6°
Path Width = 264.4 km   Duration = 07m25.4s

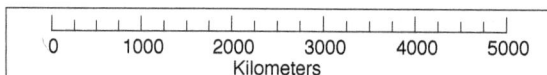

©2016 F. Espenak
www.EclipseWise.com

Partial Solar Eclipse of 2025 Mar 29

Greatest Eclipse = 10:48:36.1 TD   (= 10:47:24.2 UT1)

Eclipse Magnitude = 0.9376   Saros Series = 149
Gamma = 1.0405   Saros Member = 21 of 71

Sun at Greatest Eclipse
(Geocentric Coordinates)
R.A. = 00h33m03.1s
Dec. = +03°33'55.0"
S.D. = 00°16'01.1"
H.P. = 00°00'08.8"

Moon at Greatest Eclipse
(Geocentric Coordinates)
R.A. = 00h31m00.8s
Dec. = +04°29'34.1"
S.D. = 00°16'39.4"
H.P. = 01°01'07.8"

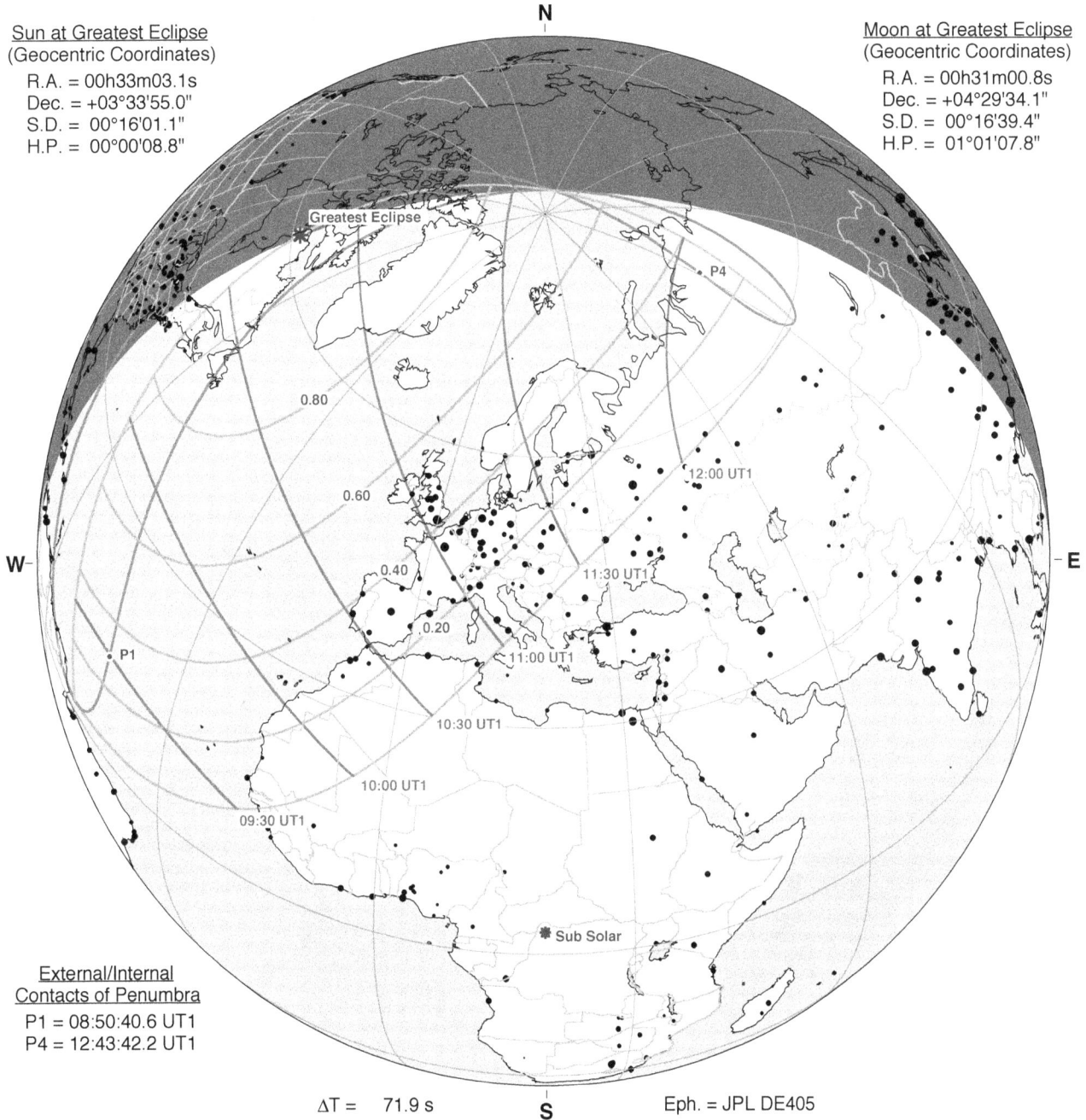

N

Greatest Eclipse

P4

0.80

0.60

0.40

0.20

12:00 UT1

11:30 UT1

11:00 UT1

10:30 UT1

10:00 UT1

09:30 UT1

P1

W

E

Sub Solar

External/Internal
Contacts of Penumbra
P1 = 08:50:40.6 UT1
P4 = 12:43:42.2 UT1

ΔT = 71.9 s

S

Eph. = JPL DE405

Circumstances at Greatest Eclipse: 10:47:24.2 UT1
Lat. = 61°06.0'N   Sun Alt. = 0.0°
Long. = 077°05.1'W   Sun Azm. = 82.6°

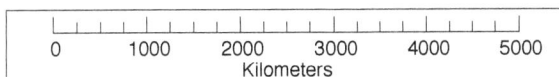

0   1000   2000   3000   4000   5000
Kilometers

©2016 F. Espenak
www.EclipseWise.com

Partial Solar Eclipse of 2025 Sep 21

Greatest Eclipse = 19:43:04.2 TD (= 19:41:52.2 UT1)

Eclipse Magnitude = 0.8550   Saros Series = 154

Gamma = -1.0651   Saros Member = 7 of 71

Sun at Greatest Eclipse
(Geocentric Coordinates)
R.A. = 11h56m36.9s
Dec. = +00°22'00.7"
S.D. = 00°15'55.9"
H.P. = 00°00'08.8"

Moon at Greatest Eclipse
(Geocentric Coordinates)
R.A. = 11h54m42.8s
Dec. = -00°29'14.7"
S.D. = 00°15'02.8"
H.P. = 00°55'13.2"

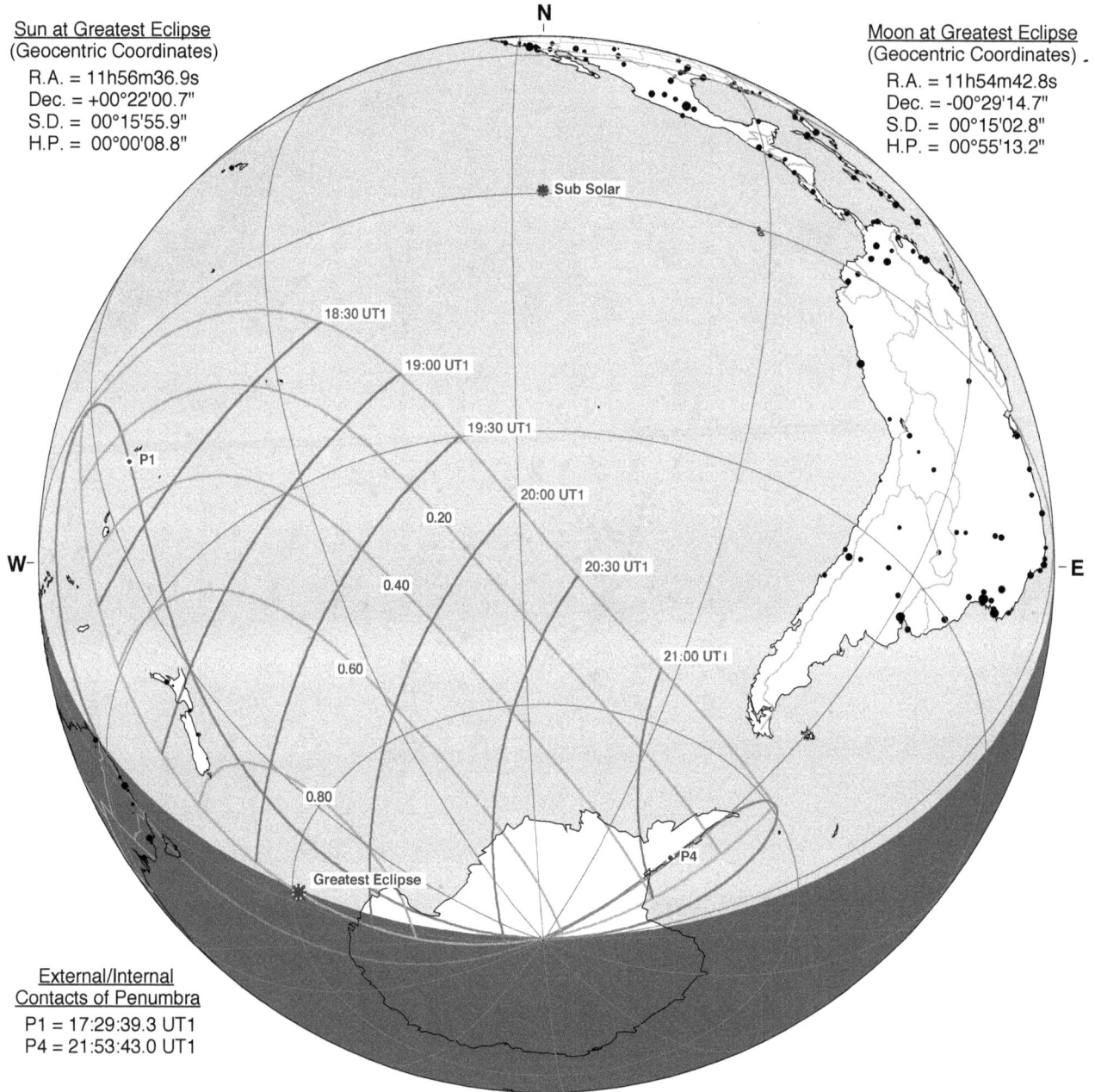

N

Sub Solar

18:30 UT1

19:00 UT1

19:30 UT1

20:00 UT1

0.20

20:30 UT1

P1

0.40

21:00 UT1

W

E

0.60

0.80

P4

Greatest Eclipse

External/Internal
Contacts of Penumbra
P1 = 17:29:39.3 UT1
P4 = 21:53:43.0 UT1

ΔT = 72.1 s   S   Eph. = JPL DE405

Circumstances at Greatest Eclipse: 19:41:52.2 UT1
Lat. = 60°54.1'S   Sun Alt. = 0.0°
Long. = 153°29.9'E   Sun Azm. = 89.2°

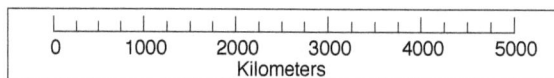

0	1000	2000	3000	4000	5000

Kilometers

©2016 F. Espenak
www.EclipseWise.com

Annular Solar Eclipse of 2026 Feb 17

Greatest Eclipse = 12:13:05.8 TD (= 12:11:53.6 UT1)

Eclipse Magnitude = 0.9630

Gamma = -0.9743

Saros Series = 121

Saros Member = 61 of 71

Sun at Greatest Eclipse
(Geocentric Coordinates)
R.A. = 22h03m54.3s
Dec. = -11°52'42.3"
S.D. = 00°16'11.1"
H.P. = 00°00'08.9"

Moon at Greatest Eclipse
(Geocentric Coordinates)
R.A. = 22h05m34.0s
Dec. = -12°42'29.5"
S.D. = 00°15'32.4"
H.P. = 00°57'02.0"

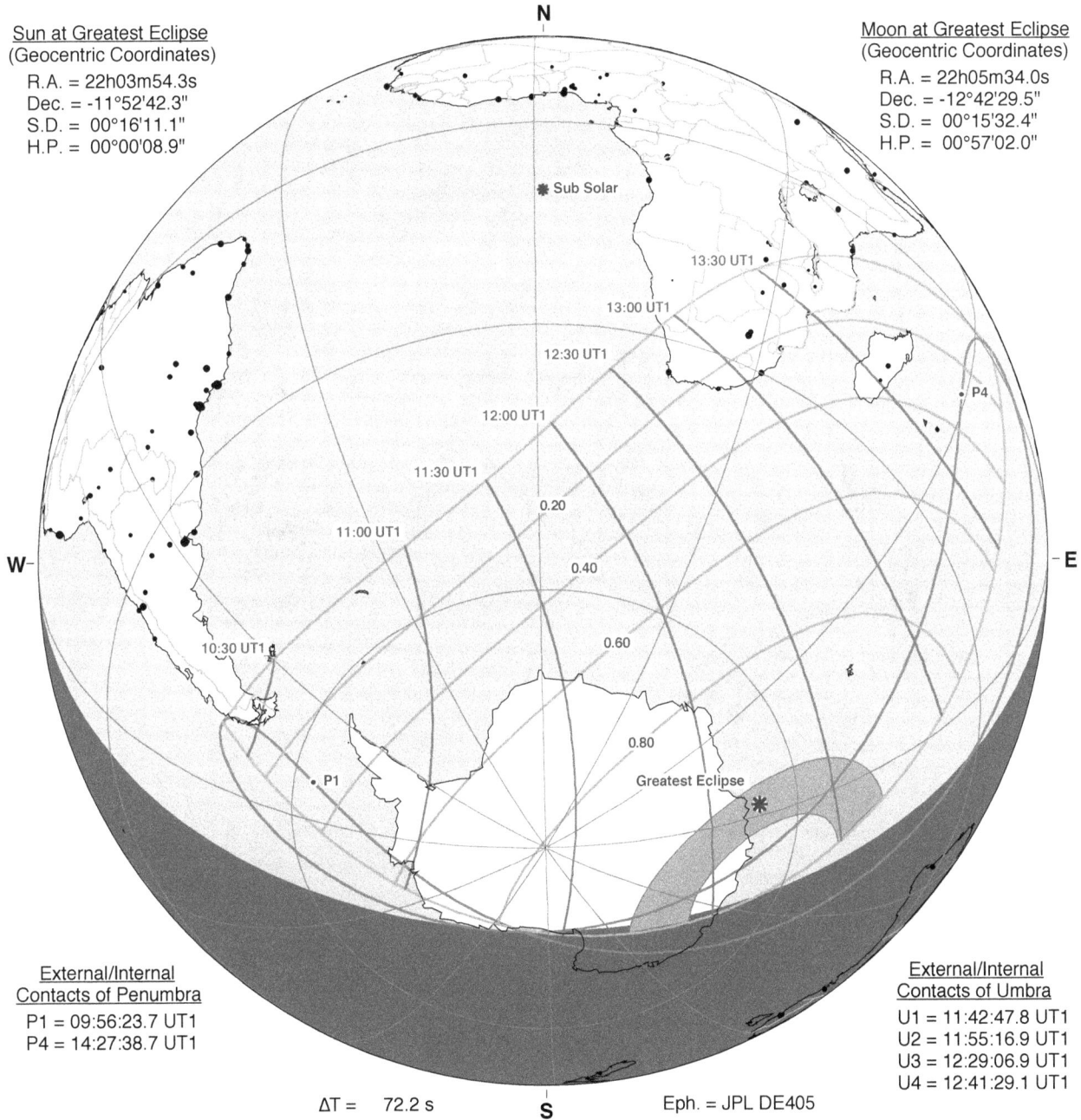

N

Sub Solar

13:30 UT1

13:00 UT1

12:30 UT1

12:00 UT1

11:30 UT1

0.20

11:00 UT1

0.40

W — — E

0.60

10:30 UT1

0.80

P1

Greatest Eclipse

P4

External/Internal
Contacts of Penumbra
P1 = 09:56:23.7 UT1
P4 = 14:27:38.7 UT1

External/Internal
Contacts of Umbra
U1 = 11:42:47.8 UT1
U2 = 11:55:16.9 UT1
U3 = 12:29:06.9 UT1
U4 = 12:41:29.1 UT1

ΔT = 72.2 s

S

Eph. = JPL DE405

Circumstances at Greatest Eclipse: 12:11:53.6 UT1

Lat. = 64°43.1'S	Sun Alt. = 12.3°
Long. = 086°44.4'E	Sun Azm. = 268.3°
Path Width = 616.2 km	Duration = 02m19.6s

Circumstances at Greatest Duration: 11:48:15.2 UT1

Lat. = 71°57.6'S	Sun Alt. = 0.0°
Long. = 136°39.8'E	Sun Azm. = 228.4°
Path Width = 765.7 km	Duration = 02m20.9s

©2016 F. Espenak
www.EclipseWise.com

0 1000 2000 3000 4000 5000
Kilometers

Total Solar Eclipse of 2026 Aug 12

Greatest Eclipse = 17:47:05.8 TD (= 17:45:53.3 UT1)

Eclipse Magnitude = 1.0386	Saros Series = 126
Gamma = 0.8977	Saros Member = 48 of 72

Sun at Greatest Eclipse
(Geocentric Coordinates)
R.A. = 09h29m47.3s
Dec. = +14°48'04.5"
S.D. = 00°15'47.0"
H.P. = 00°00'08.7"

Moon at Greatest Eclipse
(Geocentric Coordinates)
R.A. = 09h31m17.3s
Dec. = +15°36'58.5"
S.D. = 00°16'16.9"
H.P. = 00°59'45.1"

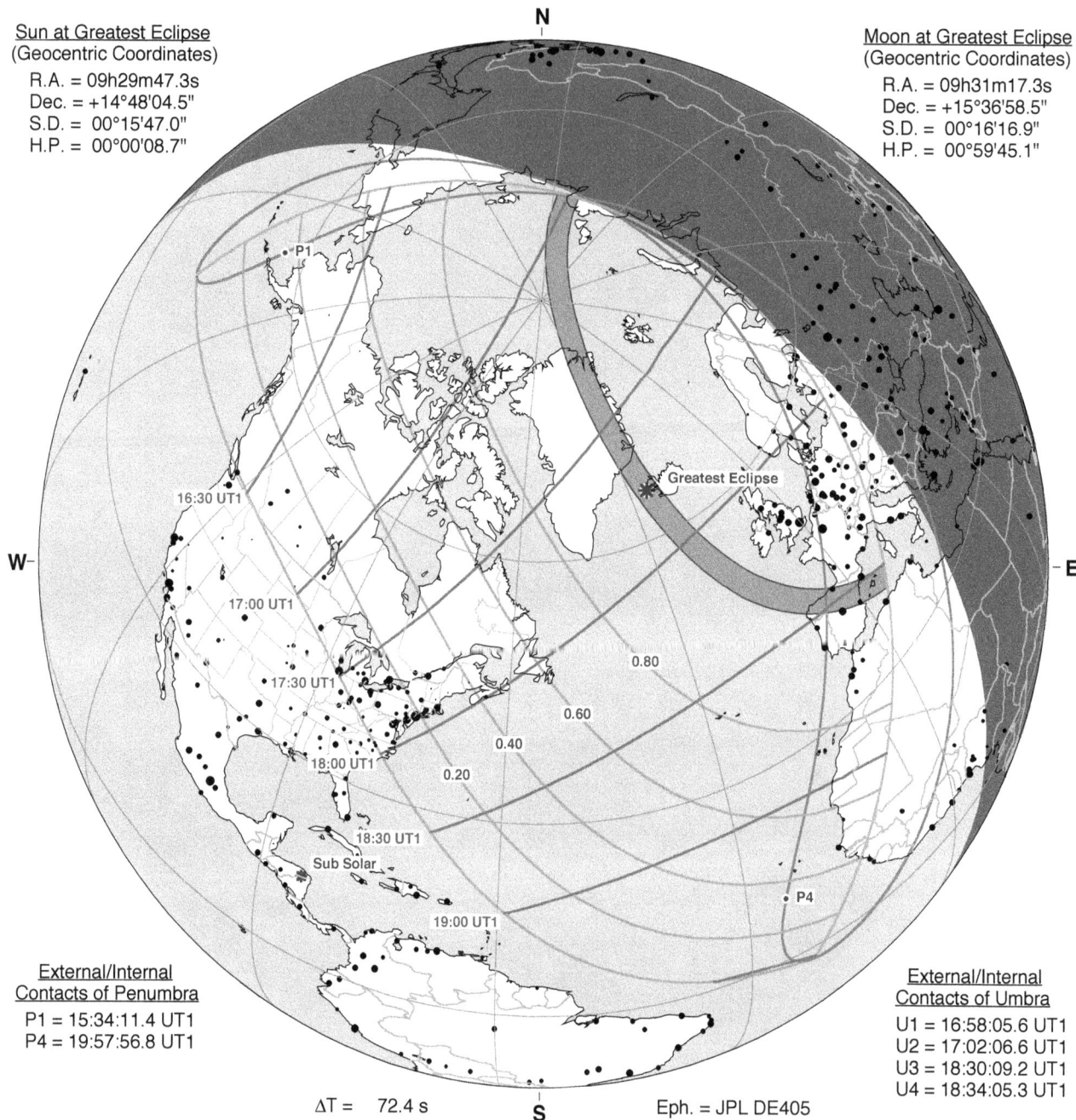

N

P1

Greatest Eclipse

16:30 UT1

W

E

17:00 UT1

17:30 UT1

0.80

0.60

18:00 UT1

0.40

0.20

18:30 UT1

Sub Solar

P4

19:00 UT1

External/Internal Contacts of Penumbra
P1 = 15:34:11.4 UT1
P4 = 19:57:56.8 UT1

External/Internal Contacts of Umbra
U1 = 16:58:05.6 UT1
U2 = 17:02:06.6 UT1
U3 = 18:30:09.2 UT1
U4 = 18:34:05.3 UT1

ΔT = 72.4 s

S

Eph. = JPL DE405

Circumstances at Greatest Eclipse: 17:45:53.3 UT1

Lat. = 65°13.4'N	Sun Alt. = 25.8°
Long. = 025°13.7'W	Sun Azm. = 248.4°
Path Width = 294.0 km	Duration = 02m18.2s

Circumstances at Greatest Duration: 17:44:41.5 UT1

Lat. = 65°49.8'N	Sun Alt. = 25.7°
Long. = 025°29.1'W	Sun Azm. = 247.6°
Path Width = 292.9 km	Duration = 02m18.2s

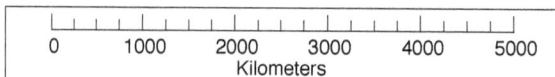

©2016 F. Espenak
www.EclipseWise.com

0   1000   2000   3000   4000   5000
Kilometers

Annular Solar Eclipse of 2027 Feb 06

Greatest Eclipse = 16:00:47.7 TD (= 15:59:35.1 UT1)

Eclipse Magnitude = 0.9281	Saros Series = 131
Gamma = -0.2952	Saros Member = 51 of 70

Sun at Greatest Eclipse
(Geocentric Coordinates)
R.A. = 21h20m17.6s
Dec. = -15°32'54.5"
S.D. = 00°16'13.1"
H.P. = 00°00'08.9"

Moon at Greatest Eclipse
(Geocentric Coordinates)
R.A. = 21h20m44.2s
Dec. = -15°47'36.0"
S.D. = 00°14'50.2"
H.P. = 00°54'27.0"

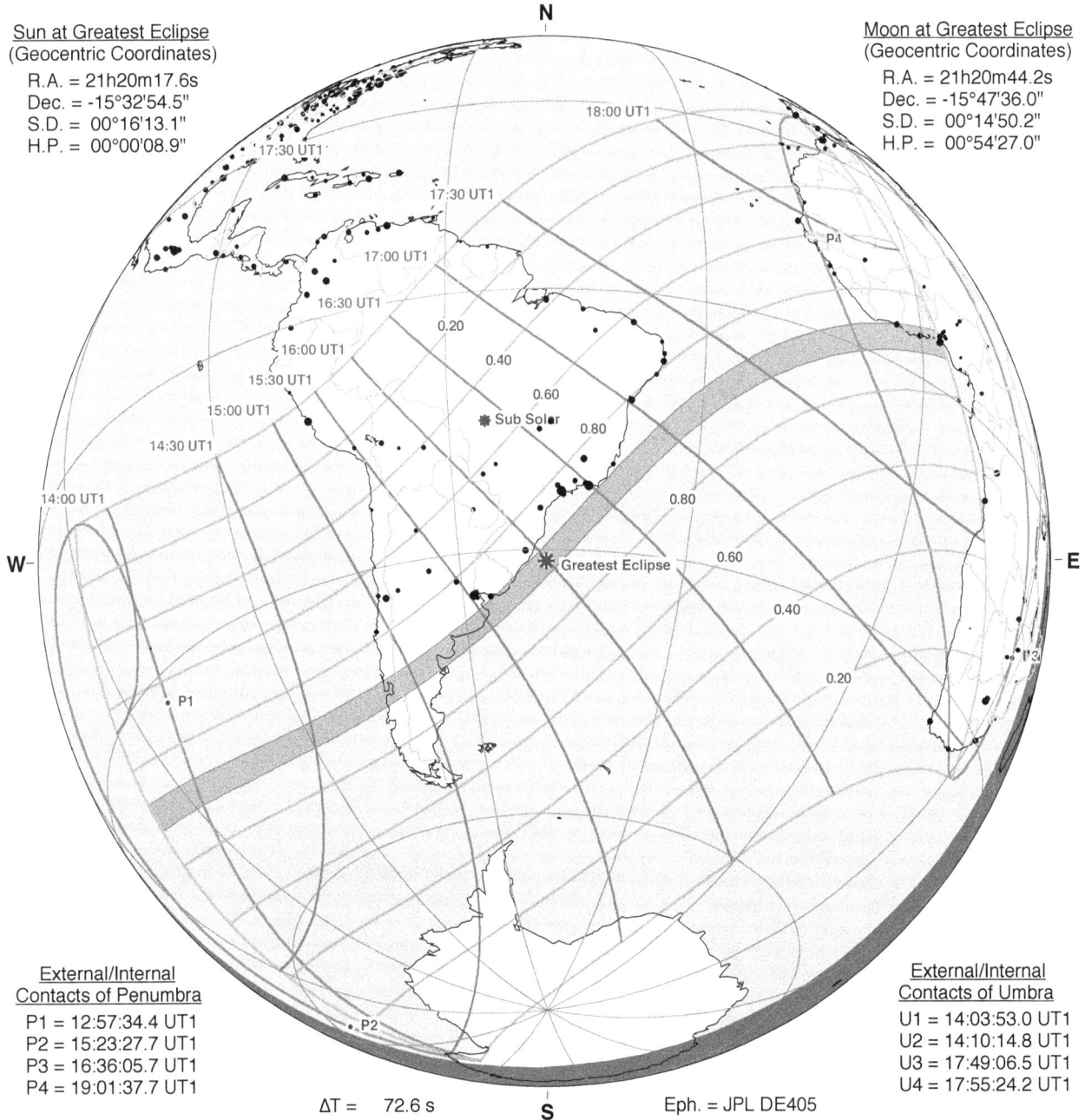

N

18:00 UT1
17:30 UT1
17:30 UT1
17:00 UT1
16:30 UT1
16:00 UT1
15:30 UT1
15:00 UT1
14:30 UT1
14:00 UT1

P4

0.20
0.40
0.60
Sub Solar
0.80

0.80
Greatest Eclipse
0.60
0.40
P3
0.20

W — E

P1

P2

External/Internal
Contacts of Penumbra
P1 = 12:57:34.4 UT1
P2 = 15:23:27.7 UT1
P3 = 16:36:05.7 UT1
P4 = 19:01:37.7 UT1

External/Internal
Contacts of Umbra
U1 = 14:03:53.0 UT1
U2 = 14:10:14.8 UT1
U3 = 17:49:06.5 UT1
U4 = 17:55:24.2 UT1

ΔT = 72.6 s

S

Eph. = JPL DE405

Circumstances at Greatest Eclipse: 15:59:35.1 UT1

Lat. = 31°18.2'S	Sun Alt. = 72.7°
Long. = 048°28.0'W	Sun Azm. = 333.5°
Path Width = 281.5 km	Duration = 07m50.9s

Circumstances at Greatest Duration: 15:41:47.6 UT1

Lat. = 35°04.2'S	Sun Alt. = 70.4°
Long. = 053°22.3'W	Sun Azm. = 4.1°
Path Width = 279.7 km	Duration = 07m53.5s

0	1000	2000	3000	4000	5000

Kilometers

©2016 F. Espenak
www.EclipseWise.com

Total Solar Eclipse of 2027 Aug 02

Greatest Eclipse = 10:07:50.2 TD   (= 10:06:37.4 UT1)

Eclipse Magnitude = 1.0790	Saros Series = 136
Gamma = 0.1421	Saros Member = 38 of 71

Sun at Greatest Eclipse
(Geocentric Coordinates)
R.A. = 08h49m26.9s
Dec. = +17°45'41.3"
S.D. = 00°15'45.5"
H.P. = 00°00'08.7"

Moon at Greatest Eclipse
(Geocentric Coordinates)
R.A. = 08h49m40.1s
Dec. = +17°53'47.8"
S.D. = 00°16'43.1"
H.P. = 01°01'21.4"

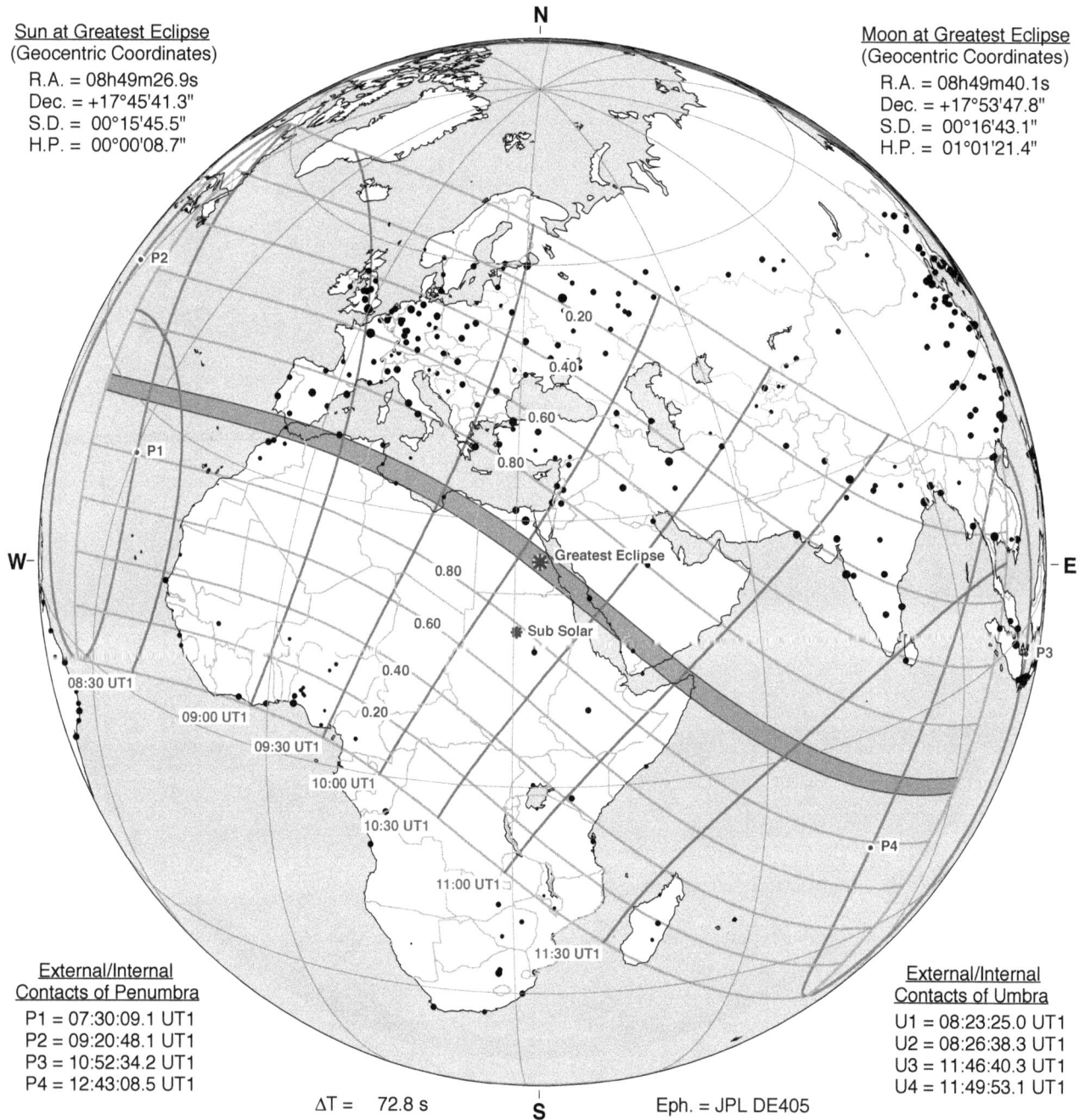

External/Internal Contacts of Penumbra
P1 = 07:30:09.1 UT1
P2 = 09:20:48.1 UT1
P3 = 10:52:34.2 UT1
P4 = 12:43:08.5 UT1

External/Internal Contacts of Umbra
U1 = 08:23:25.0 UT1
U2 = 08:26:38.3 UT1
U3 = 11:46:40.3 UT1
U4 = 11:49:53.1 UT1

ΔT = 72.8 s   Eph. = JPL DE405

Circumstances at Greatest Eclipse: 10:06:37.4 UT1

Lat. = 25°30.2'N	Sun Alt. = 81.7°
Long. = 033°11.0'E	Sun Azm. = 202.0°
Path Width = 257.7 km	Duration = 06m22.6s

Circumstances at Greatest Duration: 10:00:20.9 UT1

Lat. = 26°48.9'N	Sun Alt. = 80.9°
Long. = 031°07.9'E	Sun Azm. = 177.8°
Path Width = 257.2 km	Duration = 06m23.2s

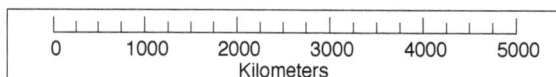

Annular Solar Eclipse of 2028 Jan 26

Greatest Eclipse = 15:08:58.8 TD (= 15:07:45.8 UT1)

Eclipse Magnitude = 0.9208   Saros Series = 141
Gamma = 0.3901   Saros Member = 24 of 70

Sun at Greatest Eclipse
(Geocentric Coordinates)
R.A. = 20h34m14.2s
Dec. = -18°43'33.0"
S.D. = 00°16'14.6"
H.P. = 00°00'08.9"

Moon at Greatest Eclipse
(Geocentric Coordinates)
R.A. = 20h33m43.7s
Dec. = -18°23'46.3"
S.D. = 00°14'45.1"
H.P. = 00°54'08.3"

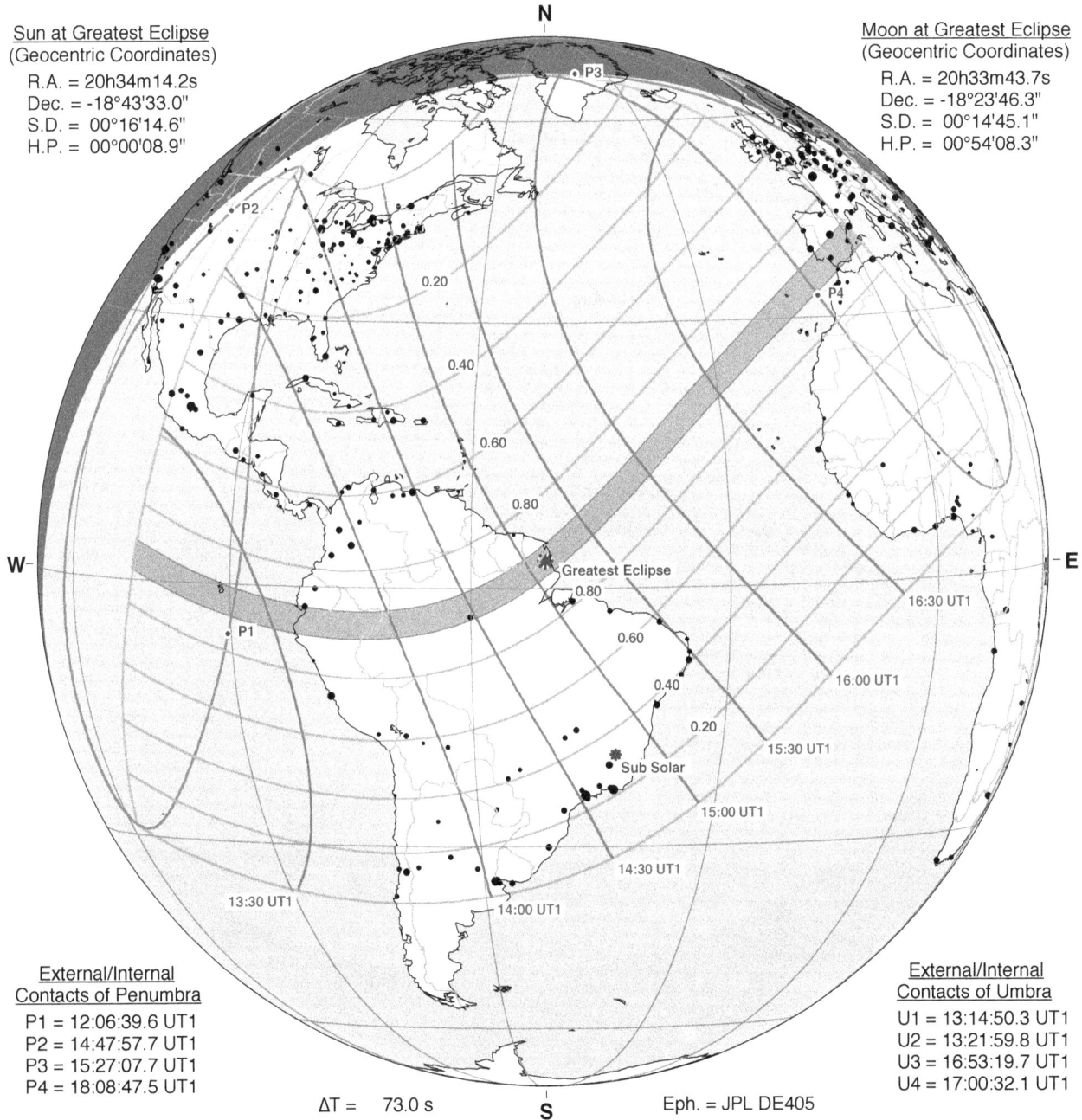

External/Internal
Contacts of Penumbra
P1 = 12:06:39.6 UT1
P2 = 14:47:57.7 UT1
P3 = 15:27:07.7 UT1
P4 = 18:08:47.5 UT1

ΔT = 73.0 s

External/Internal
Contacts of Umbra
U1 = 13:14:50.3 UT1
U2 = 13:21:59.8 UT1
U3 = 16:53:19.7 UT1
U4 = 17:00:32.1 UT1

Eph. = JPL DE405

Circumstances at Greatest Eclipse: 15:07:45.8 UT1
Lat. = 02°57.6'N   Sun Alt. = 67.0°
Long. = 051°33.6'W   Sun Azm. = 161.0°
Path Width = 323.0 km   Duration = 10m27.1s

Circumstances at Greatest Duration: 14:53:07.6 UT1
Lat. = 00°47.6'N   Sun Alt. = 65.8°
Long. = 054°39.0'W   Sun Azm. = 144.6°
Path Width = 328.7 km   Duration = 10m30.6s

©2016 F. Espenak
www.EclipseWise.com

Total Solar Eclipse of 2028 Jul 22

Greatest Eclipse = 02:56:39.6 TD (= 02:55:26.4 UT1)

Eclipse Magnitude = 1.0560   Saros Series = 146

Gamma = -0.6056   Saros Member = 28 of 76

Sun at Greatest Eclipse
(Geocentric Coordinates)
R.A. = 08h08m03.8s
Dec. = +20°10'53.0"
S.D. = 00°15'44.5"
H.P. = 00°00'08.7"

Moon at Greatest Eclipse
(Geocentric Coordinates)
R.A. = 08h07m16.7s
Dec. = +19°36'14.4"
S.D. = 00°16'24.3"
H.P. = 01°00'12.3"

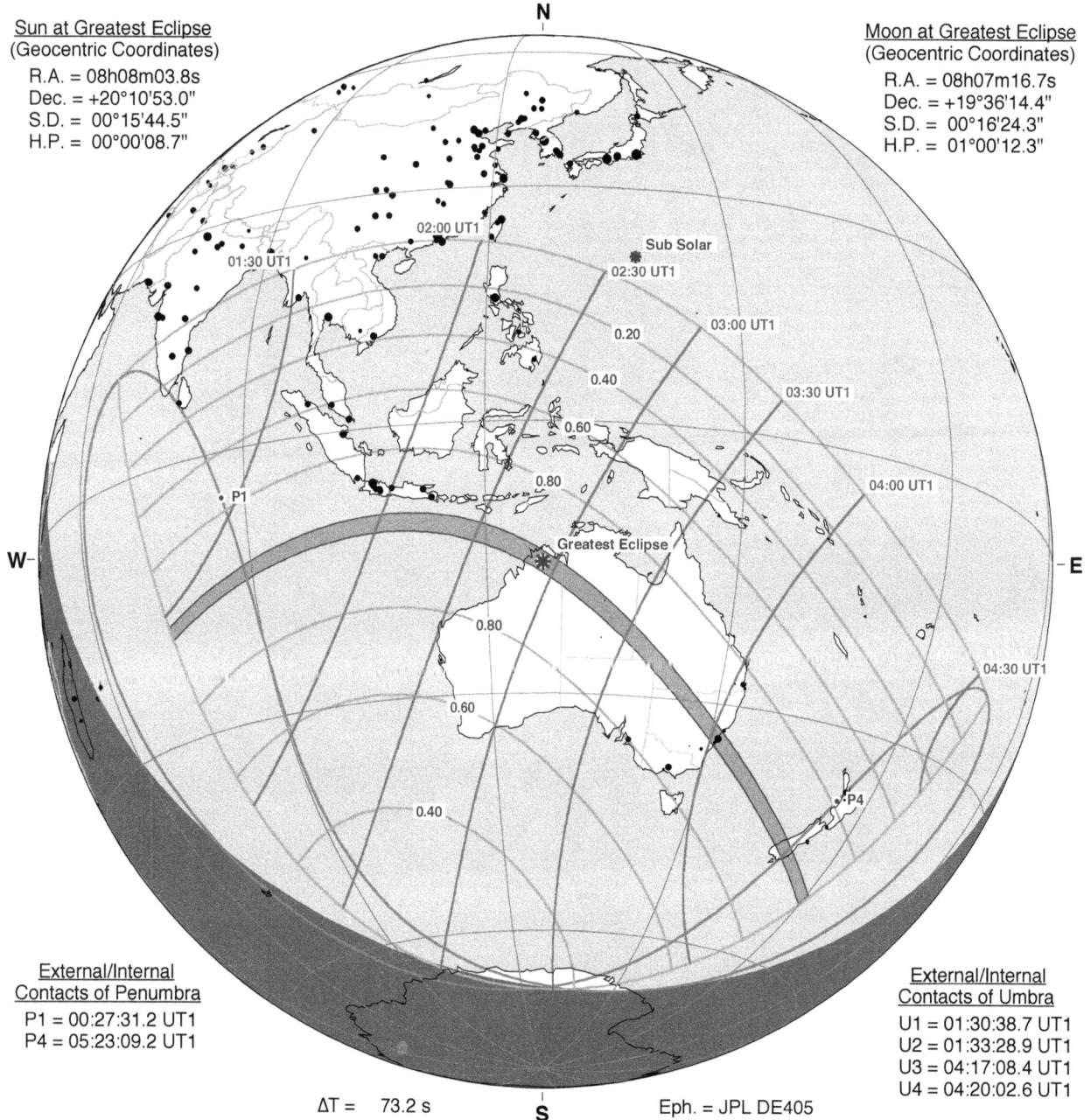

External/Internal Contacts of Penumbra
P1 = 00:27:31.2 UT1
P4 = 05:23:09.2 UT1

External/Internal Contacts of Umbra
U1 = 01:30:38.7 UT1
U2 = 01:33:28.9 UT1
U3 = 04:17:08.4 UT1
U4 = 04:20:02.6 UT1

ΔT = 73.2 s   Eph. = JPL DE405

Circumstances at Greatest Eclipse: 02:55:26.4 UT1

Lat. = 15°34.7'S	Sun Alt. = 52.6°
Long. = 126°42.4'E	Sun Azm. = 17.2°
Path Width = 230.2 km	Duration = 05m09.7s

Circumstances at Greatest Duration: 02:52:17.7 UT1

Lat. = 15°04.6'S	Sun Alt. = 52.6°
Long. = 125°47.3'E	Sun Azm. = 19.9°
Path Width = 231.4 km	Duration = 05m09.9s

0   1000   2000   3000   4000   5000
Kilometers

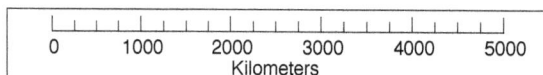

©2016 F. Espenak
www.EclipseWise.com

Partial Solar Eclipse of 2029 Jan 14

Greatest Eclipse = 17:13:47.5 TD (= 17:12:34.1 UT1)

Eclipse Magnitude = 0.8714   Saros Series = 151
Gamma = 1.0553   Saros Member = 15 of 72

Sun at Greatest Eclipse
(Geocentric Coordinates)
R.A. = 19h47m03.1s
Dec. = -21°09'31.8"
S.D. = 00°16'15.6"
H.P. = 00°00'08.9"

Moon at Greatest Eclipse
(Geocentric Coordinates)
R.A. = 19h45m53.5s
Dec. = -20°12'32.3"
S.D. = 00°15'20.6"
H.P. = 00°56'18.7"

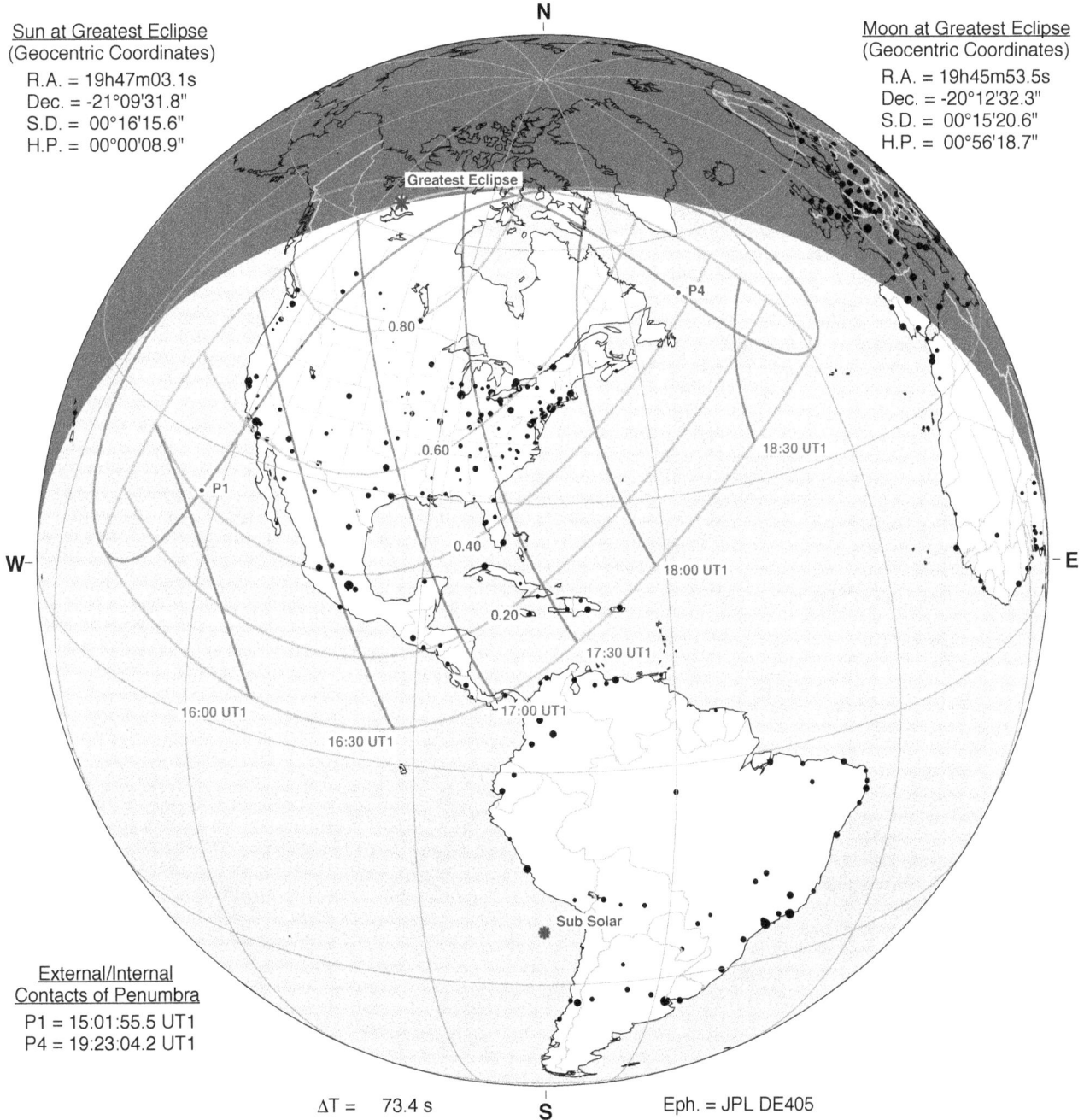

N

Greatest Eclipse

P4

0.80

0.60

0.40

18:30 UT1

18:00 UT1

0.20

17:30 UT1

17:00 UT1

16:00 UT1

16:30 UT1

P1

W

E

Sub Solar

External/Internal
Contacts of Penumbra
P1 = 15:01:55.5 UT1
P4 = 19:23:04.2 UT1

ΔT = 73.4 s

S

Eph. = JPL DE405

Circumstances at Greatest Eclipse: 17:12:34.1 UT1
Lat. = 63°42.1'N   Sun Alt. =   0.0°
Long. = 114°13.5'W   Sun Azm. = 144.6°

0   1000   2000   3000   4000   5000
Kilometers

©2016 F. Espenak
www.EclipseWise.com

Partial Solar Eclipse of 2029 Jun 12

Greatest Eclipse = 04:06:13.0 TD   (= 04:04:59.5 UT1)

Eclipse Magnitude = 0.4576   Saros Series = 118
Gamma = 1.2943   Saros Member = 69 of 72

Sun at Greatest Eclipse
(Geocentric Coordinates)
R.A. = 05h22m58.2s
Dec. = +23°09'45.7"
S.D. = 00°15'45.0"
H.P. = 00°00'08.7"

Moon at Greatest Eclipse
(Geocentric Coordinates)
R.A. = 05h23m08.9s
Dec. = +24°21'37.7"
S.D. = 00°15'10.6"
H.P. = 00°55'42.0"

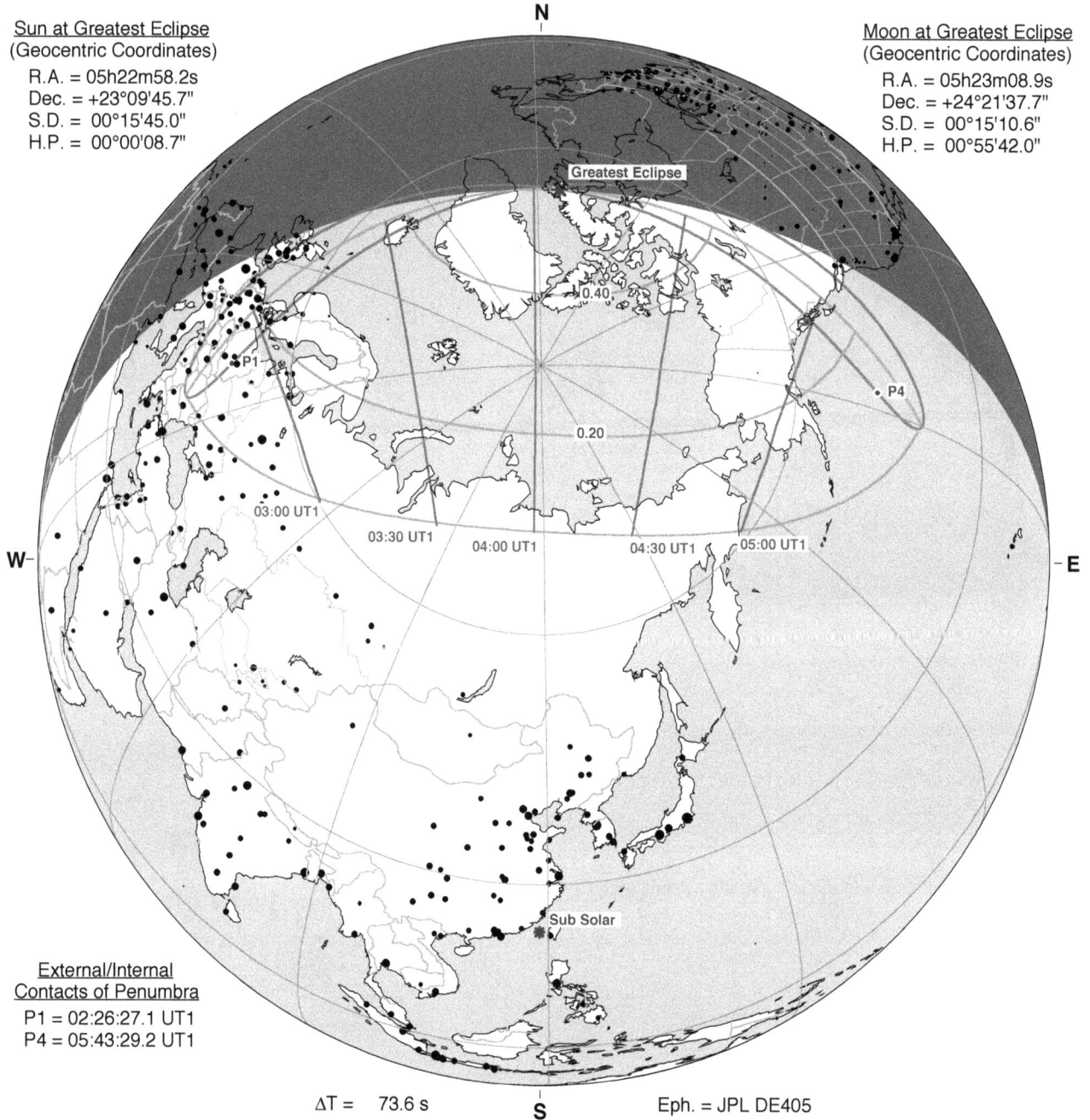

N

Greatest Eclipse

0.40

P1

P4

0.20

03:00 UT1

03:30 UT1   04:00 UT1   04:30 UT1   05:00 UT1

W   E

Sub Solar

**External/Internal
Contacts of Penumbra**
P1 = 02:26:27.1 UT1
P4 = 05:43:29.2 UT1

ΔT =   73.6 s   S   Eph. = JPL DE405

Circumstances at Greatest Eclipse: 04:04:59.5 UT1
Lat. = 66°45.9'N   Sun Alt. =   0.0°
Long. = 066°11.0'W   Sun Azm. = 355.5°

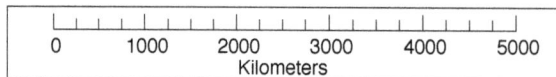

0   1000   2000   3000   4000   5000
Kilometers

©2016 F. Espenak
www.EclipseWise.com

Partial Solar Eclipse of 2029 Jul 11

Greatest Eclipse = 15:37:18.9 TD   (= 15:36:05.3 UT1)

Eclipse Magnitude = 0.2303   Saros Series = 156
Gamma = -1.4191   Saros Member = 2 of 69

Sun at Greatest Eclipse
(Geocentric Coordinates)
R.A. = 07h24m55.6s
Dec. = +22°00'04.3"
S.D. = 00°15'43.9"
H.P. = 00°00'08.7"

Moon at Greatest Eclipse
(Geocentric Coordinates)
R.A. = 07h23m33.7s
Dec. = +20°41'22.0"
S.D. = 00°15'35.3"
H.P. = 00°57'12.6"

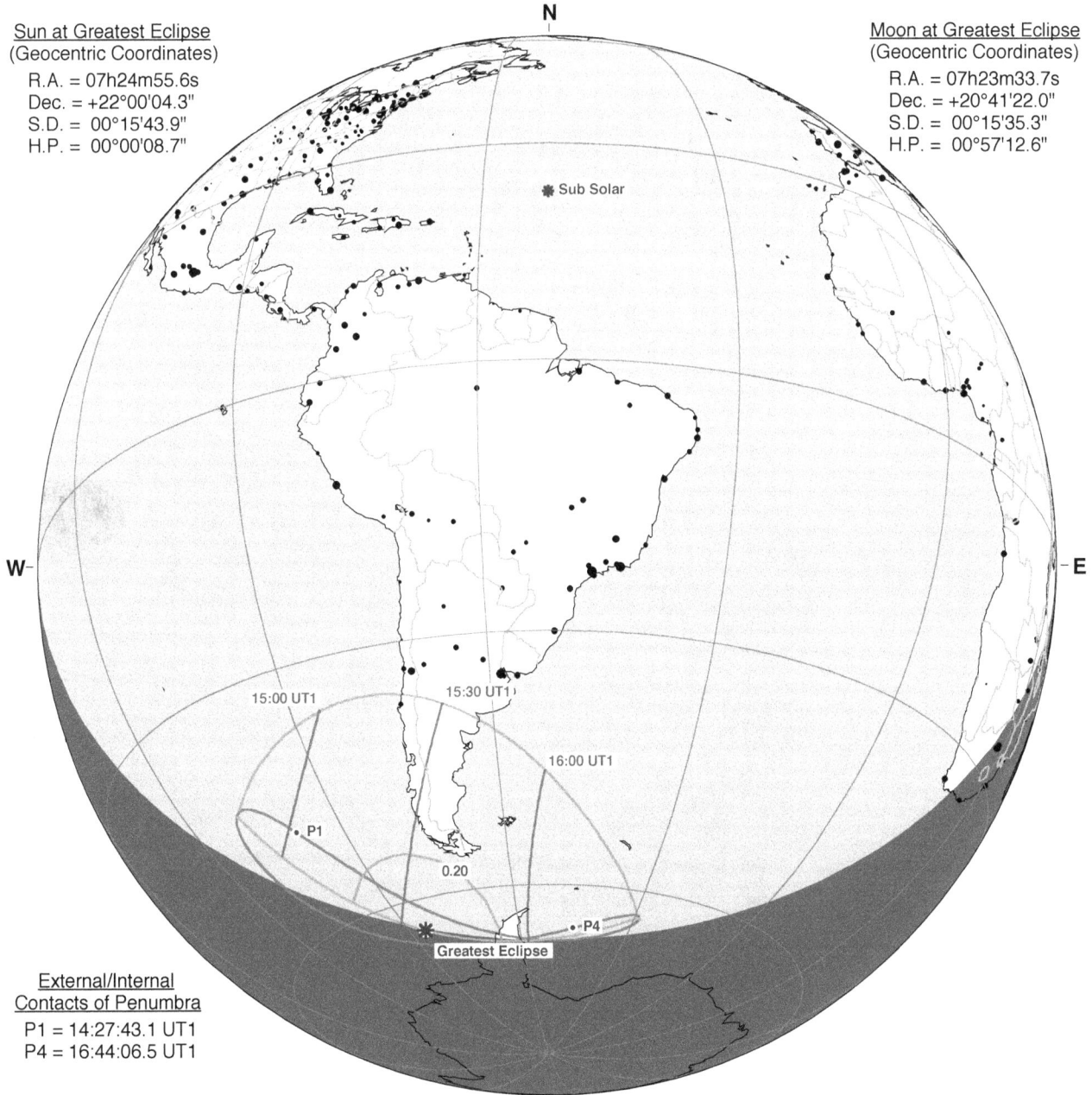

N

✳ Sub Solar

W

E

15:00 UT1   15:30 UT1
              16:00 UT1
· P1
        0.20
                    · P4
    ✳
Greatest Eclipse

External/Internal
Contacts of Penumbra
P1 = 14:27:43.1 UT1
P4 = 16:44:06.5 UT1

ΔT = 73.6 s   S   Eph. = JPL DE405

Circumstances at Greatest Eclipse: 15:36:05.3 UT1
Lat. = 64°15.8'S   Sun Alt. = 0.0°
Long. = 085°37.4'W   Sun Azm. = 30.4°

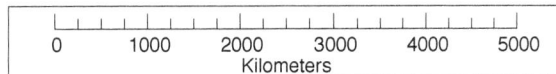

| 0 | 1000 | 2000 | 3000 | 4000 | 5000 |
Kilometers

©2016 F. Espenak
www.EclipseWise.com

Partial Solar Eclipse of 2029 Dec 05

Greatest Eclipse = 15:03:58.0 TD   (= 15:02:44.2 UT1)

Eclipse Magnitude = 0.8911   Saros Series = 123
Gamma = -1.0609   Saros Member = 54 of 70

Sun at Greatest Eclipse
(Geocentric Coordinates)
R.A. = 16h49m34.2s
Dec. = -22°26'54.3"
S.D. = 00°16'13.8"
H.P. = 00°00'08.9"

Moon at Greatest Eclipse
(Geocentric Coordinates)
R.A. = 16h49m27.4s
Dec. = -23°31'15.0"
S.D. = 00°16'34.3"
H.P. = 01°00'49.1"

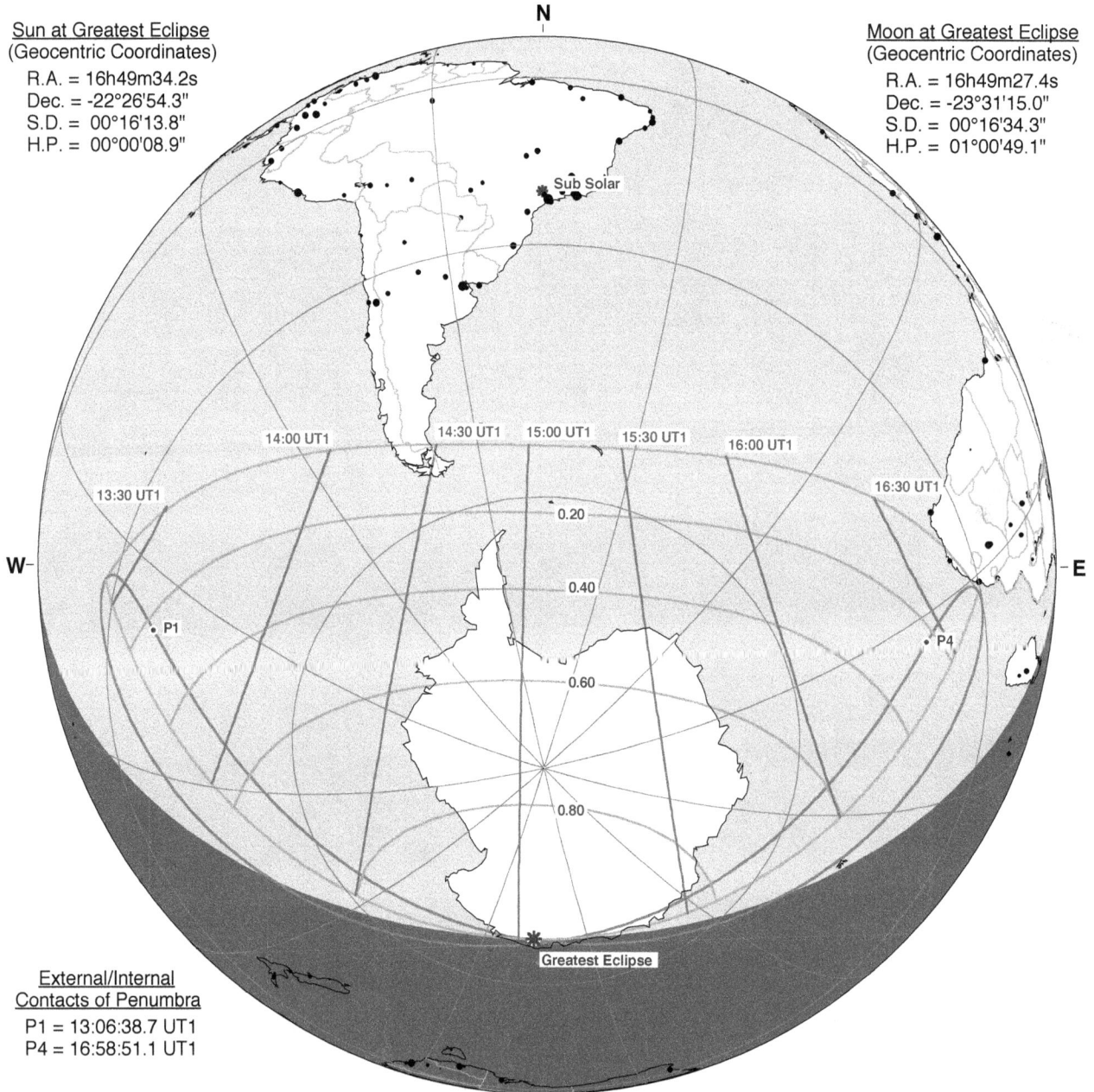

N

Sub Solar

14:00 UT1   14:30 UT1   15:00 UT1   15:30 UT1   16:00 UT1

13:30 UT1   16:30 UT1

W   E

0.20

P1   P4

0.40

0.60

0.80

Greatest Eclipse

External/Internal
Contacts of Penumbra
P1 = 13:06:38.7 UT1
P4 = 16:58:51.1 UT1

ΔT = 73.8 s   S   Eph. = JPL DE405

Circumstances at Greatest Eclipse: 15:02:44.2 UT1
Lat. = 67°30.8'S   Sun Alt. =   0.0°
Long. = 135°38.7'E   Sun Azm. = 176.6°

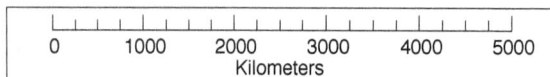

```
0    1000   2000   3000   4000   5000
            Kilometers
```

©2016 F. Espenak
www.EclipseWise.com

Annular Solar Eclipse of 2030 Jun 01

Greatest Eclipse = 06:29:12.9 TD (= 06:27:58.8 UT1)

Eclipse Magnitude = 0.9443   Saros Series = 128
Gamma = 0.5626   Saros Member = 59 of 73

Sun at Greatest Eclipse
(Geocentric Coordinates)
R.A. = 04h37m01.2s
Dec. = +22°03'55.3"
S.D. = 00°15'46.4"
H.P. = 00°00'08.7"

Moon at Greatest Eclipse
(Geocentric Coordinates)
R.A. = 04h36m55.8s
Dec. = +22°34'11.5"
S.D. = 00°14'42.7"
H.P. = 00°53'59.6"

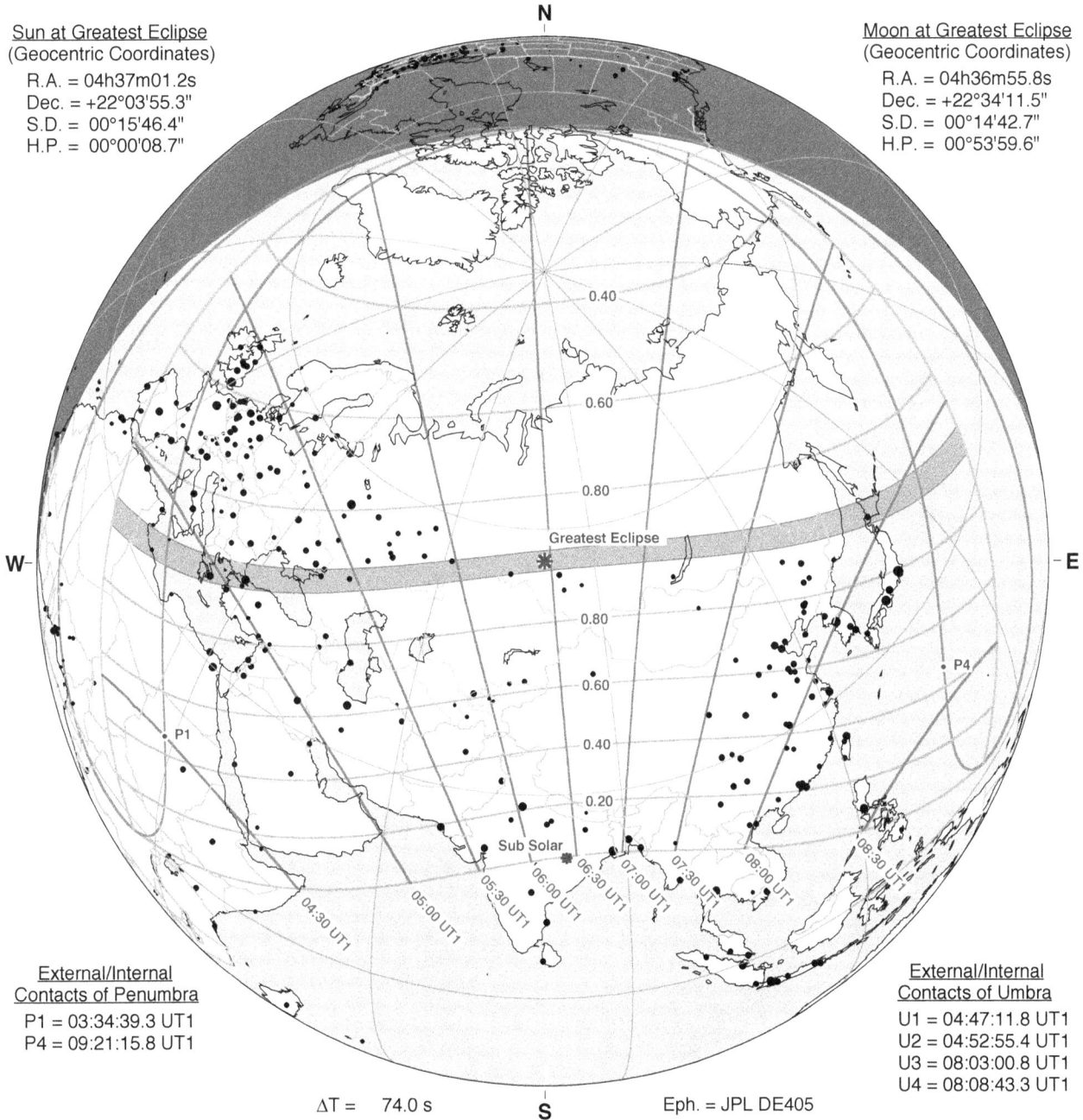

N

0.40
0.60
0.80
Greatest Eclipse
0.80
0.60
0.40
0.20

P1
P4

Sub Solar

04:30 UT1 05:00 UT1 05:30 UT1 06:00 UT1 06:30 UT1 07:00 UT1 07:30 UT1 08:00 UT1 08:30 UT1

W

E

External/Internal
Contacts of Penumbra
P1 = 03:34:39.3 UT1
P4 = 09:21:15.8 UT1

External/Internal
Contacts of Umbra
U1 = 04:47:11.8 UT1
U2 = 04:52:55.4 UT1
U3 = 08:03:00.8 UT1
U4 = 08:08:43.3 UT1

ΔT = 74.0 s

S

Eph. = JPL DE405

Circumstances at Greatest Eclipse: 06:27:58.8 UT1
Lat. = 56°31.5'N   Sun Alt. = 55.5°
Long. = 080°03.7'E   Sun Azm. = 176.1°
Path Width = 249.6 km   Duration = 05m20.8s

Circumstances at Greatest Duration: 06:28:41.2 UT1
Lat. = 56°32.9'N   Sun Alt. = 55.5°
Long. = 080°30.1'E   Sun Azm. = 177.1°
Path Width = 249.6 km   Duration = 05m20.8s

©2016 F. Espenak
www.EclipseWise.com

0 1000 2000 3000 4000 5000
Kilometers

Total Solar Eclipse of 2030 Nov 25

Greatest Eclipse = 06:51:36.9 TD   (= 06:50:22.7 UT1)

Eclipse Magnitude = 1.0468   Saros Series = 133
Gamma = -0.3867   Saros Member = 46 of 72

Sun at Greatest Eclipse
(Geocentric Coordinates)
R.A. = 16h03m58.7s
Dec. = -20°45'39.0"
S.D. = 00°16'12.1"
H.P. = 00°00'08.9"

Moon at Greatest Eclipse
(Geocentric Coordinates)
R.A. = 16h03m49.1s
Dec. = -21°09'10.6"
S.D. = 00°16'41.7"
H.P. = 01°01'16.4"

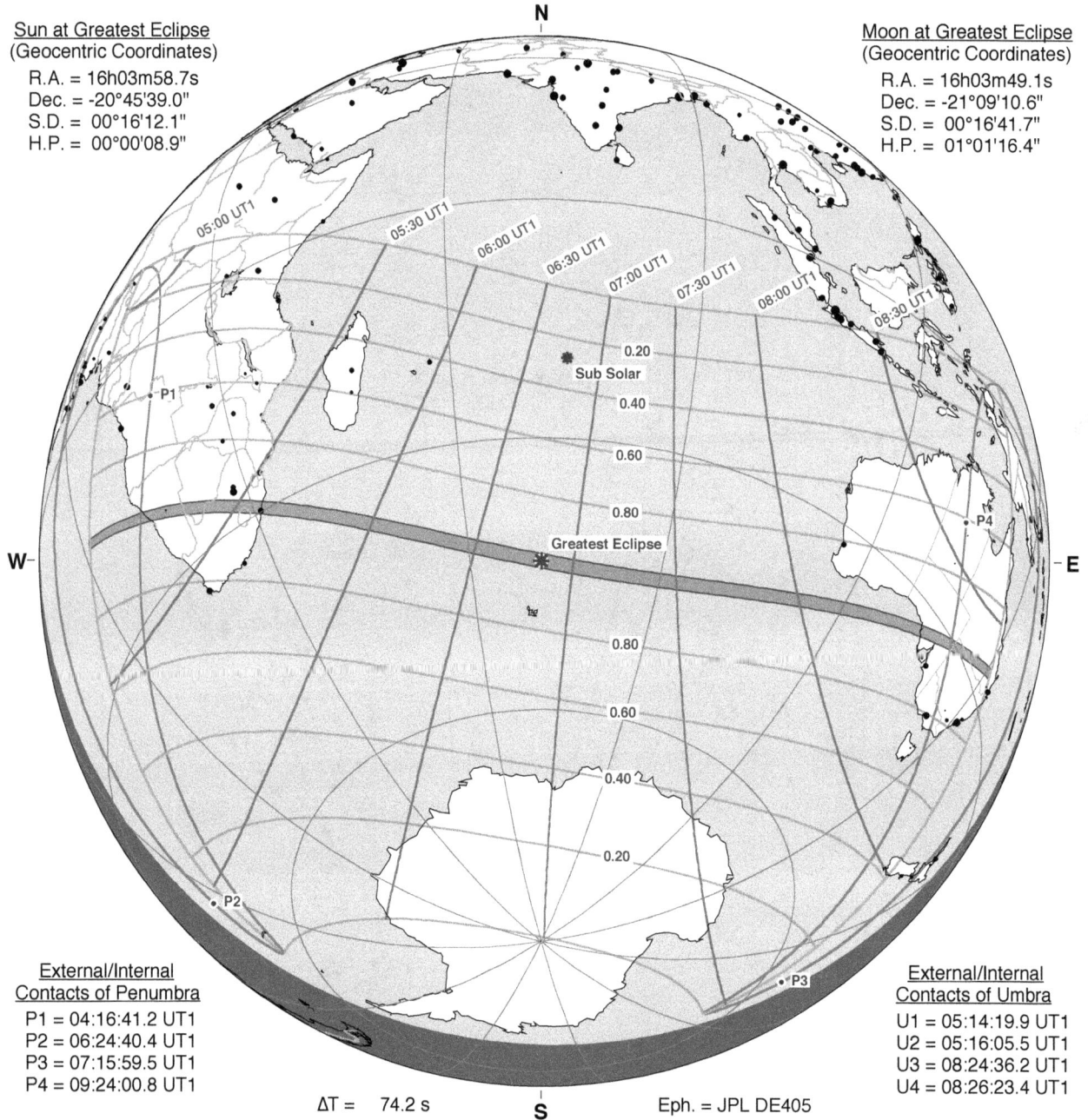

N

05:00 UT1   05:30 UT1   06:00 UT1   06:30 UT1   07:00 UT1   07:30 UT1   08:00 UT1   08:30 UT1

P1

0.20
Sub Solar
0.40
0.60
0.80
Greatest Eclipse
0.80
0.60
0.40
0.20

W ─   ─ E

P4

P2

P3

External/Internal
Contacts of Penumbra
P1 = 04:16:41.2 UT1
P2 = 06:24:40.4 UT1
P3 = 07:15:59.5 UT1
P4 = 09:24:00.8 UT1

External/Internal
Contacts of Umbra
U1 = 05:14:19.9 UT1
U2 = 05:16:05.5 UT1
U3 = 08:24:36.2 UT1
U4 = 08:26:23.4 UT1

ΔT = 74.2 s   S   Eph. = JPL DE405

Circumstances at Greatest Eclipse: 06:50:22.7 UT1

Lat. = 43°36.6'S	Sun Alt. = 67.0°
Long. = 071°13.7'E	Sun Azm. = 7.0°
Path Width = 169.3 km	Duration = 03m43.5s

Circumstances at Greatest Duration: 06:51:56.0 UT1

Lat. = 43°42.7'S	Sun Alt. = 67.0°
Long. = 072°01.4'E	Sun Azm. = 4.1°
Path Width = 169.2 km	Duration = 03m43.6s

0   1000   2000   3000   4000   5000
Kilometers

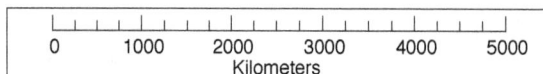

©2016 F. Espenak
www.EclipseWise.com

Annular Solar Eclipse of 2031 May 21

Greatest Eclipse = 07:16:04.3 TD   (= 07:14:49.9 UT1)

Eclipse Magnitude = 0.9589   Saros Series = 138
Gamma = -0.1970   Saros Member = 32 of 70

Sun at Greatest Eclipse
(Geocentric Coordinates)
R.A. = 03h51m34.6s
Dec. = +20°09'39.2"
S.D. = 00°15'48.2"
H.P. = 00°00'08.7"

Moon at Greatest Eclipse
(Geocentric Coordinates)
R.A. = 03h51m39.8s
Dec. = +19°58'57.5"
S.D. = 00°14'55.8"
H.P. = 00°54'47.5"

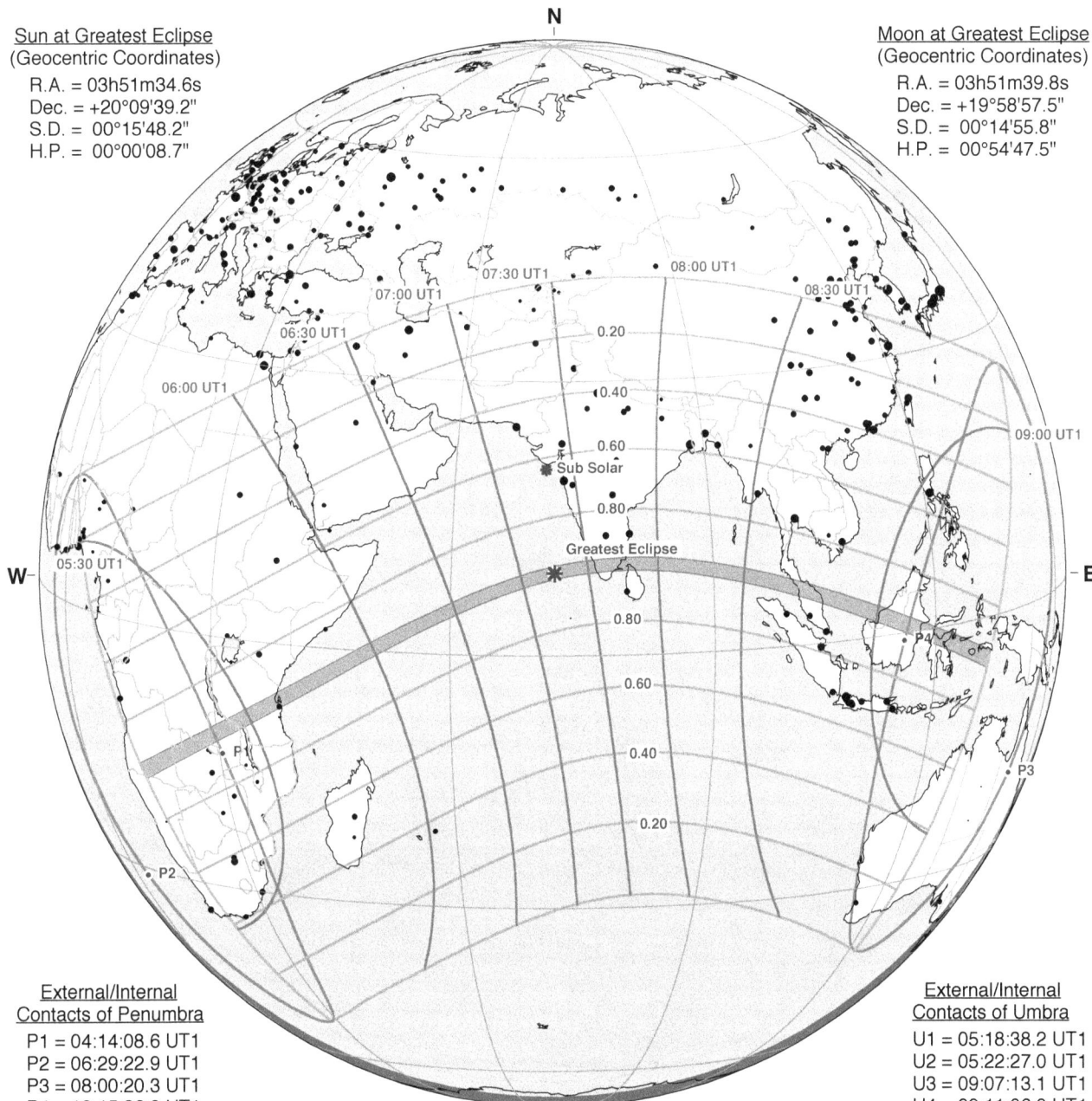

N

07:30 UT1
08:00 UT1
07:00 UT1
08:30 UT1
06:30 UT1
06:00 UT1
0.20
0.40
0.60
Sub Solar
0.80
09:00 UT1
05:30 UT1
W
Greatest Eclipse
E
0.80
P4
0.60
P1
0.40
P2
0.20
P3

External/Internal Contacts of Penumbra
P1 = 04:14:08.6 UT1
P2 = 06:29:22.9 UT1
P3 = 08:00:20.3 UT1
P4 = 10:15:36.9 UT1

External/Internal Contacts of Umbra
U1 = 05:18:38.2 UT1
U2 = 05:22:27.0 UT1
U3 = 09:07:13.1 UT1
U4 = 09:11:06.0 UT1

ΔT = 74.4 s

S

Eph. = JPL DE405

Circumstances at Greatest Eclipse: 07:14:49.9 UT1
Lat. = 08°55.6'N   Sun Alt. = 78.7°
Long. = 071°43.6'E   Sun Azm. = 353.8°
Path Width = 152.2 km Duration = 05m25.5s

Circumstances at Greatest Duration: 07:23:14.4 UT1
Lat. = 09°19.1'N   Sun Alt. = 77.9°
Long. = 073°47.6'E   Sun Azm. = 334.7°
Path Width = 152.7 km Duration = 05m26.1s

0 1000 2000 3000 4000 5000
Kilometers

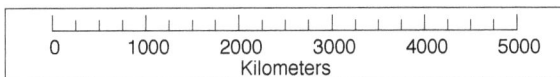

©2016 F. Espenak
www.EclipseWise.com

Hybrid Solar Eclipse of 2031 Nov 14

Greatest Eclipse = 21:07:30.7 TD   (= 21:06:16.0 UT1)

Eclipse Magnitude = 1.0106	Saros Series = 143
Gamma = 0.3078	Saros Member = 24 of 72

Sun at Greatest Eclipse
(Geocentric Coordinates)
R.A. = 15h19m31.2s
Dec. = -18°20'14.5"
S.D. = 00°16'09.9"
H.P. = 00°00'08.9"

Moon at Greatest Eclipse
(Geocentric Coordinates)
R.A. = 15h19m43.3s
Dec. = -18°02'21.3"
S.D. = 00°16'05.0"
H.P. = 00°59'01.4"

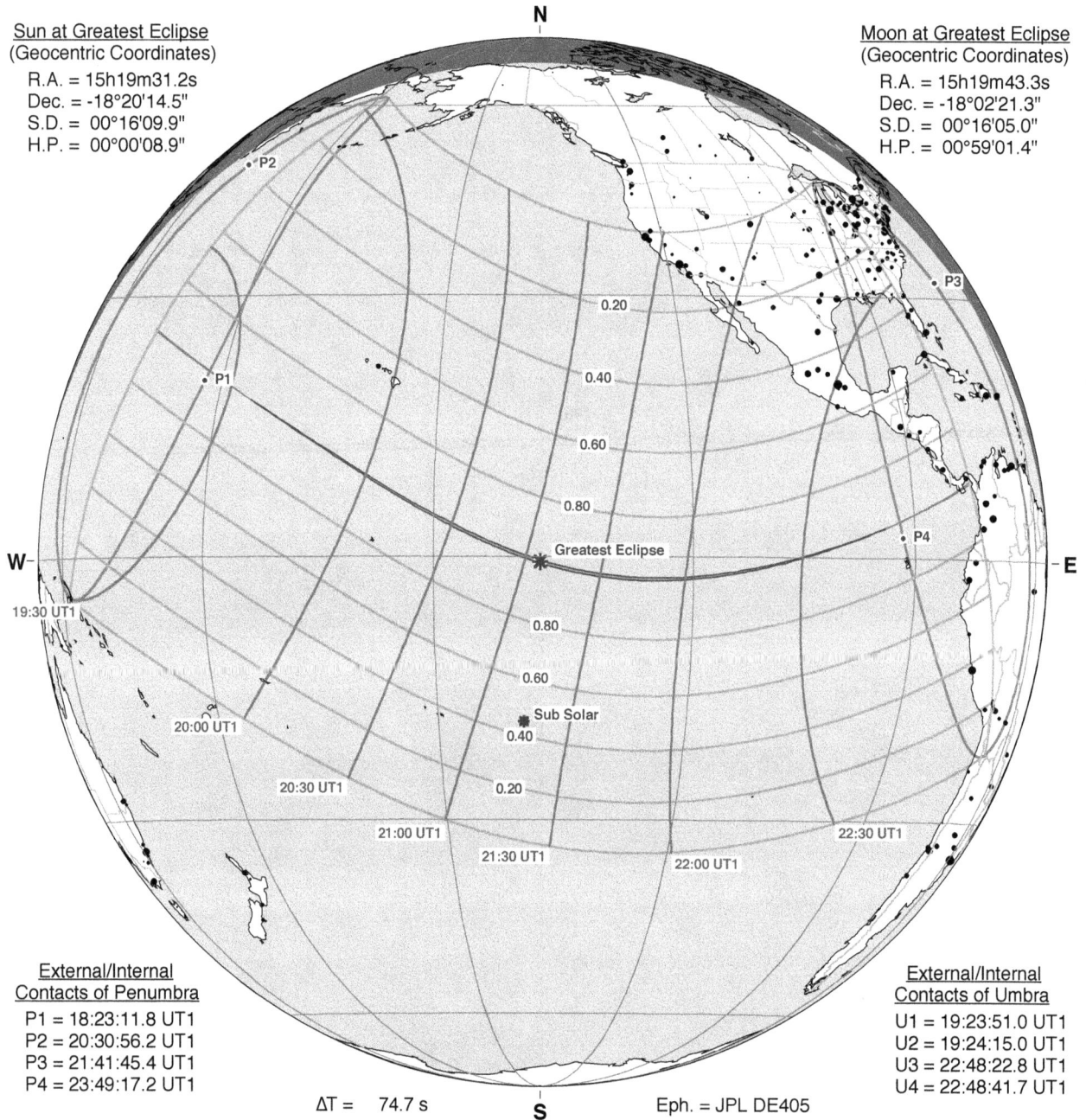

N

P2

P1

0.20
0.40
0.60
0.80

P3

Greatest Eclipse

W— —E

19:30 UT1

P4

0.80
0.60

Sub Solar
0.40

20:00 UT1
20:30 UT1
0.20
21:00 UT1
21:30 UT1
22:00 UT1
22:30 UT1

**External/Internal
Contacts of Penumbra**
P1 = 18:23:11.8 UT1
P2 = 20:30:56.2 UT1
P3 = 21:41:45.4 UT1
P4 = 23:49:17.2 UT1

ΔT = 74.7 s

S

Eph. = JPL DE405

**External/Internal
Contacts of Umbra**
U1 = 19:23:51.0 UT1
U2 = 19:24:15.0 UT1
U3 = 22:48:22.8 UT1
U4 = 22:48:41.7 UT1

Circumstances at Greatest Eclipse: 21:06:16.0 UT1

Lat. = 00°37.9'S	Sun Alt. = 72.1°
Long. = 137°38.7'W	Sun Azm. = 188.7°
Path Width = 38.3 km	Duration = 01m08.3s

Circumstances at Greatest Duration: 21:10:29.2 UT1

Lat. = 00°57.7'S	Sun Alt. = 71.9°
Long. = 136°28.4'W	Sun Azm. = 195.6°
Path Width = 38.4 km	Duration = 01m08.4s

0	1000	2000	3000	4000	5000

Kilometers

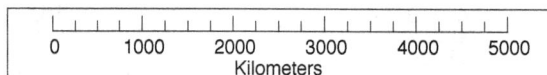

©2016 F. Espenak
www.EclipseWise.com

Annular Solar Eclipse of 2032 May 09

Greatest Eclipse = 13:26:42.4 TD   (= 13:25:27.5 UT1)

Eclipse Magnitude = 0.9957   Saros Series = 148

Gamma = -0.9375   Saros Member = 22 of 75

Sun at Greatest Eclipse
(Geocentric Coordinates)

R.A. = 03h08m06.7s
Dec. = +17°35'43.5"
S.D. = 00°15'50.4"
H.P. = 00°00'08.7"

Moon at Greatest Eclipse
(Geocentric Coordinates)

R.A. = 03h08m46.1s
Dec. = +16°42'42.0"
S.D. = 00°15'41.5"
H.P. = 00°57'35.4"

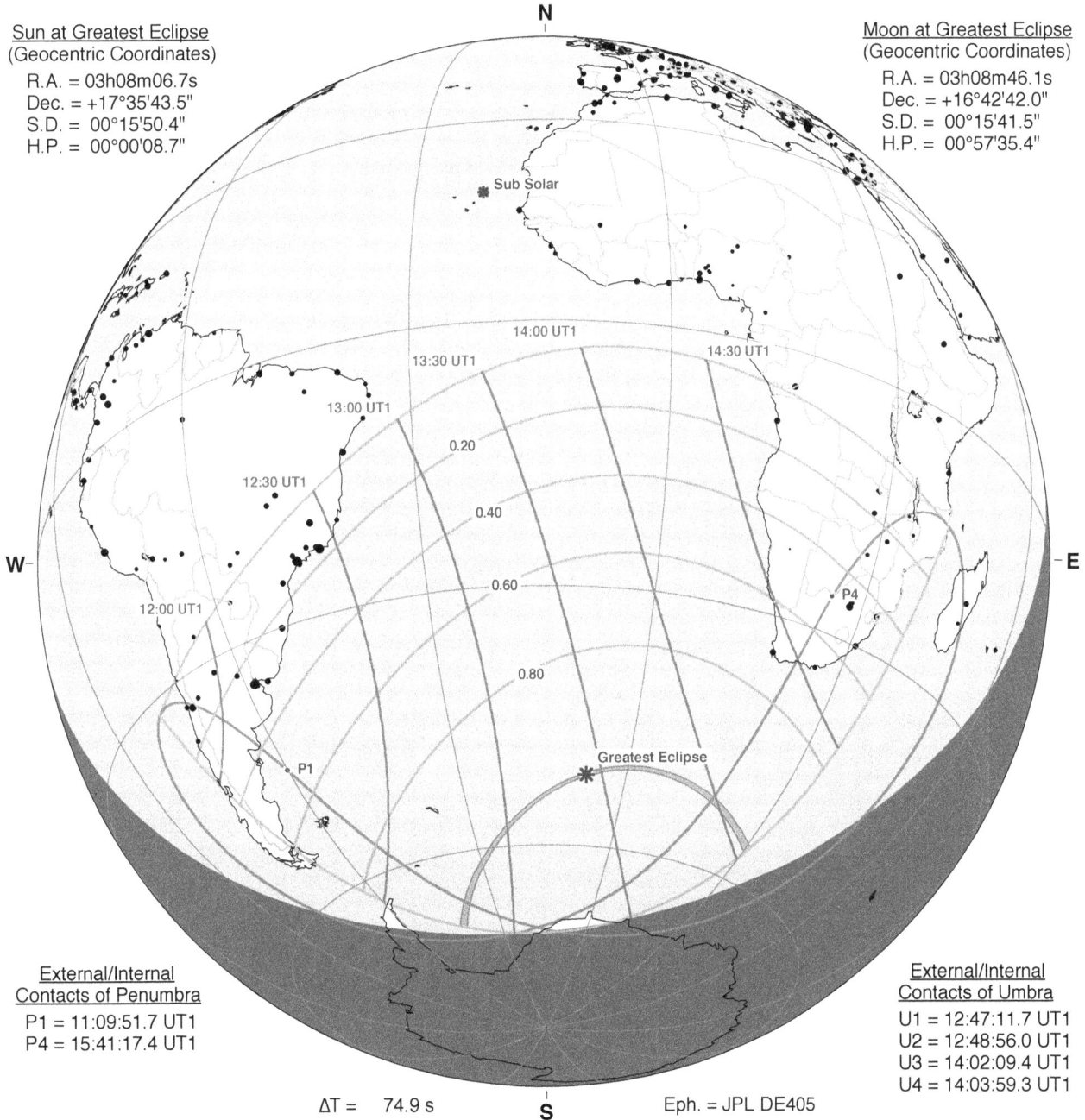

N

Sub Solar

14:00 UT1

13:30 UT1   14:30 UT1

13:00 UT1

0.20

12:30 UT1   0.40

W   0.60   E

12:00 UT1

0.80

P4

P1

Greatest Eclipse

External/Internal Contacts of Penumbra
P1 = 11:09:51.7 UT1
P4 = 15:41:17.4 UT1

External/Internal Contacts of Umbra
U1 = 12:47:11.7 UT1
U2 = 12:48:56.0 UT1
U3 = 14:02:09.4 UT1
U4 = 14:03:59.3 UT1

ΔT = 74.9 s   S   Eph. = JPL DE405

Circumstances at Greatest Eclipse: 13:25:27.5 UT1

Lat. = 51°15.6'S	Sun Alt. = 19.9°
Long. = 007°04.4'W	Sun Azm. = 344.6°
Path Width = 43.7 km	Duration = 00m22.0s

Circumstances at Greatest Duration: 12:48:03.3 UT1

Lat. = 69°57.7'S	Sun Alt. = 0.0°
Long. = 042°31.9'W	Sun Azm. = 28.1°
Path Width = 86.7 km	Duration = 00m40.8s

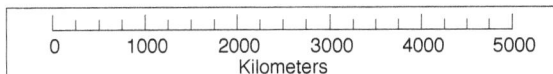

0 1000 2000 3000 4000 5000
Kilometers

Partial Solar Eclipse of 2032 Nov 03

Greatest Eclipse = 05:34:12.9 TD   (= 05:32:57.8 UT1)

Eclipse Magnitude = 0.8554	Saros Series = 153
Gamma = 1.0643	Saros Member = 10 of 70

Sun at Greatest Eclipse
(Geocentric Coordinates)
R.A. = 14h35m40.9s
Dec. = -15°13'54.9"
S.D. = 00°16'07.4"
H.P. = 00°00'08.9"

Moon at Greatest Eclipse
(Geocentric Coordinates)
R.A. = 14h36m33.6s
Dec. = -14°16'01.1"
S.D. = 00°15'13.0"
H.P. = 00°55'50.8"

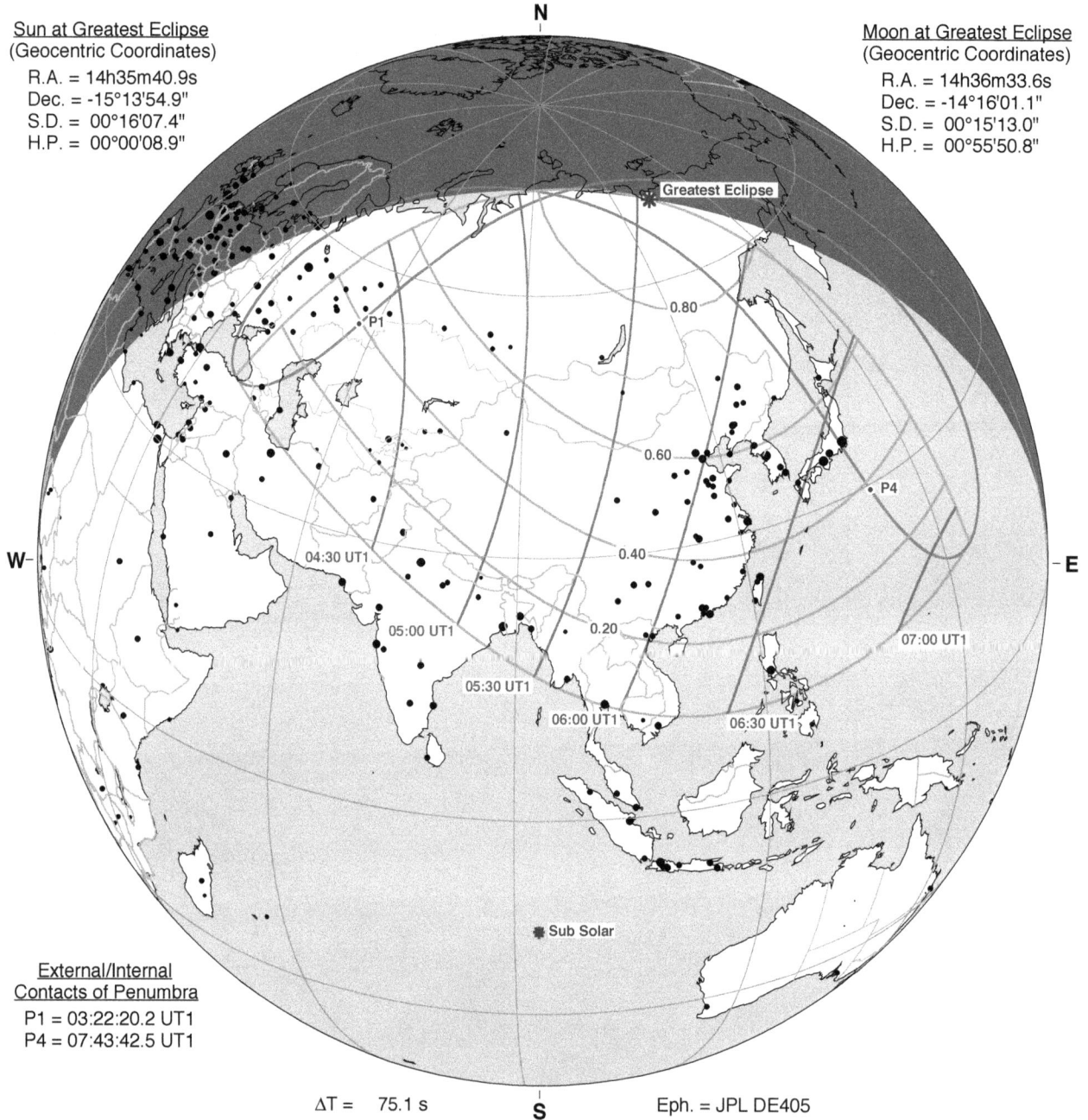

External/Internal
Contacts of Penumbra
P1 = 03:22:20.2 UT1
P4 = 07:43:42.5 UT1

ΔT = 75.1 s   Eph. = JPL DE405

Circumstances at Greatest Eclipse: 05:32:57.8 UT1

Lat. = 70°25.7'N	Sun Alt. = 0.0°
Long. = 132°36.5'E	Sun Azm. = 218.3°

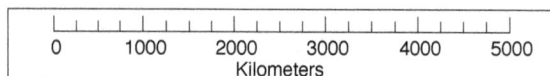

©2016 F. Espenak
www.EclipseWise.com

Total Solar Eclipse of 2033 Mar 30

Greatest Eclipse = 18:02:35.7 TD   (= 18:01:20.4 UT1)

Eclipse Magnitude = 1.0462   Saros Series = 120
Gamma = 0.9778   Saros Member = 62 of 71

Sun at Greatest Eclipse
(Geocentric Coordinates)
R.A. = 00h38m02.8s
Dec. = +04°05'47.8"
S.D. = 00°16'00.8"
H.P. = 00°00'08.8"

Moon at Greatest Eclipse
(Geocentric Coordinates)
R.A. = 00h36m50.4s
Dec. = +05°02'48.6"
S.D. = 00°16'42.2"
H.P. = 01°01'18.3"

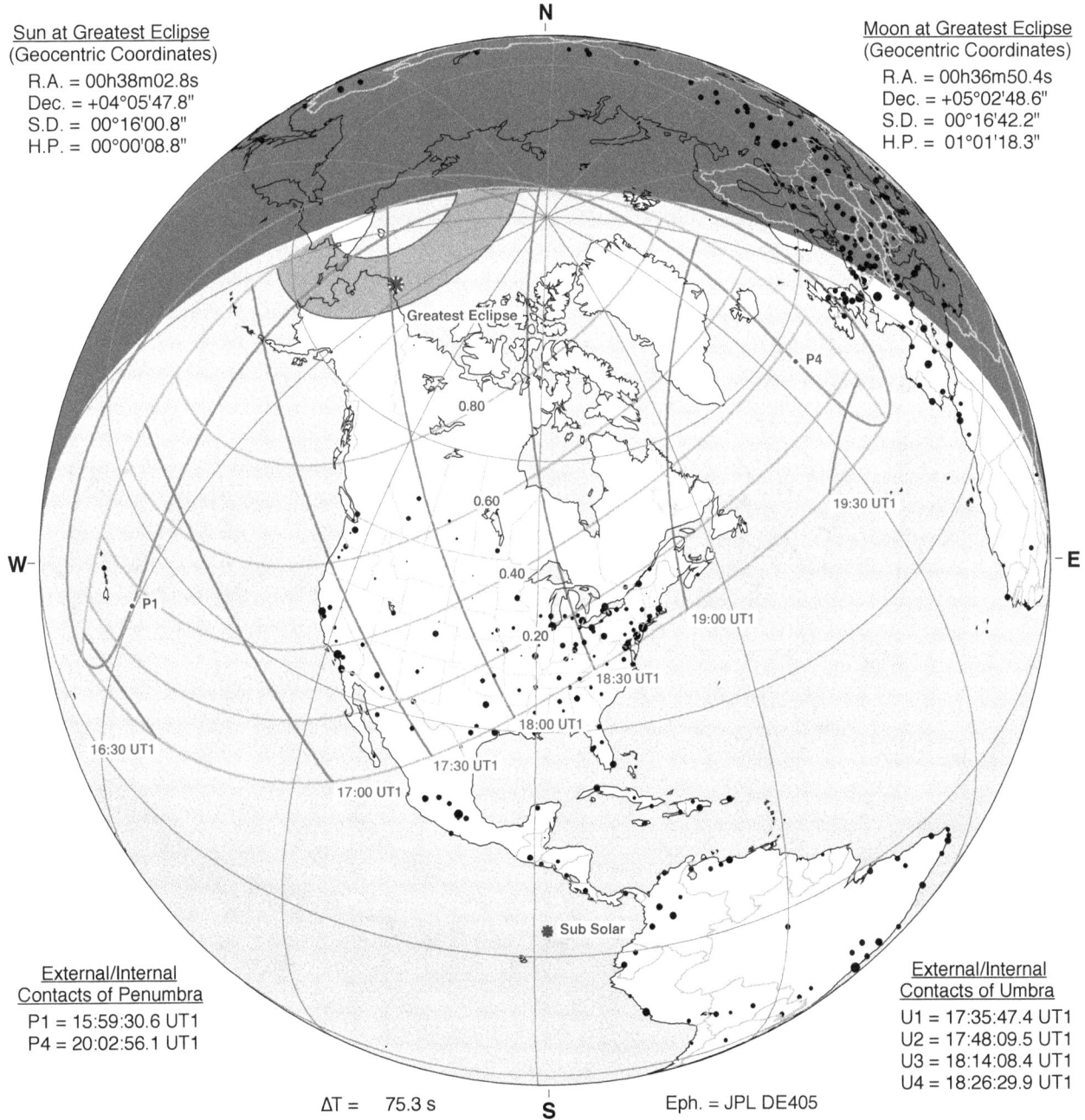

N

Greatest Eclipse

P4

0.80

0.60

19:30 UT1

W

0.40

P1

19:00 UT1

0.20

18:30 UT1

18:00 UT1

16:30 UT1

17:30 UT1

17:00 UT1

E

Sub Solar

External/Internal
Contacts of Penumbra
P1 = 15:59:30.6 UT1
P4 = 20:02:56.1 UT1

External/Internal
Contacts of Umbra
U1 = 17:35:47.4 UT1
U2 = 17:48:09.5 UT1
U3 = 18:14:08.4 UT1
U4 = 18:26:29.9 UT1

ΔT = 75.3 s

S

Eph. = JPL DE405

Circumstances at Greatest Eclipse: 18:01:20.4 UT1
Lat. = 71°18.9'N   Sun Alt. = 11.2°
Long. = 155°50.1'W   Sun Azm. = 111.1°
Path Width = 781.2 km   Duration = 02m37.0s

Circumstances at Greatest Duration: 18:01:04.2 UT1
Lat. = 71°10.5'N   Sun Alt. = 11.2°
Long. = 155°55.9'W   Sun Azm. = 110.9°
Path Width = 782.6 km   Duration = 02m37.0s

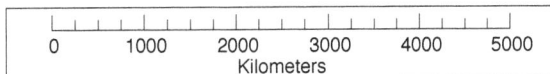

©2016 F. Espenak
www.EclipseWise.com

0   1000   2000   3000   4000   5000
Kilometers

Partial Solar Eclipse of 2033 Sep 23

Greatest Eclipse = 13:54:31.2 TD   (= 13:53:15.7 UT1)

Eclipse Magnitude = 0.6890   Saros Series = 125
Gamma = -1.1583   Saros Member = 55 of 73

Sun at Greatest Eclipse
(Geocentric Coordinates)
R.A. = 12h03m08.9s
Dec. = -00°20'27.7"
S.D. = 00°15'56.3"
H.P. = 00°00'08.8"

Moon at Greatest Eclipse
(Geocentric Coordinates)
R.A. = 12h01m52.5s
Dec. = -01°19'54.7"
S.D. = 00°14'43.6"
H.P. = 00°54'03.0"

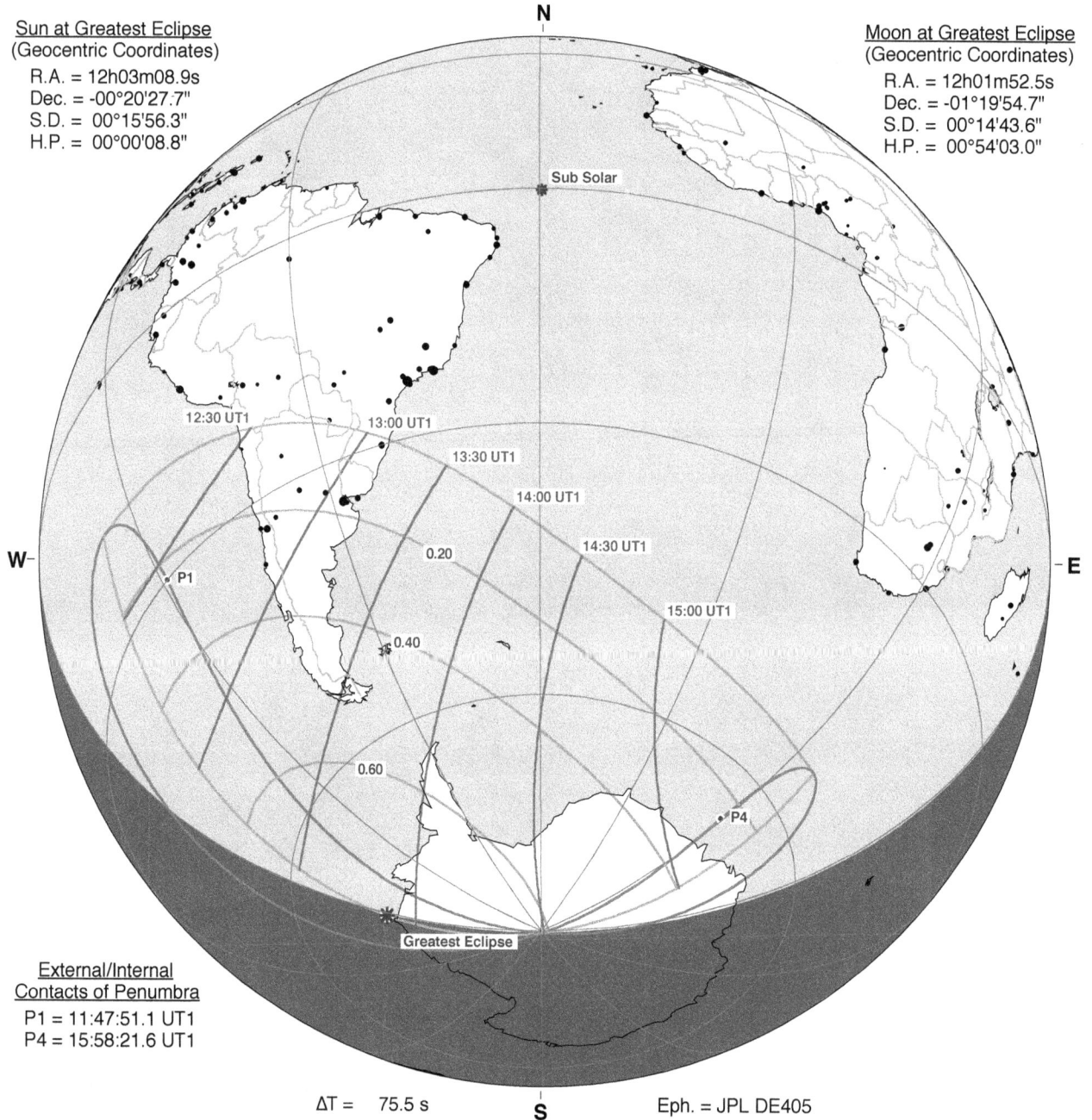

N

Sub Solar

12:30 UT1
13:00 UT1
13:30 UT1
14:00 UT1
14:30 UT1
0.20
15:00 UT1
W
P1
0.40
E
0.60
P4
Greatest Eclipse

External/Internal Contacts of Penumbra
P1 = 11:47:51.1 UT1
P4 = 15:58:21.6 UT1

ΔT = 75.5 s   S   Eph. = JPL DE405

Circumstances at Greatest Eclipse: 13:53:15.7 UT1
Lat. = 72°12.2'S   Sun Alt. = 0.0°
Long. = 121°15.2'W   Sun Azm. = 91.1°

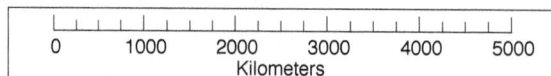

0 1000 2000 3000 4000 5000
Kilometers

©2016 F. Espenak
www.EclipseWise.com

Total Solar Eclipse of 2034 Mar 20

Greatest Eclipse = 10:18:45.2 TD   (= 10:17:29.5 UT1)

Eclipse Magnitude = 1.0458   Saros Series = 130
Gamma = 0.2894   Saros Member = 53 of 73

Sun at Greatest Eclipse
(Geocentric Coordinates)
R.A. = 23h59m32.7s
Dec. = -00°02'58.0"
S.D. = 00°16'03.7"
H.P. = 00°00'08.8"

Moon at Greatest Eclipse
(Geocentric Coordinates)
R.A. = 23h59m11.3s
Dec. = +00°13'42.6"
S.D. = 00°16'31.6"
H.P. = 01°00'39.3"

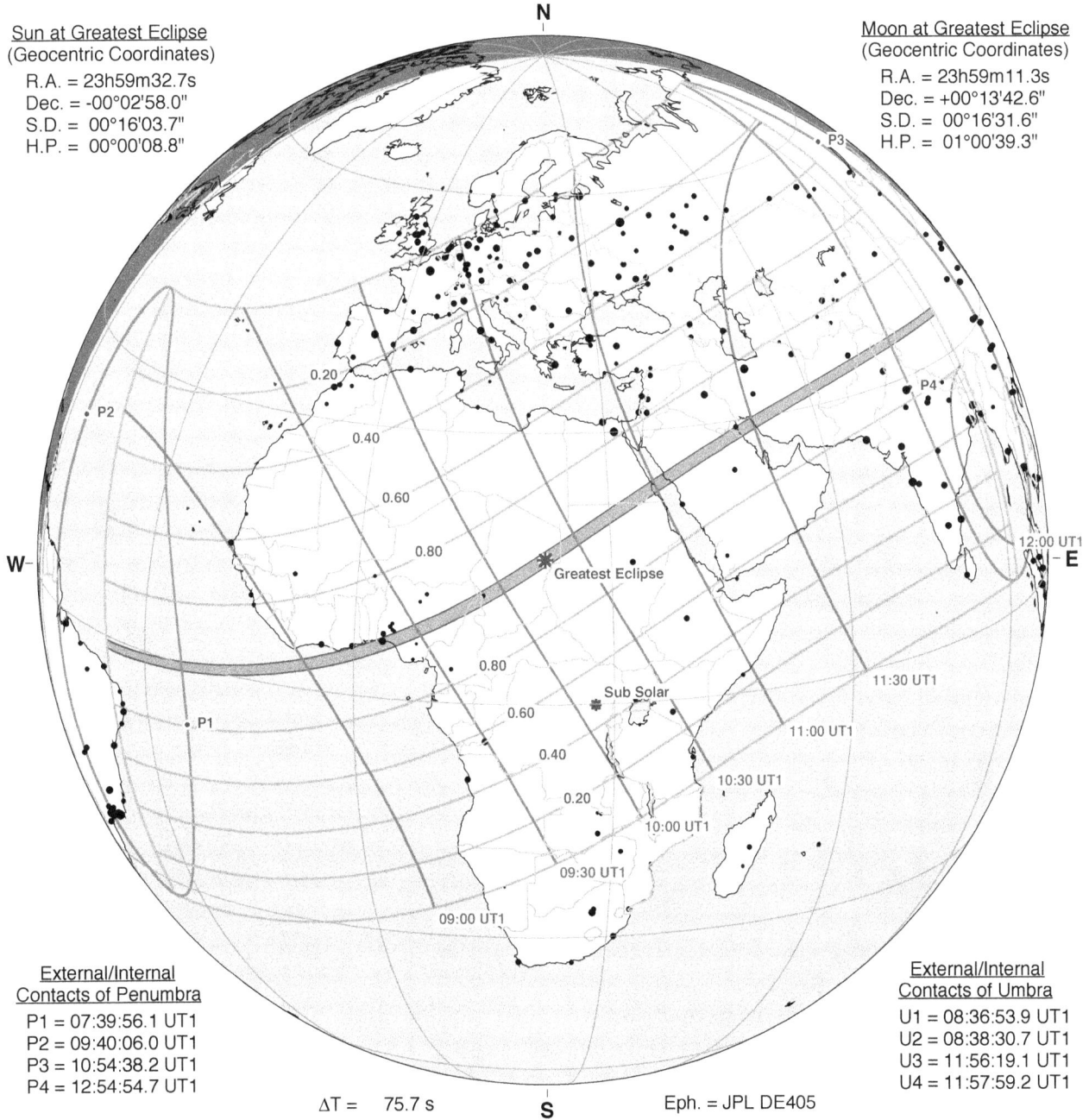

External/Internal
Contacts of Penumbra
P1 = 07:39:56.1 UT1
P2 = 09:40:06.0 UT1
P3 = 10:54:38.2 UT1
P4 = 12:54:54.7 UT1

External/Internal
Contacts of Umbra
U1 = 08:36:53.9 UT1
U2 = 08:38:30.7 UT1
U3 = 11:56:19.1 UT1
U4 = 11:57:59.2 UT1

ΔT = 75.7 s   Eph. = JPL DE405

Circumstances at Greatest Eclipse: 10:17:29.5 UT1
Lat. = 16°03.3'N   Sun Alt. = 73.1°
Long. = 022°13.1'E   Sun Azm. = 161.6°
Path Width = 159.1 km   Duration = 04m09.3s

Circumstances at Greatest Duration: 10:18:26.1 UT1
Lat. = 16°13.3'N   Sun Alt. = 73.1°
Long. = 022°29.6'E   Sun Azm. = 163.5°
Path Width = 159.0 km   Duration = 04m09.4s

©2016 F. Espenak
www.EclipseWise.com

0   1000   2000   3000   4000   5000
Kilometers

Annular Solar Eclipse of 2034 Sep 12

Greatest Eclipse = 16:19:27.5 TD (= 16:18:11.6 UT1)

Eclipse Magnitude = 0.9736	Saros Series = 135
Gamma = -0.3936	Saros Member = 40 of 71

Sun at Greatest Eclipse
(Geocentric Coordinates)
R.A. = 11h23m10.9s
Dec. = +03°57'57.5"
S.D. = 00°15'53.5"
H.P. = 00°00'08.7"

Moon at Greatest Eclipse
(Geocentric Coordinates)
R.A. = 11h22m44.5s
Dec. = +03°36'59.6"
S.D. = 00°15'15.1"
H.P. = 00°55'58.6"

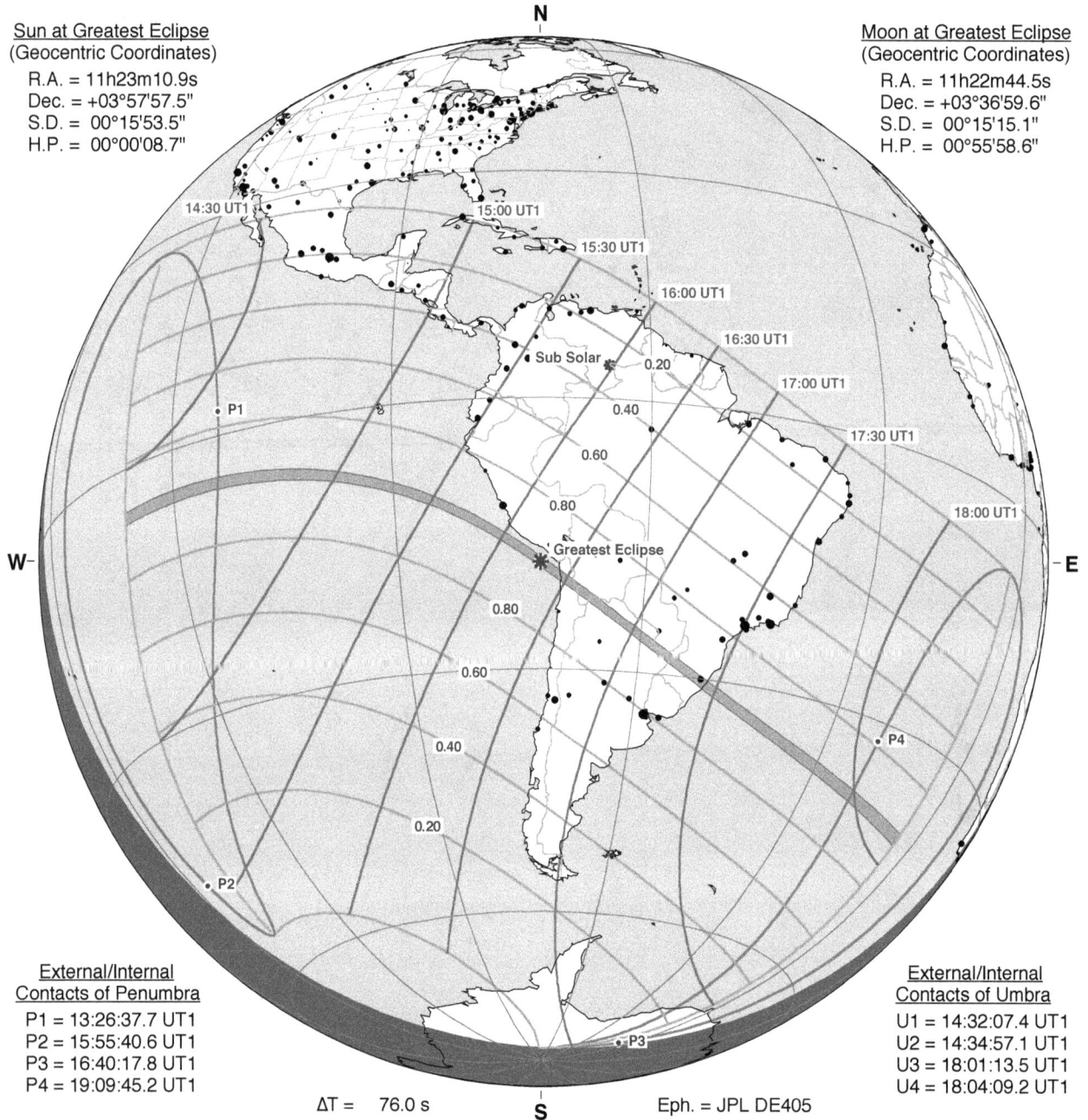

N

14:30 UT1
15:00 UT1
15:30 UT1
16:00 UT1
16:30 UT1
17:00 UT1
17:30 UT1
18:00 UT1

Sub Solar
0.20
0.40
0.60
0.80
Greatest Eclipse
0.80
0.60
0.40
0.20

P1
P2
P3
P4

W — — E

S

External/Internal
Contacts of Penumbra
P1 = 13:26:37.7 UT1
P2 = 15:55:40.6 UT1
P3 = 16:40:17.8 UT1
P4 = 19:09:45.2 UT1

ΔT = 76.0 s

Eph. = JPL DE405

External/Internal
Contacts of Umbra
U1 = 14:32:07.4 UT1
U2 = 14:34:57.1 UT1
U3 = 18:01:13.5 UT1
U4 = 18:04:09.2 UT1

Circumstances at Greatest Eclipse: 16:18:11.6 UT1

Lat. = 18°14.5'S	Sun Alt. = 66.7°
Long. = 072°37.4'W	Sun Azm. = 18.2°
Path Width = 102.1 km	Duration = 02m57.7s

Circumstances at Greatest Duration: 16:29:28.2 UT1

Lat. = 20°04.3'S	Sun Alt. = 65.9°
Long. = 069°44.5'W	Sun Azm. = 3.5°
Path Width = 101.7 km	Duration = 02m57.8s

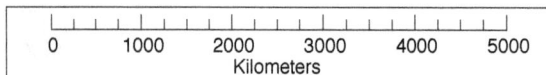

©2016 F. Espenak
www.EclipseWise.com

0   1000   2000   3000   4000   5000
Kilometers

Annular Solar Eclipse of 2035 Mar 09

Greatest Eclipse = 23:05:53.6 TD (= 23:04:37.4 UT1)

Eclipse Magnitude = 0.9919   Saros Series = 140

Gamma = -0.4368   Saros Member = 30 of 71

Sun at Greatest Eclipse
(Geocentric Coordinates)
R.A. = 23h20m17.6s
Dec. = -04°16'22.2"
S.D. = 00°16'06.5"
H.P. = 00°00'08.9"

Moon at Greatest Eclipse
(Geocentric Coordinates)
R.A. = 23h20m47.9s
Dec. = -04°40'23.8"
S.D. = 00°15'44.9"
H.P. = 00°57'47.9"

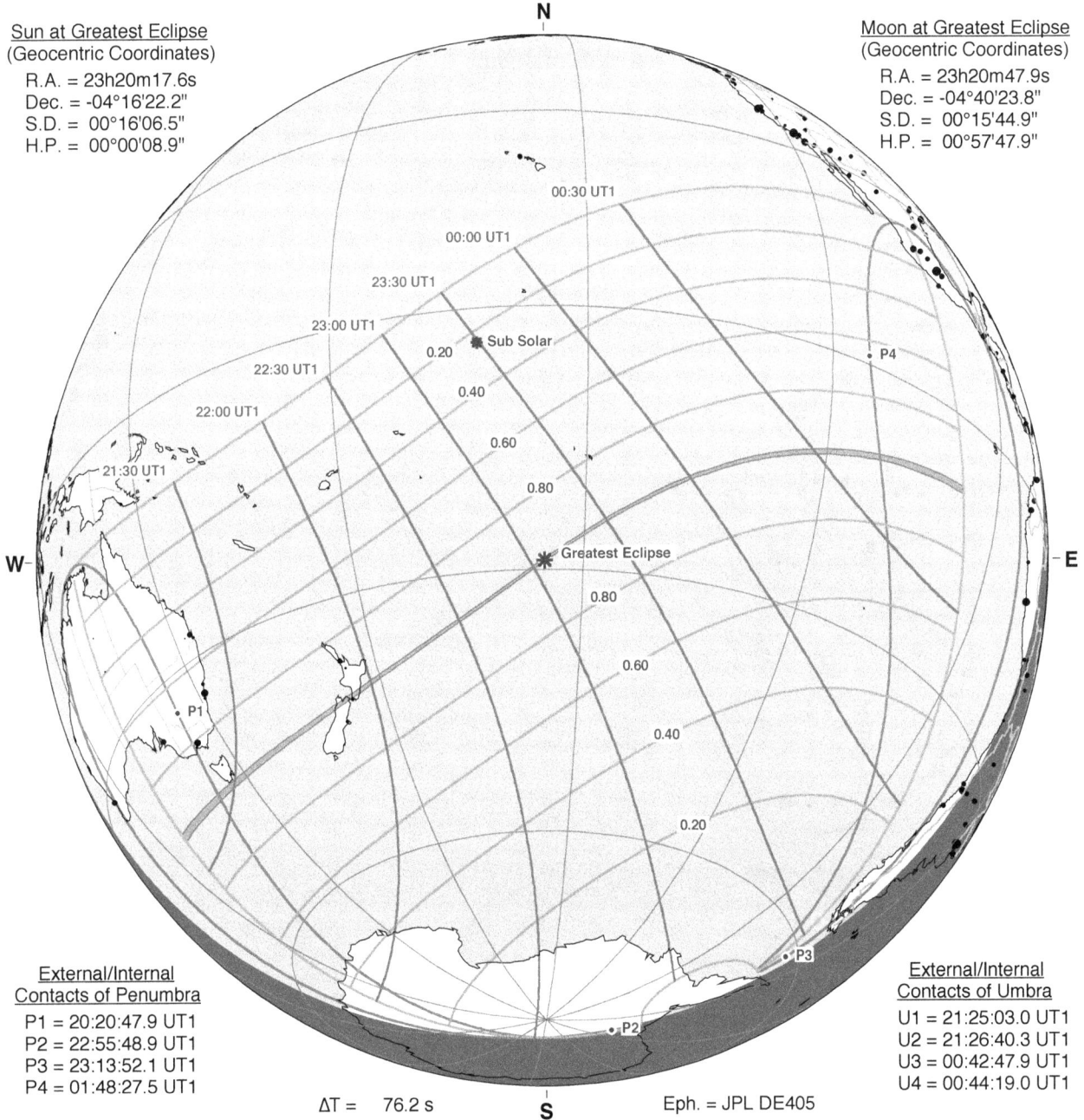

External/Internal
Contacts of Penumbra
P1 = 20:20:47.9 UT1
P2 = 22:55:48.9 UT1
P3 = 23:13:52.1 UT1
P4 = 01:48:27.5 UT1

External/Internal
Contacts of Umbra
U1 = 21:25:03.0 UT1
U2 = 21:26:40.3 UT1
U3 = 00:42:47.9 UT1
U4 = 00:44:19.0 UT1

ΔT = 76.2 s   Eph. = JPL DE405

Circumstances at Greatest Eclipse: 23:04:37.4 UT1
Lat. = 29°02.6'S   Sun Alt. = 63.9°
Long. = 154°58.1'W   Sun Azm. = 340.2°
Path Width = 31.5 km   Duration = 00m47.5s

Circumstances at Greatest Duration: 21:25:51.6 UT1
Lat. = 43°22.2'S   Sun Alt. = 0.0°
Long. = 127°04.5'E   Sun Azm. = 95.9°
Path Width = 95.5 km   Duration = 01m25.9s

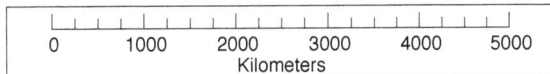

©2016 F. Espenak
www.EclipseWise.com

0 1000 2000 3000 4000 5000
Kilometers

Total Solar Eclipse of 2035 Sep 02

Greatest Eclipse = 01:56:46.3 TD   (= 01:55:29.9 UT1)

Eclipse Magnitude = 1.0320	Saros Series = 145
Gamma = 0.3727	Saros Member = 23 of 77

Sun at Greatest Eclipse
(Geocentric Coordinates)
R.A. = 10h44m07.3s
Dec. = +08°01'09.8"
S.D. = 00°15'50.9"
H.P. = 00°00'08.7"

Moon at Greatest Eclipse
(Geocentric Coordinates)
R.A. = 10h44m32.4s
Dec. = +08°22'14.7"
S.D. = 00°16'06.4"
H.P. = 00°59'06.9"

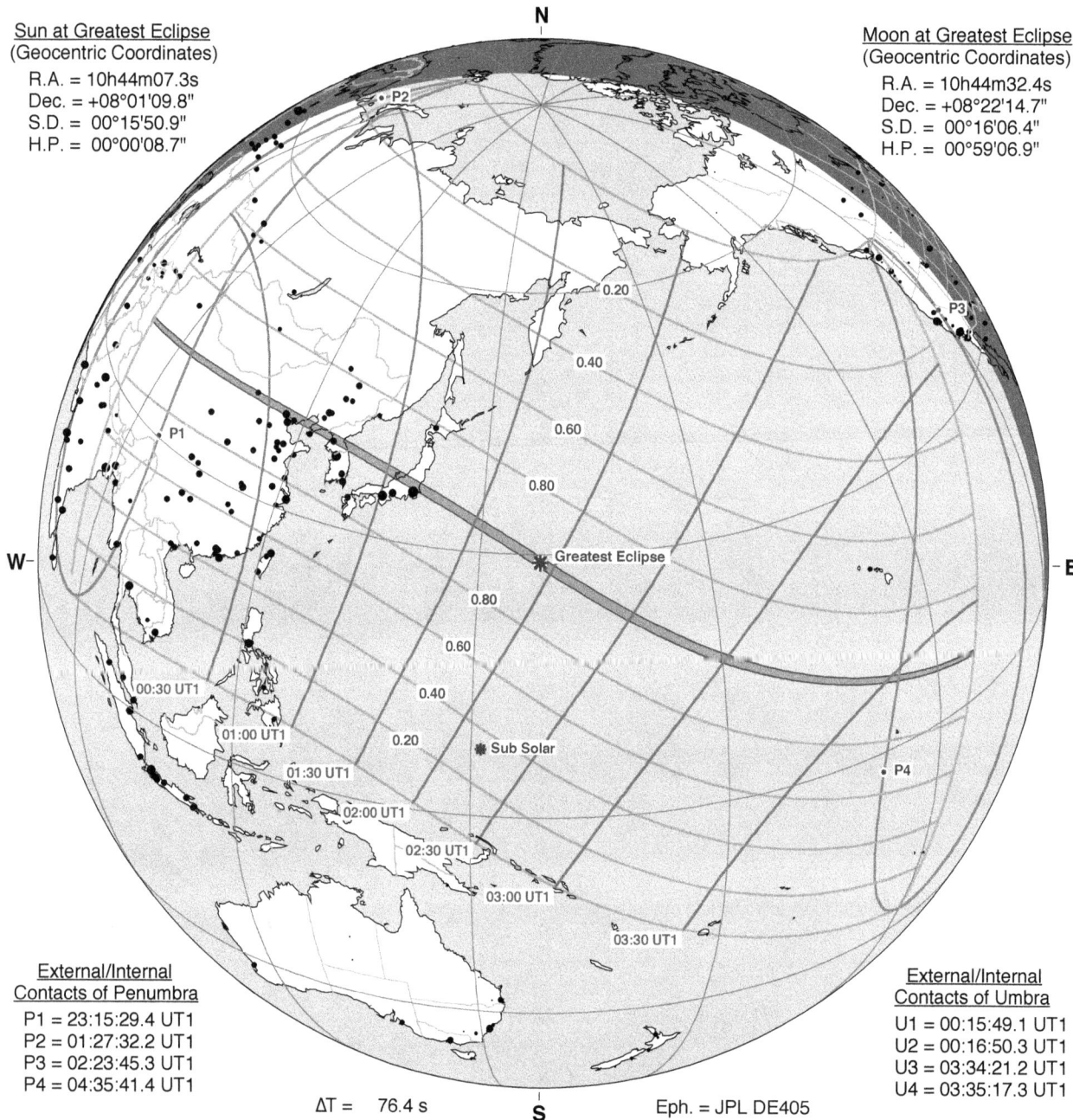

N

P2

0.20
0.40
0.60
0.80

P3

W

Greatest Eclipse

E

0.80
0.60
0.40
0.20

Sub Solar

P1

00:30 UT1
01:00 UT1
01:30 UT1
02:00 UT1
02:30 UT1
03:00 UT1
03:30 UT1

P4

**External/Internal
Contacts of Penumbra**
P1 = 23:15:29.4 UT1
P2 = 01:27:32.2 UT1
P3 = 02:23:45.3 UT1
P4 = 04:35:41.4 UT1

ΔT =   76.4 s

S

Eph. = JPL DE405

**External/Internal
Contacts of Umbra**
U1 = 00:15:49.1 UT1
U2 = 00:16:50.3 UT1
U3 = 03:34:21.2 UT1
U4 = 03:35:17.3 UT1

Circumstances at Greatest Eclipse: 01:55:29.9 UT1

Lat. = 29°05.6'N	Sun Alt. = 67.9°
Long. = 158°00.9'E	Sun Azm. = 198.5°
Path Width = 116.3 km	Duration = 02m54.2s

Circumstances at Greatest Duration: 01:52:01.1 UT1

Lat. = 29°41.4'N	Sun Alt. = 67.8°
Long. = 156°51.6'E	Sun Azm. = 192.9°
Path Width = 116.1 km	Duration = 02m54.4s

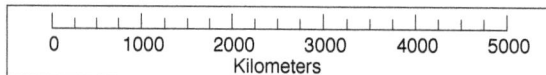

0   1000   2000   3000   4000   5000
Kilometers

©2016 F. Espenak
www.EclipseWise.com

Partial Solar Eclipse of 2036 Feb 27

Greatest Eclipse = 04:46:49.0 TD (= 04:45:32.4 UT1)

Eclipse Magnitude = 0.6286	Saros Series = 150
Gamma = -1.1942	Saros Member = 18 of 71

Sun at Greatest Eclipse
(Geocentric Coordinates)
R.A. = 22h39m15.4s
Dec. = -08°30'21.2"
S.D. = 00°16'09.1"
H.P. = 00°00'08.9"

Moon at Greatest Eclipse
(Geocentric Coordinates)
R.A. = 22h40m29.9s
Dec. = -09°33'05.6"
S.D. = 00°14'57.5"
H.P. = 00°54'53.9"

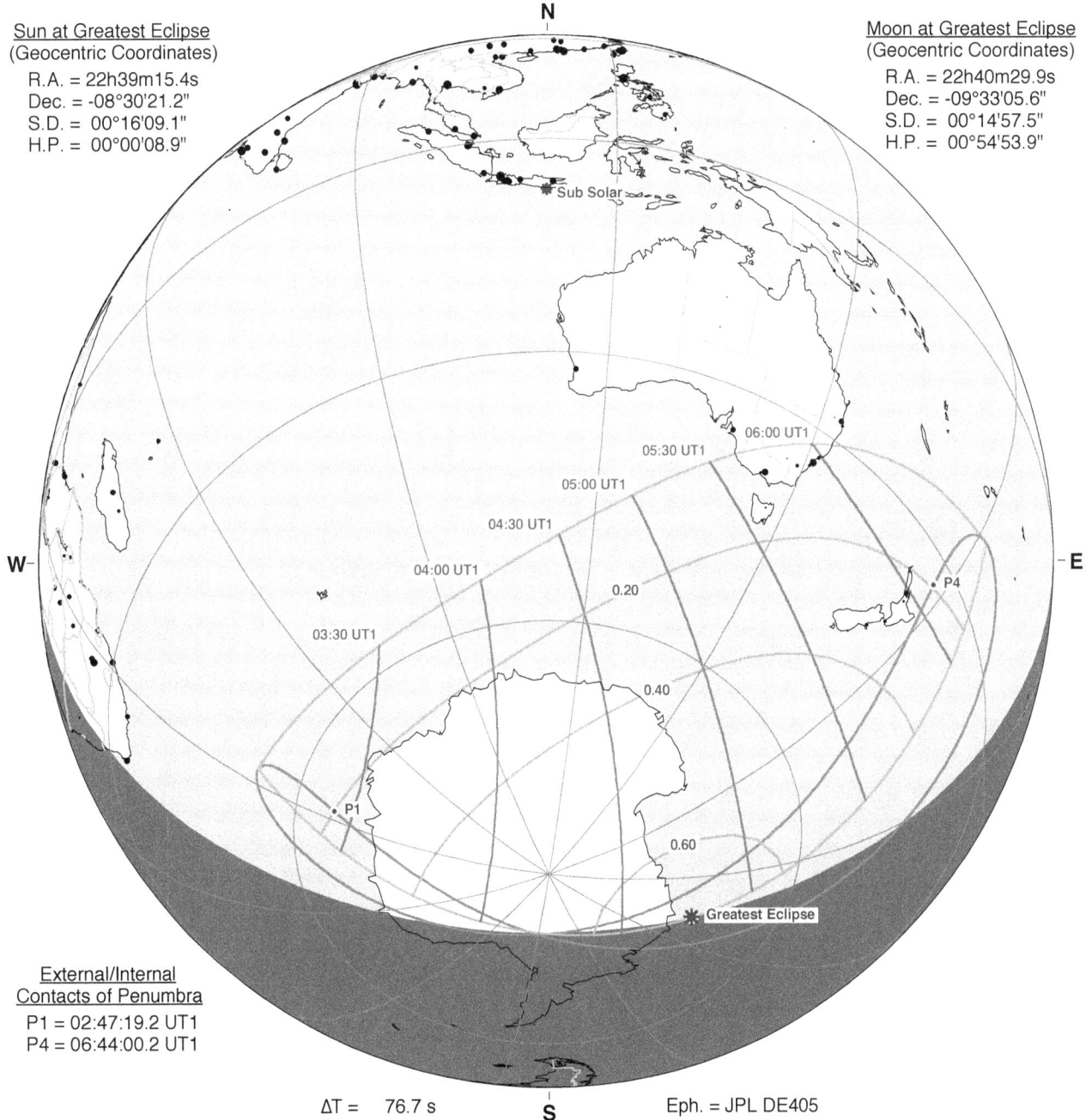

N

Sub Solar

06:00 UT1
05:30 UT1
05:00 UT1
04:30 UT1
04:00 UT1
0.20
03:30 UT1
P4
0.40
P1
0.60

W

E

* Greatest Eclipse

External/Internal
Contacts of Penumbra
P1 = 02:47:19.2 UT1
P4 = 06:44:00.2 UT1

ΔT = 76.7 s

S

Eph. = JPL DE405

Circumstances at Greatest Eclipse: 04:45:32.4 UT1
Lat. = 71°38.9'S   Sun Alt. =   0.0°
Long. = 131°27.3'W   Sun Azm. = 242.0°

0	1000	2000	3000	4000	5000

Kilometers

©2016 F. Espenak
www.EclipseWise.com

Partial Solar Eclipse of 2036 Jul 23

Greatest Eclipse = 10:32:06.5 TD (= 10:30:49.6 UT1)

Eclipse Magnitude = 0.1992	Saros Series = 117
Gamma = -1.4250	Saros Member = 70 of 71

Sun at Greatest Eclipse
(Geocentric Coordinates)
R.A. = 08h13m32.5s
Dec. = +19°53'41.2"
S.D. = 00°15'44.6"
H.P. = 00°00'08.7"

Moon at Greatest Eclipse
(Geocentric Coordinates)
R.A. = 08h12m46.3s
Dec. = +18°27'12.2"
S.D. = 00°16'42.4"
H.P. = 01°01'18.7"

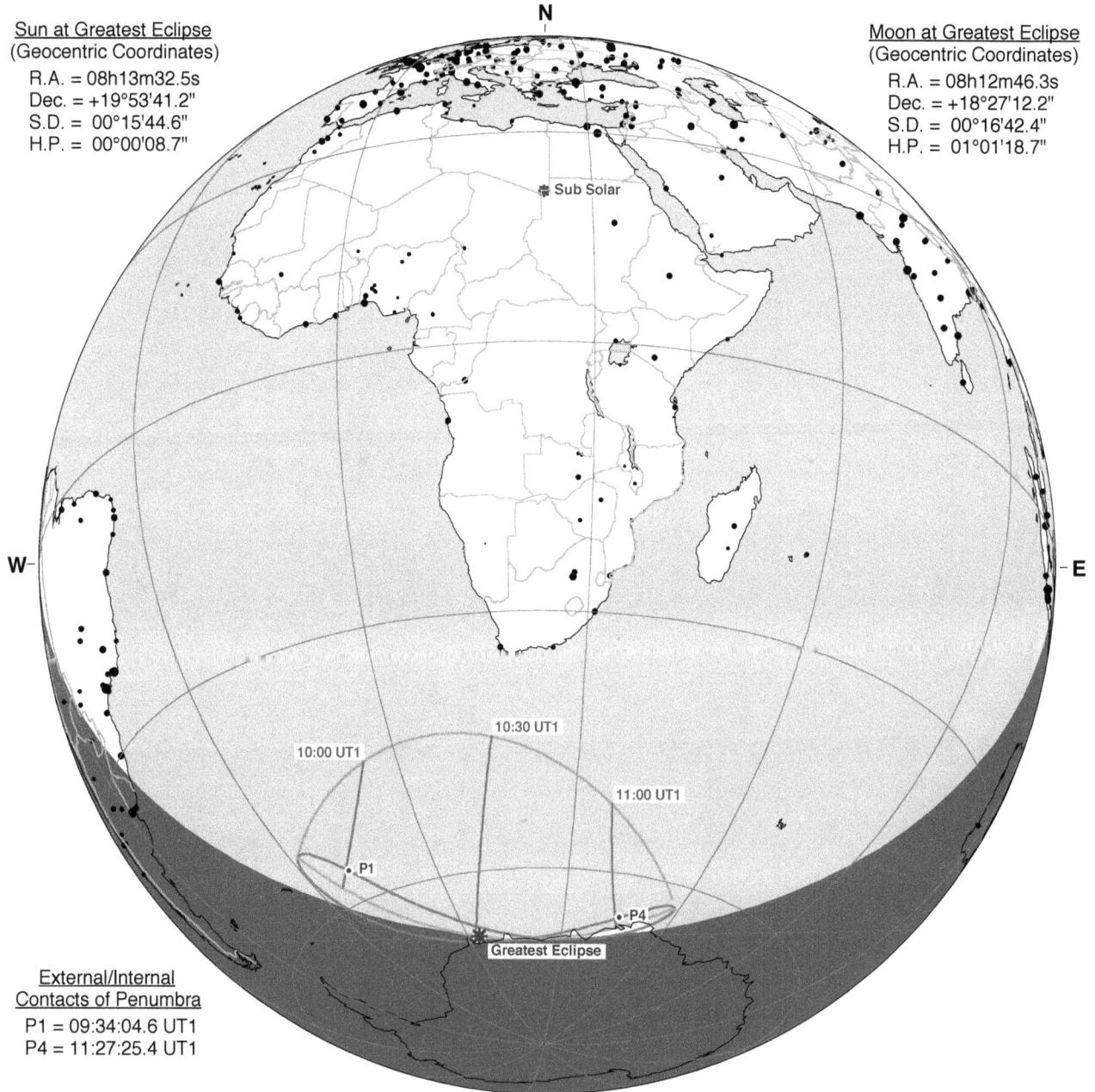

N

Sub Solar

W

E

10:00 UT1

10:30 UT1

11:00 UT1

P1

P4

Greatest Eclipse

External/Internal
Contacts of Penumbra
P1 = 09:34:04.6 UT1
P4 = 11:27:25.4 UT1

ΔT = 76.8 s

S

Eph. = JPL DE405

Circumstances at Greatest Eclipse: 10:30:49.6 UT1

Lat. = 68°53.0'S	Sun Alt. = 0.0°
Long. = 003°32.3'E	Sun Azm. = 19.1°

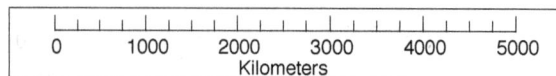

0 1000 2000 3000 4000 5000
Kilometers

Partial Solar Eclipse of 2036 Aug 21

Greatest Eclipse = 17:25:45.4 TD   (= 17:24:28.5 UT1)

Eclipse Magnitude = 0.8622   Saros Series = 155
           Gamma = 1.0825   Saros Member = 7 of 71

Sun at Greatest Eclipse
(Geocentric Coordinates)
R.A. = 10h05m24.9s
Dec. = +11°44'16.4"
S.D. = 00°15'48.7"
H.P. = 00°00'08.7"

Moon at Greatest Eclipse
(Geocentric Coordinates)
R.A. = 10h06m34.6s
Dec. = +12°48'10.2"
S.D. = 00°16'41.1"
H.P. = 01°01'14.1"

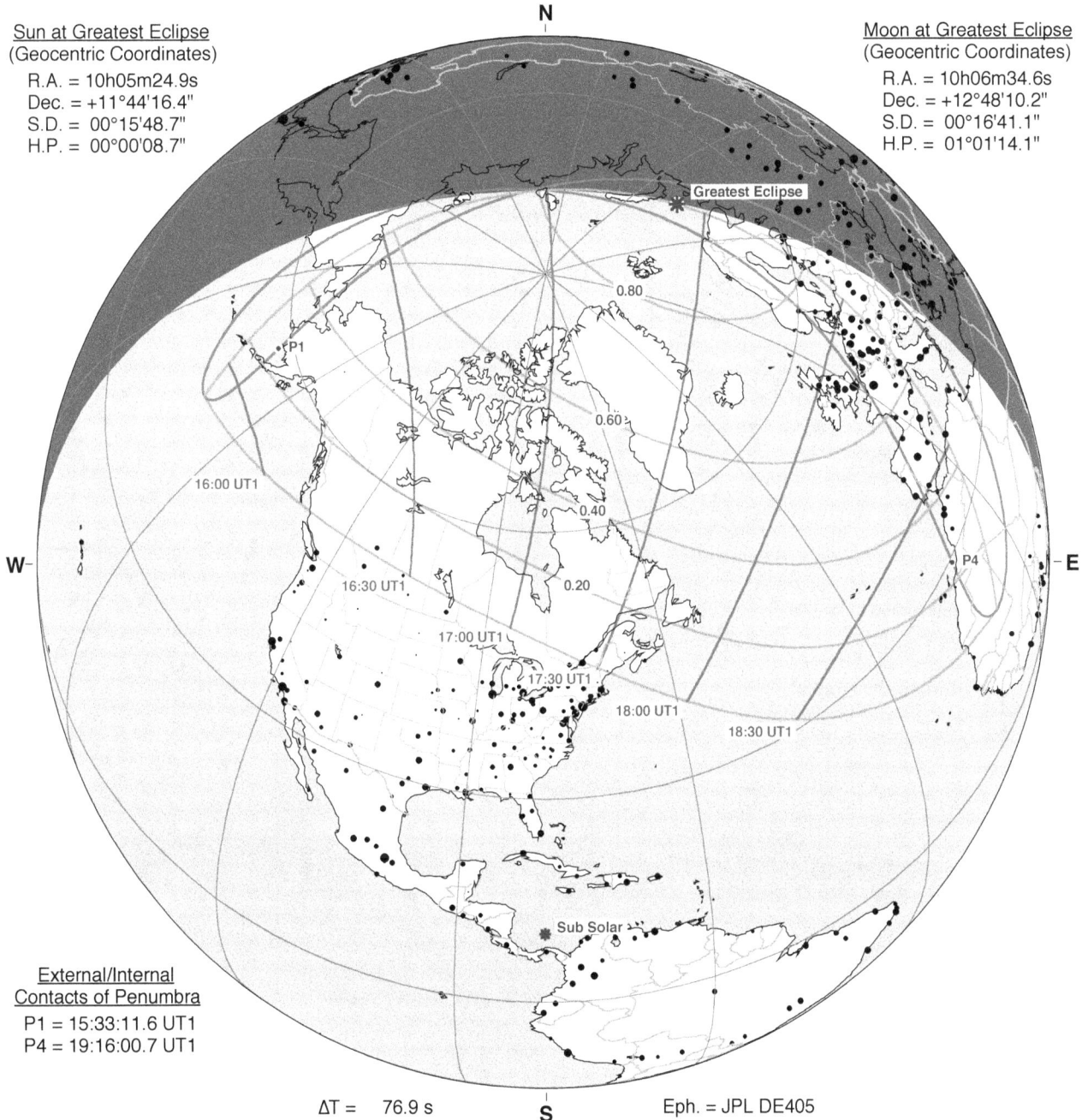

N

Greatest Eclipse

0.80
0.60
0.40
0.20

16:00 UT1
16:30 UT1
17:00 UT1
17:30 UT1
18:00 UT1
18:30 UT1

P1
P4

W

E

Sub Solar

External/Internal
Contacts of Penumbra
P1 = 15:33:11.6 UT1
P4 = 19:16:00.7 UT1

ΔT =   76.9 s   S   Eph. = JPL DE405

Circumstances at Greatest Eclipse: 17:24:28.5 UT1
     Lat. = 71°07.2'N   Sun Alt. =   0.0°
     Long. = 046°59.1'E   Sun Azm. = 308.9°

0   1000   2000   3000   4000   5000
            Kilometers

©2016 F. Espenak
www.EclipseWise.com

101

Partial Solar Eclipse of 2037 Jan 16

Greatest Eclipse = 09:48:55.1 TD (= 09:47:38.0 UT1)

Eclipse Magnitude = 0.7049   Saros Series = 122
Gamma = 1.1477   Saros Member = 59 of 70

Sun at Greatest Eclipse
(Geocentric Coordinates)
R.A. = 19h54m30.1s
Dec. = -20°49'43.5"
S.D. = 00°16'15.5"
H.P. = 00°00'08.9"

Moon at Greatest Eclipse
(Geocentric Coordinates)
R.A. = 19h54m05.4s
Dec. = -19°47'33.7"
S.D. = 00°14'51.8"
H.P. = 00°54'32.8"

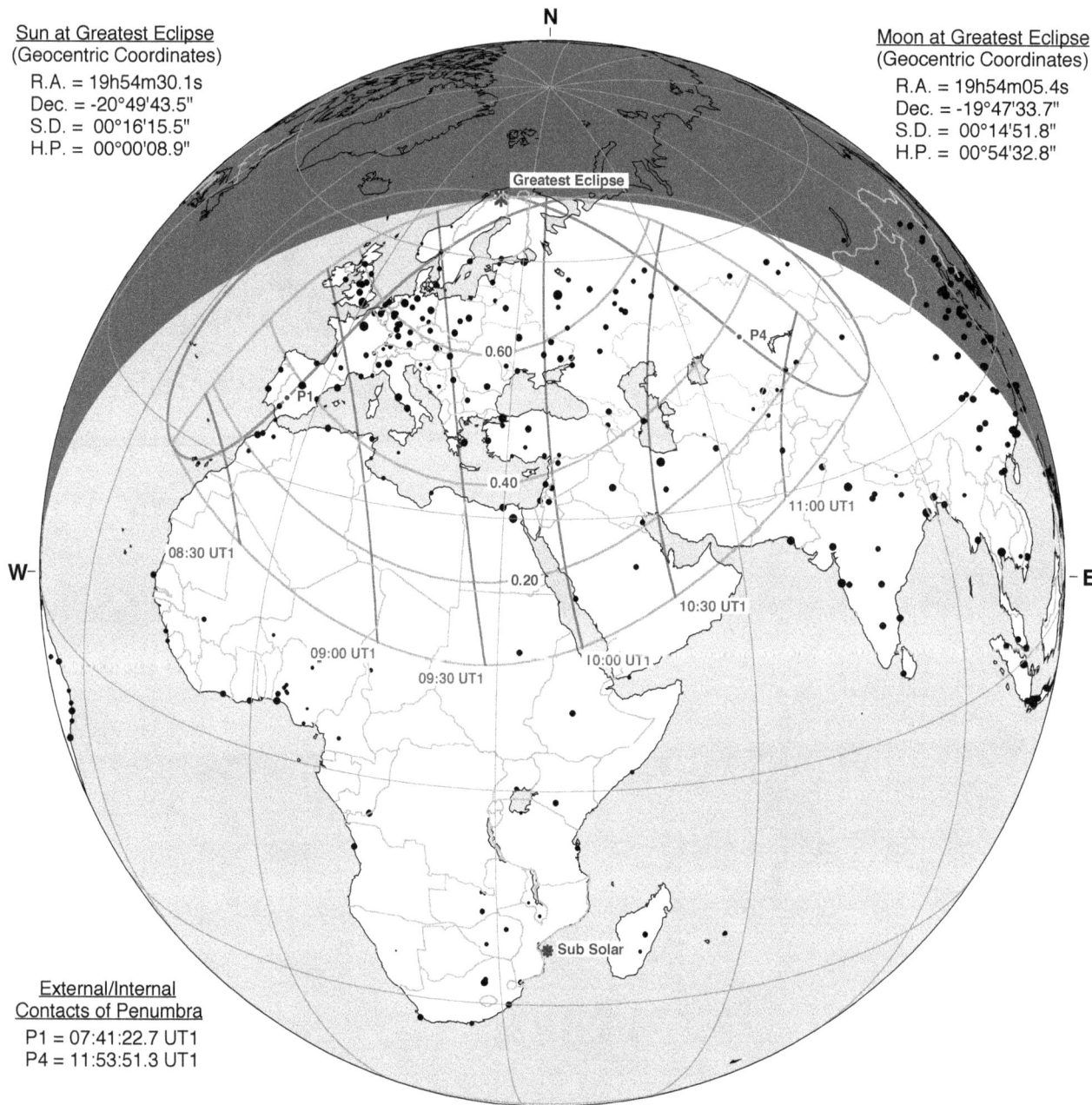

N

Greatest Eclipse

P4

P1

0.60

0.40

0.20

11:00 UT1

08:30 UT1

10:30 UT1

09:00 UT1

09:30 UT1

10:00 UT1

W

E

Sub Solar

**External/Internal
Contacts of Penumbra**
P1 = 07:41:22.7 UT1
P4 = 11:53:51.3 UT1

ΔT = 77.1 s   S   Eph. = JPL DE405

Circumstances at Greatest Eclipse: 09:47:38.0 UT1
Lat. = 68°31.3'N   Sun Alt. =   0.0°
Long. = 020°47.6'E   Sun Azm. = 166.2°

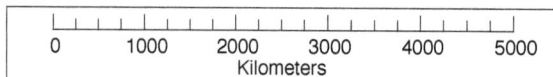

0 1000 2000 3000 4000 5000
Kilometers

Total Solar Eclipse of 2037 Jul 13

Greatest Eclipse = 02:40:35.9 TD   (= 02:39:18.5 UT1)

Eclipse Magnitude = 1.0413   Saros Series = 127
Gamma = -0.7246   Saros Member = 59 of 82

Sun at Greatest Eclipse
(Geocentric Coordinates)
R.A. = 07h31m06.7s
Dec. = +21°46'57.5"
S.D. = 00°15'44.0"
H.P. = 00°00'08.7"

Moon at Greatest Eclipse
(Geocentric Coordinates)
R.A. = 07h30m56.4s
Dec. = +21°04'03.1"
S.D. = 00°16'12.0"
H.P. = 00°59'27.3"

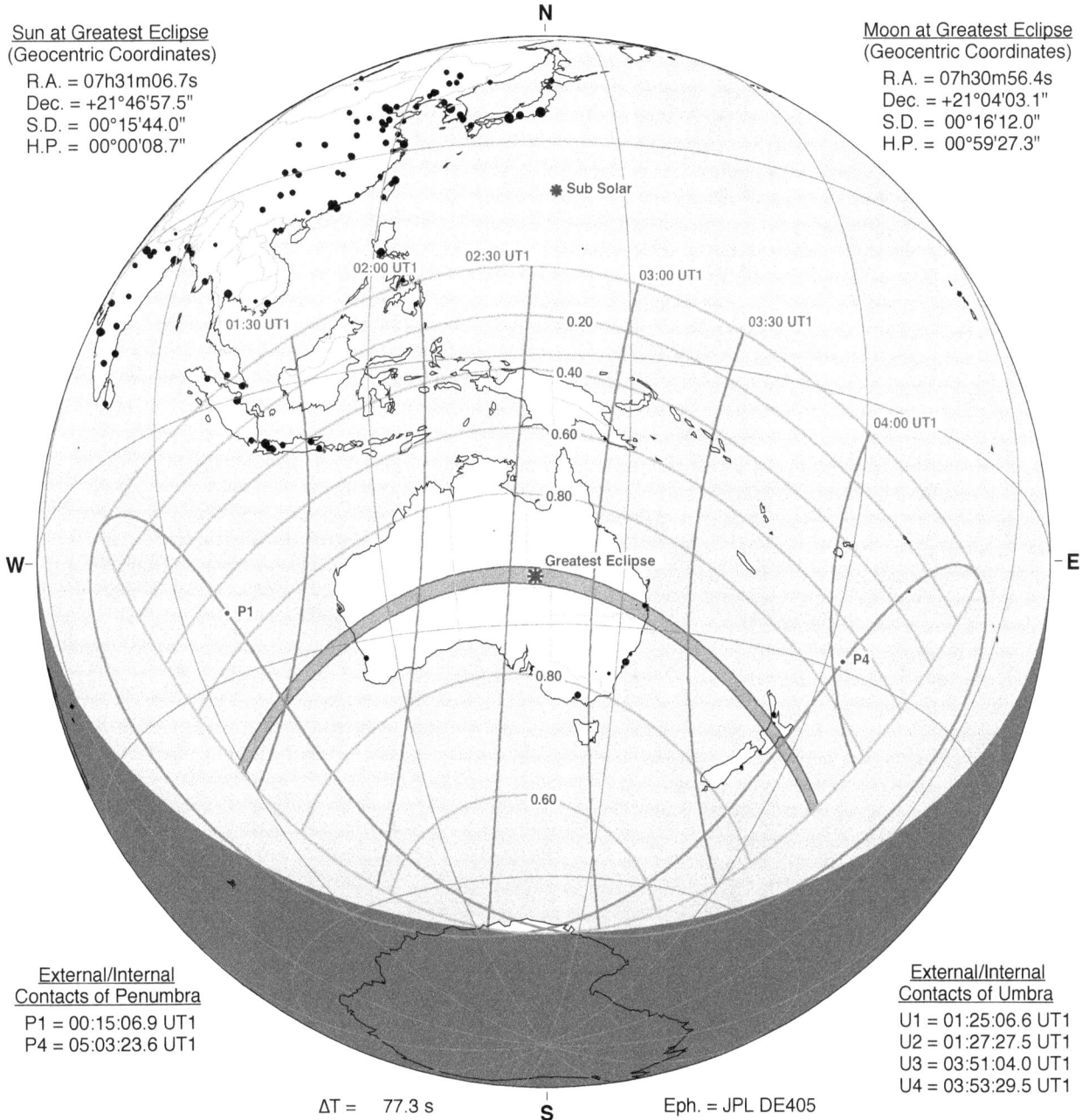

N

Sub Solar

02:00 UT1   02:30 UT1
01:30 UT1   03:00 UT1
                              0.20   03:30 UT1
                        0.40   04:00 UT1
                  0.60

W   E

Greatest Eclipse

P1

0.80

P4

0.80

0.60

S

External/Internal Contacts of Penumbra
P1 = 00:15:06.9 UT1
P4 = 05:03:23.6 UT1

External/Internal Contacts of Umbra
U1 = 01:25:06.6 UT1
U2 = 01:27:27.5 UT1
U3 = 03:51:04.0 UT1
U4 = 03:53:29.5 UT1

ΔT = 77.3 s   Eph. = JPL DE405

Circumstances at Greatest Eclipse: 02:39:18.5 UT1

Lat. = 24°45.8'S	Sun Alt. = 43.4°
Long. = 139°03.2'E	Sun Azm. = 3.3°
Path Width = 200.7 km	Duration = 03m58.4s

Circumstances at Greatest Duration: 02:39:45.9 UT1

Lat. = 24°46.7'S	Sun Alt. = 43.4°
Long. = 139°12.5'E	Sun Azm. = 2.9°
Path Width = 200.6 km	Duration = 03m58.4s

0   1000   2000   3000   4000   5000
Kilometers

Annular Solar Eclipse of 2038 Jan 05

Greatest Eclipse = 13:47:10.9 TD   (= 13:45:53.4 UT1)

Eclipse Magnitude = 0.9728   Saros Series = 132
Gamma = 0.4169   Saros Member = 47 of 71

Sun at Greatest Eclipse
(Geocentric Coordinates)
R.A. = 19h06m27.4s
Dec. = -22°33'17.3"
S.D. = 00°16'15.9"
H.P. = 00°00'08.9"

Moon at Greatest Eclipse
(Geocentric Coordinates)
R.A. = 19h06m25.9s
Dec. = -22°09'29.7"
S.D. = 00°15'35.7"
H.P. = 00°57'13.9"

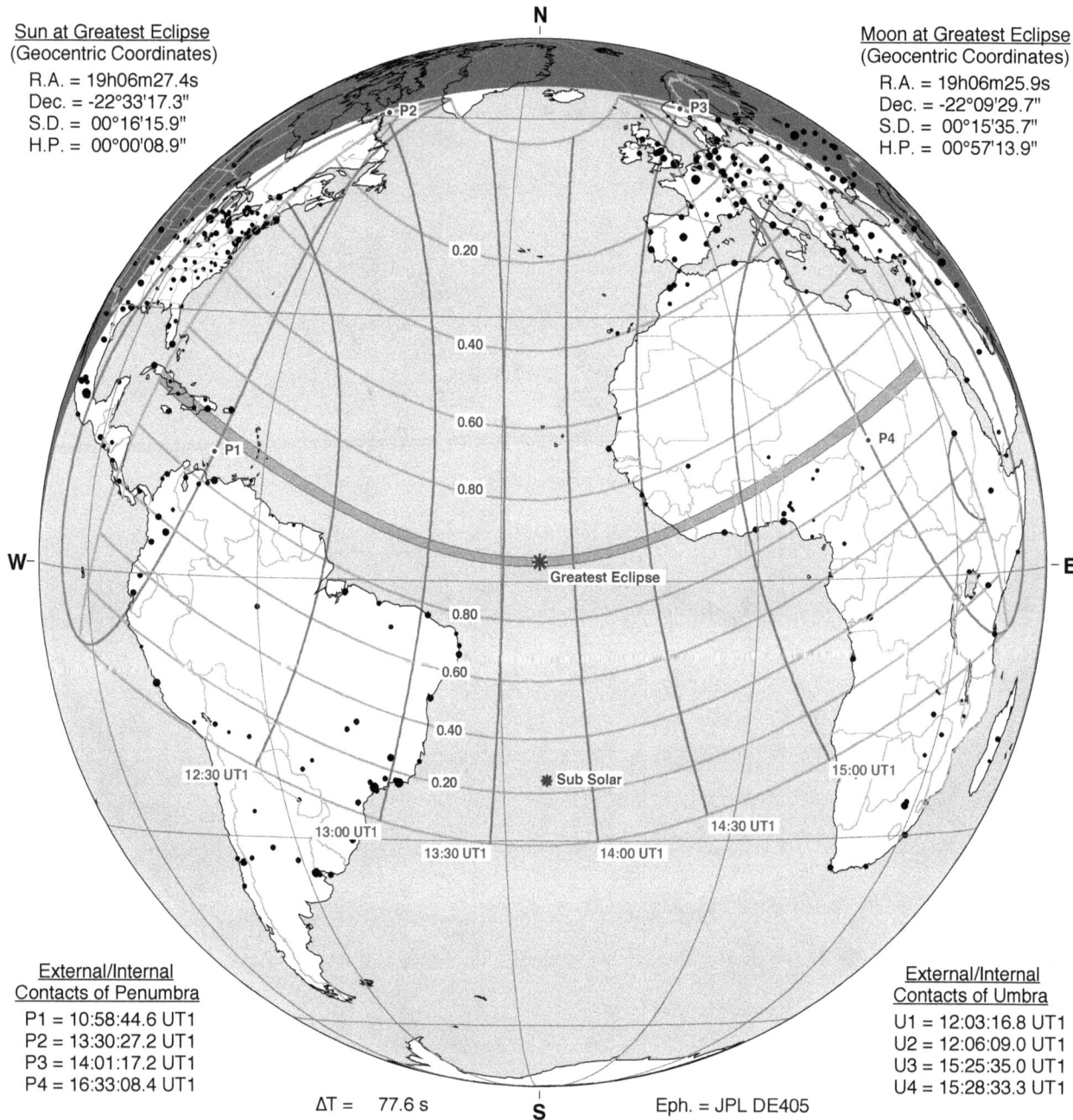

N

P2   P3

0.20

0.40

0.60

P1

0.80

W — Greatest Eclipse — E

0.80

Sub Solar

0.60

P4

0.40

0.20

12:30 UT1   15:00 UT1

13:00 UT1   14:30 UT1

13:30 UT1   14:00 UT1

External/Internal
Contacts of Penumbra
P1 = 10:58:44.6 UT1
P2 = 13:30:27.2 UT1
P3 = 14:01:17.2 UT1
P4 = 16:33:08.4 UT1

External/Internal
Contacts of Umbra
U1 = 12:03:16.8 UT1
U2 = 12:06:09.0 UT1
U3 = 15:25:35.0 UT1
U4 = 15:28:33.3 UT1

ΔT =   77.6 s   S   Eph. = JPL DE405

Circumstances at Greatest Eclipse: 13:45:53.4 UT1
Lat. = 02°05.7'N   Sun Alt. = 65.3°
Long. = 025°28.0'W   Sun Azm. = 179.2°
Path Width = 107.2 km   Duration = 03m18.5s

Circumstances at Greatest Duration: 13:52:36.3 UT1
Lat. = 02°11.2'N   Sun Alt. = 65.1°
Long. = 023°41.3'W   Sun Azm. = 186.8°
Path Width = 107.4 km   Duration = 03m18.6s

| 0 | 1000 | 2000 | 3000 | 4000 | 5000 |
Kilometers

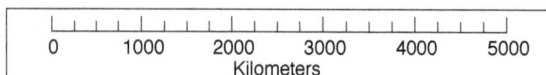

©2016 F. Espenak
www.EclipseWise.com

Annular Solar Eclipse of 2038 Jul 02

Greatest Eclipse = 13:32:55.0 TD   (= 13:31:37.2 UT1)

Eclipse Magnitude = 0.9911   Saros Series = 137
Gamma = 0.0398   Saros Member = 37 of 70

Sun at Greatest Eclipse
(Geocentric Coordinates)

R.A. = 06h46m55.4s
Dec. = +22°59'44.2"
S.D. = 00°15'43.9"
H.P. = 00°00'08.6"

Moon at Greatest Eclipse
(Geocentric Coordinates)

R.A. = 06h46m55.2s
Dec. = +23°01'58.2"
S.D. = 00°15'20.9"
H.P. = 00°56'19.9"

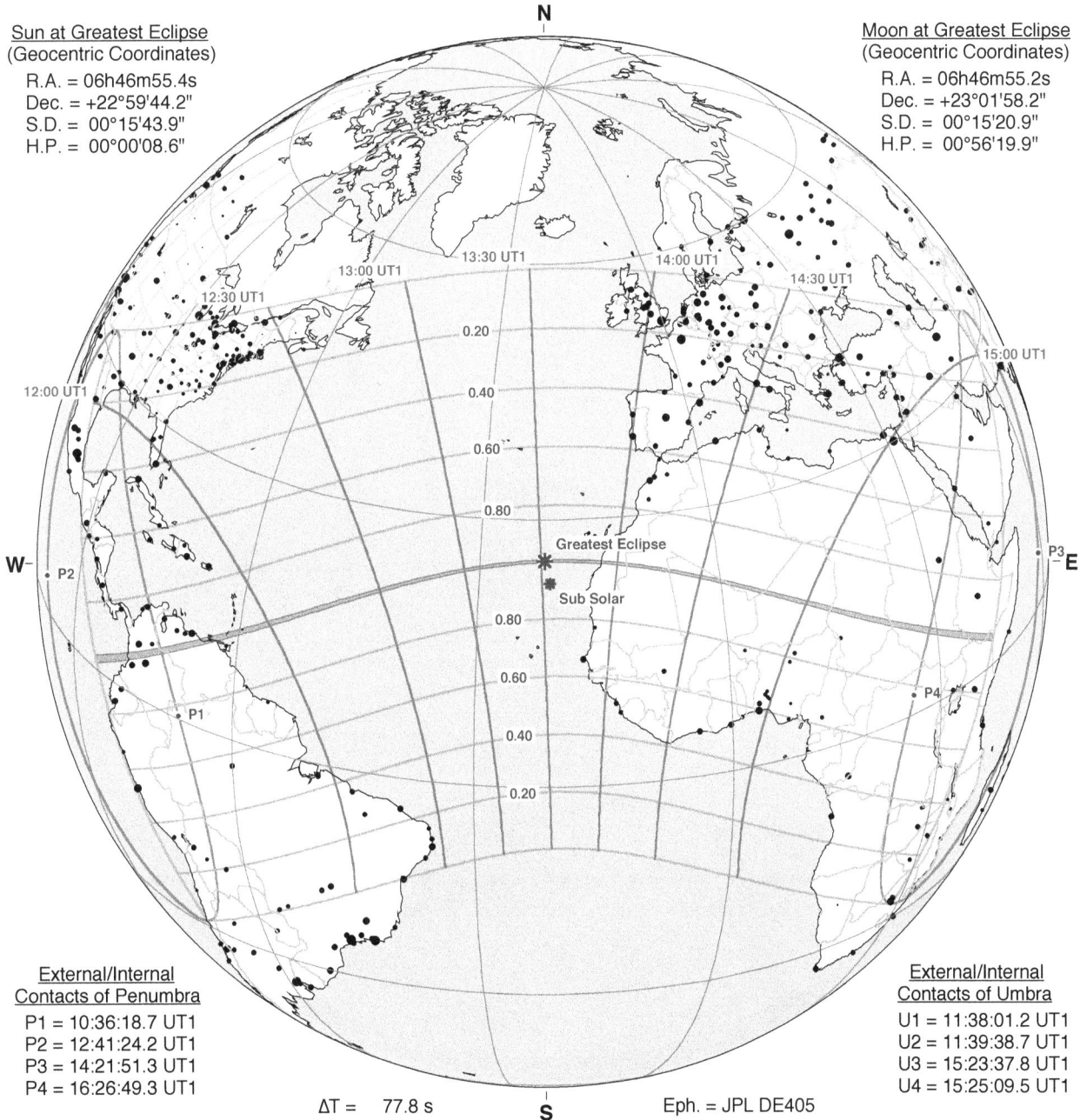

External/Internal Contacts of Penumbra

P1 = 10:36:18.7 UT1
P2 = 12:41:24.2 UT1
P3 = 14:21:51.3 UT1
P4 = 16:26:49.3 UT1

External/Internal Contacts of Umbra

U1 = 11:38:01.2 UT1
U2 = 11:39:38.7 UT1
U3 = 15:23:37.8 UT1
U4 = 15:25:09.5 UT1

ΔT = 77.8 s   Eph. = JPL DE405

Circumstances at Greatest Eclipse: 13:31:37.2 UT1

Lat. = 25°25.4'N   Sun Alt. = 87.6°
Long. = 021°54.8'W   Sun Azm. = 179.0°
Path Width = 31.2 km   Duration = 00m59.7s

Circumstances at Greatest Duration: 11:38:50.0 UT1

Lat. = 01°05.8'N   Sun Alt. = 0.0°
Long. = 084°08.2'W   Sun Azm. = 67.0°
Path Width = 94.2 km   Duration = 01m34.9s

0   1000   2000   3000   4000   5000
Kilometers

©2016 F. Espenak
www.EclipseWise.com

105

Total Solar Eclipse of 2038 Dec 26

Greatest Eclipse = 01:00:09.7 TD (= 00:58:51.6 UT1)

Eclipse Magnitude = 1.0269   Saros Series = 142
Gamma = -0.2881   Saros Member = 24 of 72

Sun at Greatest Eclipse
(Geocentric Coordinates)
R.A. = 18h18m51.7s
Dec. = -23°21'47.8"
S.D. = 00°16'15.7"
H.P. = 00°00'08.9"

Moon at Greatest Eclipse
(Geocentric Coordinates)
R.A. = 18h18m46.7s
Dec. = -23°39'05.4"
S.D. = 00°16'25.8"
H.P. = 01°00'18.1"

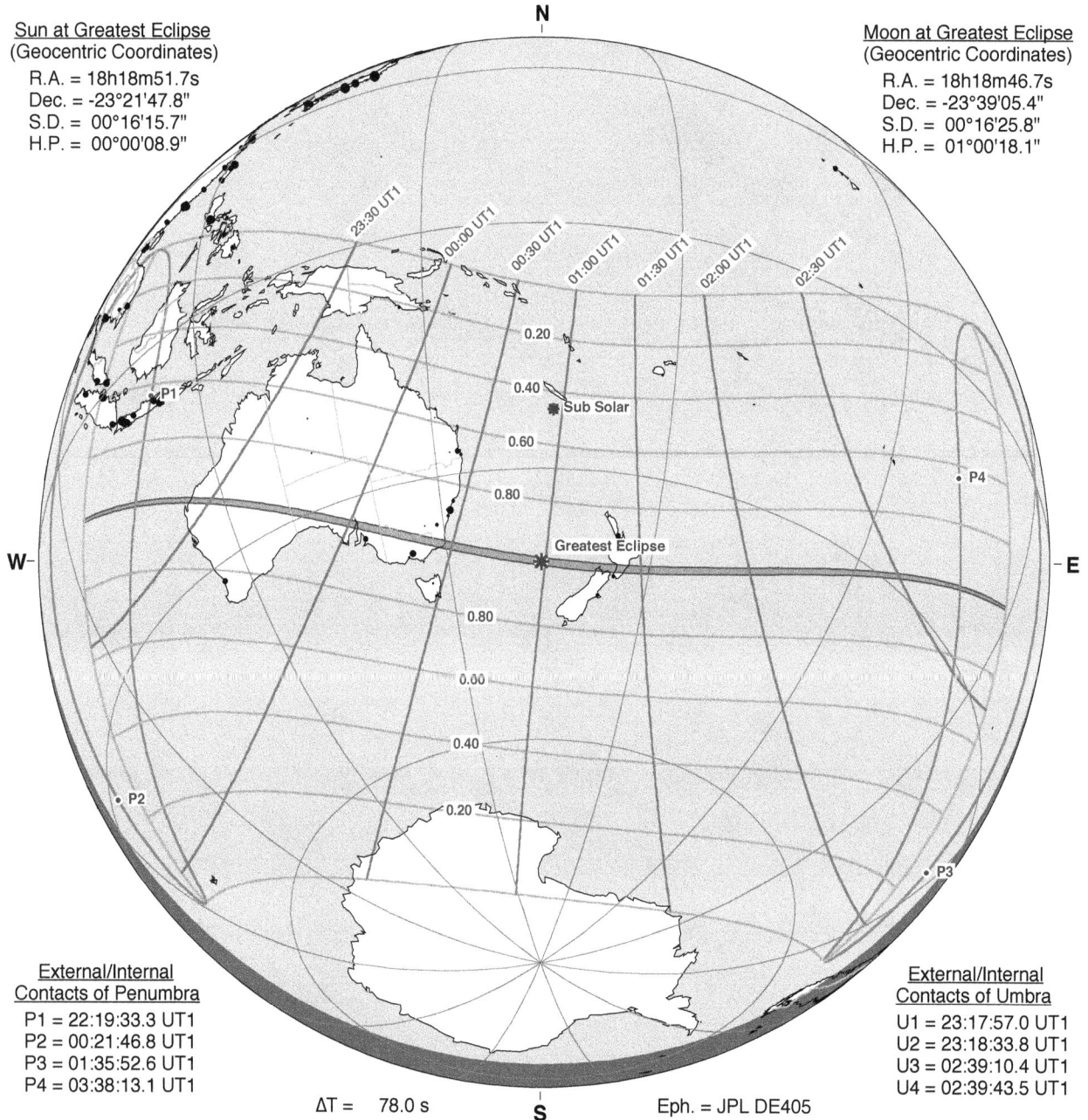

N

23:30 UT1
00:00 UT1
00:30 UT1
01:00 UT1
01:30 UT1
02:00 UT1
02:30 UT1

0.20
0.40
Sub Solar
0.60
0.80
Greatest Eclipse
P1
P4
W — — E
0.80
0.00
0.40
P2
0.20
P3

External/Internal Contacts of Penumbra
P1 = 22:19:33.3 UT1
P2 = 00:21:46.8 UT1
P3 = 01:35:52.6 UT1
P4 = 03:38:13.1 UT1

ΔT = 78.0 s   S   Eph. = JPL DE405

External/Internal Contacts of Umbra
U1 = 23:17:57.0 UT1
U2 = 23:18:33.8 UT1
U3 = 02:39:10.4 UT1
U4 = 02:39:43.5 UT1

Circumstances at Greatest Eclipse: 00:58:51.6 UT1
Lat. = 40°16.7'S   Sun Alt. = 73.0°
Long. = 163°55.6'E   Sun Azm. = 4.5°
Path Width = 95.3 km   Duration = 02m18.2s

Circumstances at Greatest Duration: 00:58:08.2 UT1
Lat. = 40°14.8'S   Sun Alt. = 73.0°
Long. = 163°35.6'E   Sun Azm. = 6.2°
Path Width = 95.3 km   Duration = 02m18.2s

0   1000   2000   3000   4000   5000
Kilometers

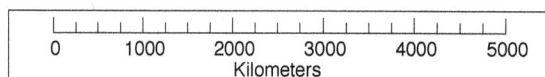

©2016 F. Espenak
www.EclipseWise.com

Annular Solar Eclipse of 2039 Jun 21

Greatest Eclipse = 17:12:53.8 TD (= 17:11:35.5 UT1)

Eclipse Magnitude = 0.9454   Saros Series = 147
Gamma = 0.8312   Saros Member = 24 of 80

Sun at Greatest Eclipse
(Geocentric Coordinates)
R.A. = 06h00m54.5s
Dec. = +23°26'03.6"
S.D. = 00°15'44.3"
H.P. = 00°00'08.7"

Moon at Greatest Eclipse
(Geocentric Coordinates)
R.A. = 06h00m35.3s
Dec. = +24°10'44.9"
S.D. = 00°14'45.6"
H.P. = 00°54'10.2"

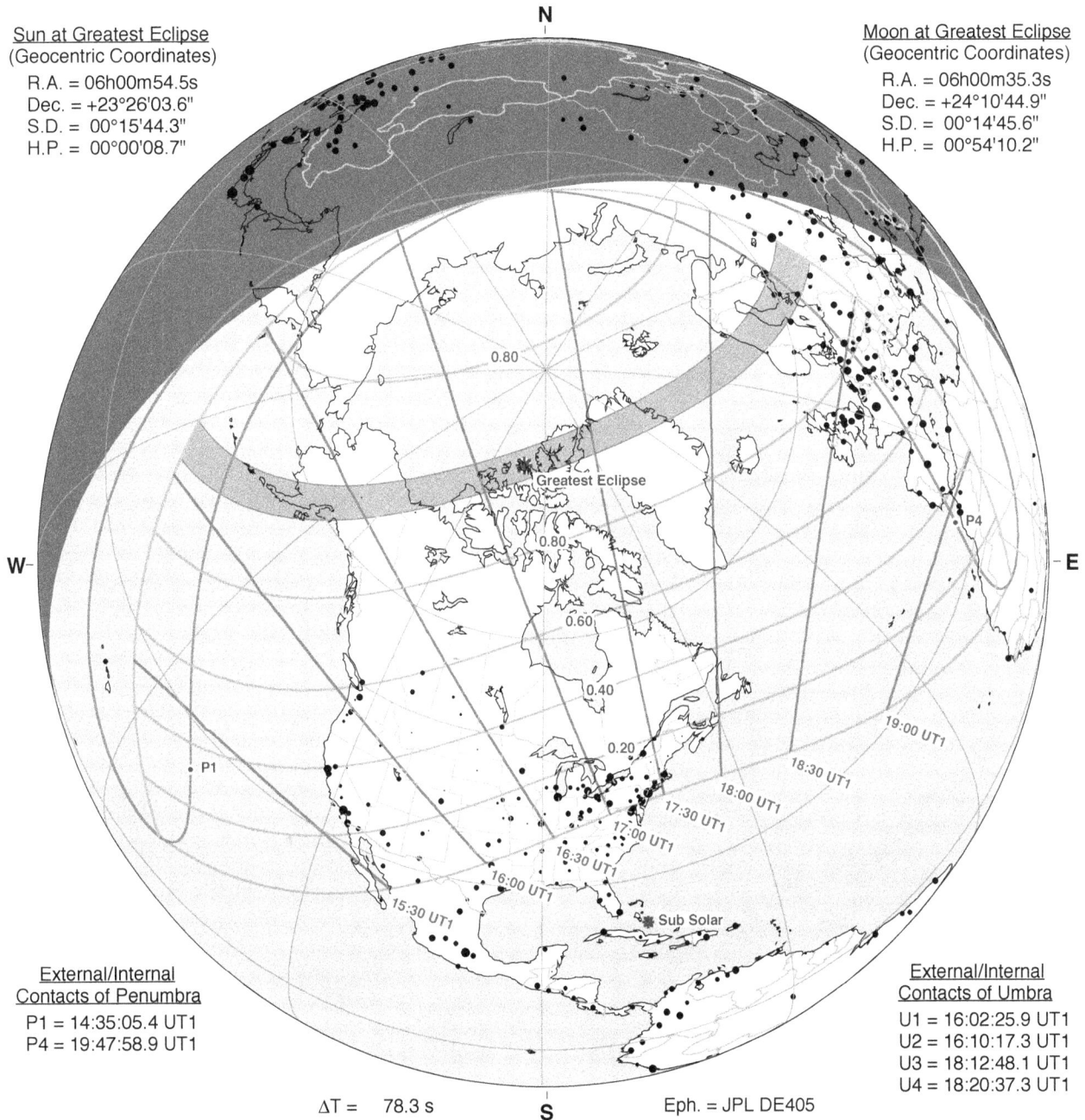

N

W — — E

S

External/Internal
Contacts of Penumbra
P1 = 14:35:05.4 UT1
P4 = 19:47:58.9 UT1

External/Internal
Contacts of Umbra
U1 = 16:02:25.9 UT1
U2 = 16:10:17.3 UT1
U3 = 18:12:48.1 UT1
U4 = 18:20:37.3 UT1

ΔT = 78.3 s   Eph. = JPL DE405

Circumstances at Greatest Eclipse: 17:11:35.5 UT1

Lat. = 78°52.5'N	Sun Alt. = 33.4°
Long. = 102°07.9'W	Sun Azm. = 152.6°
Path Width = 365.3 km	Duration = 04m04.9s

Circumstances at Greatest Duration: 17:11:45.6 UT1

Lat. = 78°54.8'N	Sun Alt. = 33.4°
Long. = 101°47.9'W	Sun Azm. = 153.1°
Path Width = 365.2 km	Duration = 04m04.9s

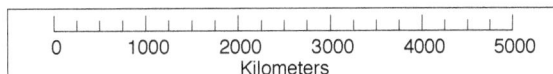

0   1000   2000   3000   4000   5000
Kilometers

Total Solar Eclipse of 2039 Dec 15

Greatest Eclipse = 16:23:45.9 TD (= 16:22:27.4 UT1)

Eclipse Magnitude = 1.0356	Saros Series = 152
Gamma = -0.9458	Saros Member = 14 of 70

Sun at Greatest Eclipse
(Geocentric Coordinates)
R.A. = 17h31m51.4s
Dec. = -23°16'37.6"
S.D. = 00°16'14.9"
H.P. = 00°00'08.9"

Moon at Greatest Eclipse
(Geocentric Coordinates)
R.A. = 17h31m14.4s
Dec. = -24°13'58.8"
S.D. = 00°16'44.6"
H.P. = 01°01'26.8"

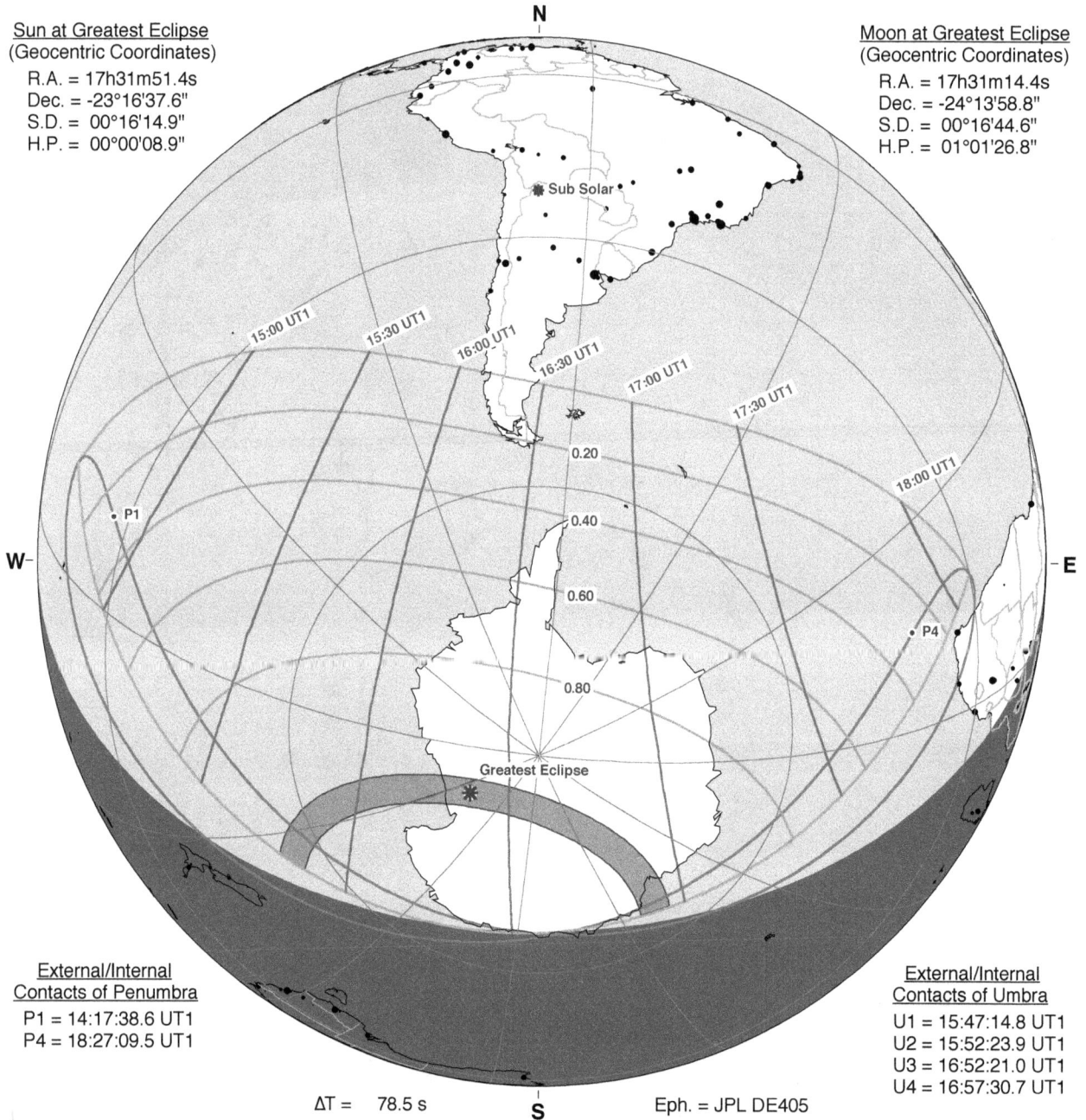

N

15:00 UT1
15:30 UT1
16:00 UT1
16:30 UT1
17:00 UT1
17:30 UT1
18:00 UT1

Sub Solar

0.20
0.40
0.60
0.80

P1

W

E

P4

Greatest Eclipse

External/Internal
Contacts of Penumbra
P1 = 14:17:38.6 UT1
P4 = 18:27:09.5 UT1

External/Internal
Contacts of Umbra
U1 = 15:47:14.8 UT1
U2 = 15:52:23.9 UT1
U3 = 16:52:21.0 UT1
U4 = 16:57:30.7 UT1

ΔT = 78.5 s

S

Eph. = JPL DE405

Circumstances at Greatest Eclipse: 16:22:27.4 UT1

Lat. = 80°51.4'S	Sun Alt. = 18.4°
Long. = 172°44.9'E	Sun Azm. = 123.4°
Path Width = 379.8 km	Duration = 01m51.4s

Circumstances at Greatest Duration: 16:22:32.9 UT1

Lat. = 80°54.0'S	Sun Alt. = 18.4°
Long. = 172°32.7'E	Sun Azm. = 123.5°
Path Width = 379.7 km	Duration = 01m51.4s

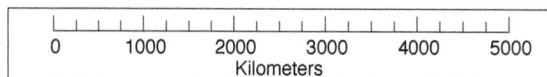

©2016 F. Espenak
www.EclipseWise.com

0   1000   2000   3000   4000   5000
Kilometers

Partial Solar Eclipse of 2040 May 11

Greatest Eclipse = 03:43:02.1 TD   (= 03:41:43.4 UT1)

Eclipse Magnitude = 0.5306   Saros Series = 119
Gamma = -1.2529   Saros Member = 67 of 71

Sun at Greatest Eclipse
(Geocentric Coordinates)
R.A. = 03h14m33.6s
Dec. = +18°01'19.7"
S.D. = 00°15'50.1"
H.P. = 00°00'08.7"

Moon at Greatest Eclipse
(Geocentric Coordinates)
R.A. = 03h16m16.3s
Dec. = +16°56'30.8"
S.D. = 00°15'06.4"
H.P. = 00°55'26.7"

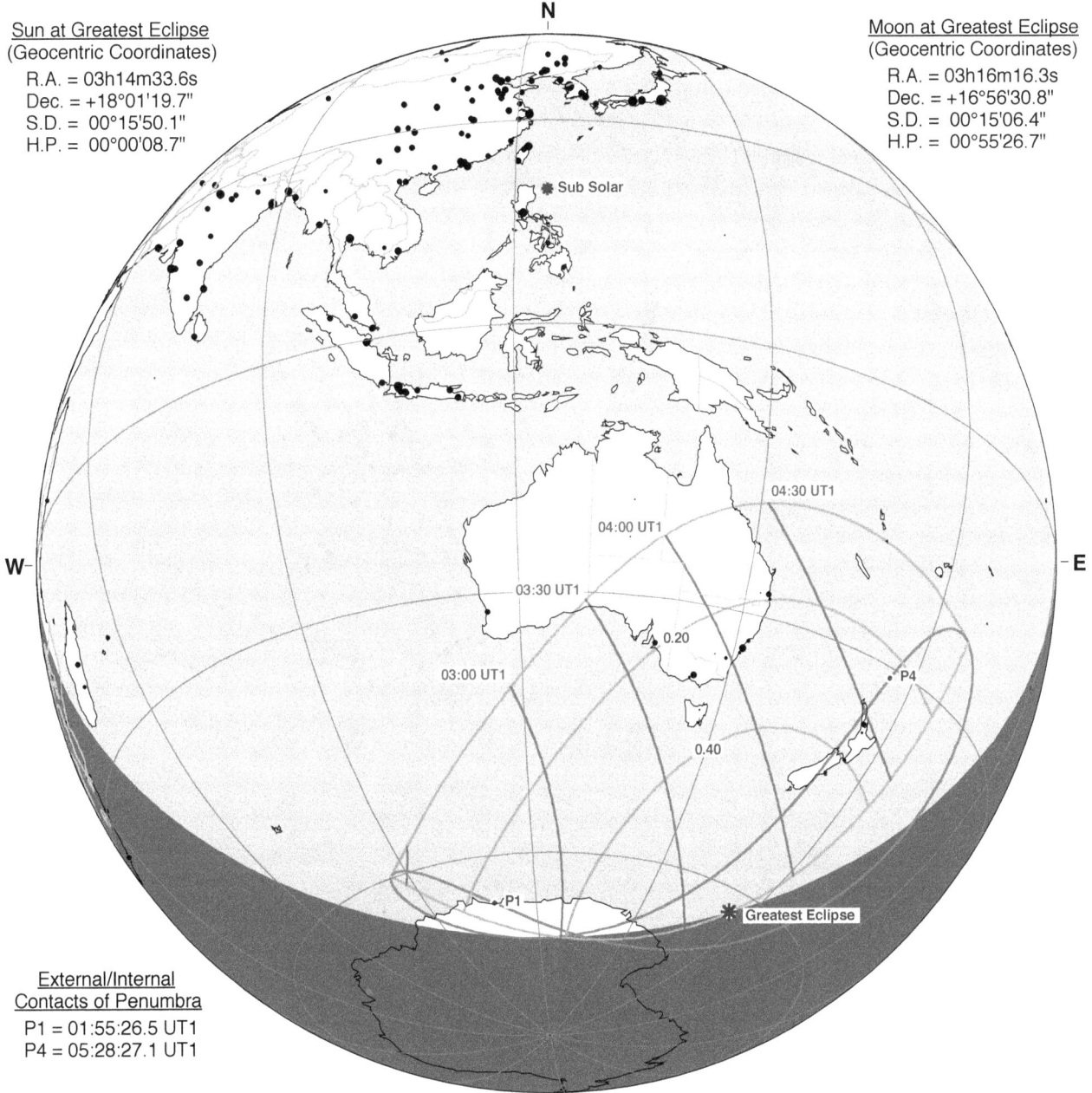

N

Sub Solar

04:30 UT1
04:00 UT1
03:30 UT1
03:00 UT1
0.20
0.40

W

E

P4

P1

Greatest Eclipse

External/Internal
Contacts of Penumbra
P1 = 01:55:26.5 UT1
P4 = 05:28:27.1 UT1

ΔT =   78.8 s   S   Eph. = JPL DE405

Circumstances at Greatest Eclipse: 03:41:43.4 UT1
Lat. = 62°46.1'S   Sun Alt. =   0.0°
Long. = 174°23.3'E   Sun Azm. = 312.5°

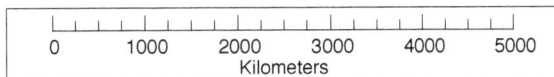

0	1000	2000	3000	4000	5000

Kilometers

©2016 F. Espenak
www.EclipseWise.com

Partial Solar Eclipse of 2040 Nov 04

Greatest Eclipse = 19:09:02.0 TD (= 19:07:43.0 UT1)

Eclipse Magnitude = 0.8074	Saros Series = 124
Gamma = 1.0993	Saros Member = 56 of 73

Sun at Greatest Eclipse
(Geocentric Coordinates)
R.A. = 14h42m06.9s
Dec. = -15°43'53.8"
S.D. = 00°16'07.7"
H.P. = 00°00'08.9"

Moon at Greatest Eclipse
(Geocentric Coordinates)
R.A. = 14h43m50.8s
Dec. = -14°45'19.8"
S.D. = 00°15'49.8"
H.P. = 00°58'05.7"

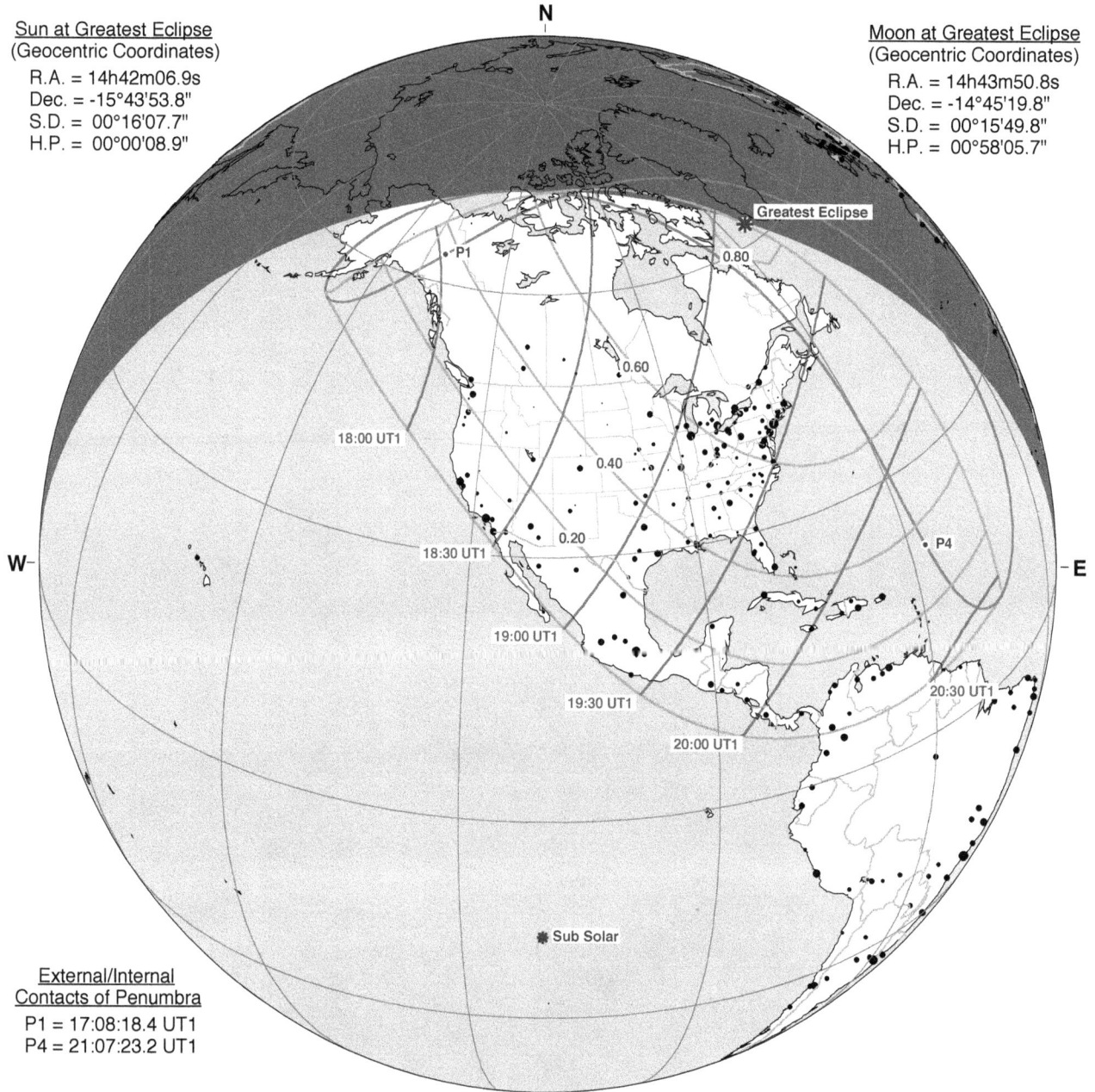

N

Greatest Eclipse

P1

0.80

0.60

18:00 UT1

0.40

W

0.20

18:30 UT1

P4

E

19:00 UT1

19:30 UT1

20:30 UT1

20:00 UT1

Sub Solar

**External/Internal
Contacts of Penumbra**
P1 = 17:08:18.4 UT1
P4 = 21:07:23.2 UT1

ΔT = 79.0 s

S

Eph. = JPL DE405

Circumstances at Greatest Eclipse: 19:07:43.0 UT1

Lat. = 62°12.1'N	Sun Alt. = 0.0°
Long. = 053°24.2'W	Sun Azm. = 234.4°

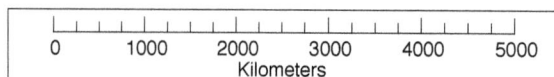

0   1000   2000   3000   4000   5000
Kilometers

©2016 F. Espenak
www.EclipseWise.com

Total Solar Eclipse of 2041 Apr 30

Greatest Eclipse = 11:52:20.8 TD (= 11:51:01.5 UT1)

Eclipse Magnitude = 1.0189   Saros Series = 129

Gamma = -0.4492   Saros Member = 53 of 80

Sun at Greatest Eclipse
(Geocentric Coordinates)

R.A. = 02h32m22.2s
Dec. = +14°58'18.8"
S.D. = 00°15'52.6"
H.P. = 00°00'08.7"

Moon at Greatest Eclipse
(Geocentric Coordinates)

R.A. = 02h33m06.0s
Dec. = +14°34'20.1"
S.D. = 00°15'56.6"
H.P. = 00°58'30.8"

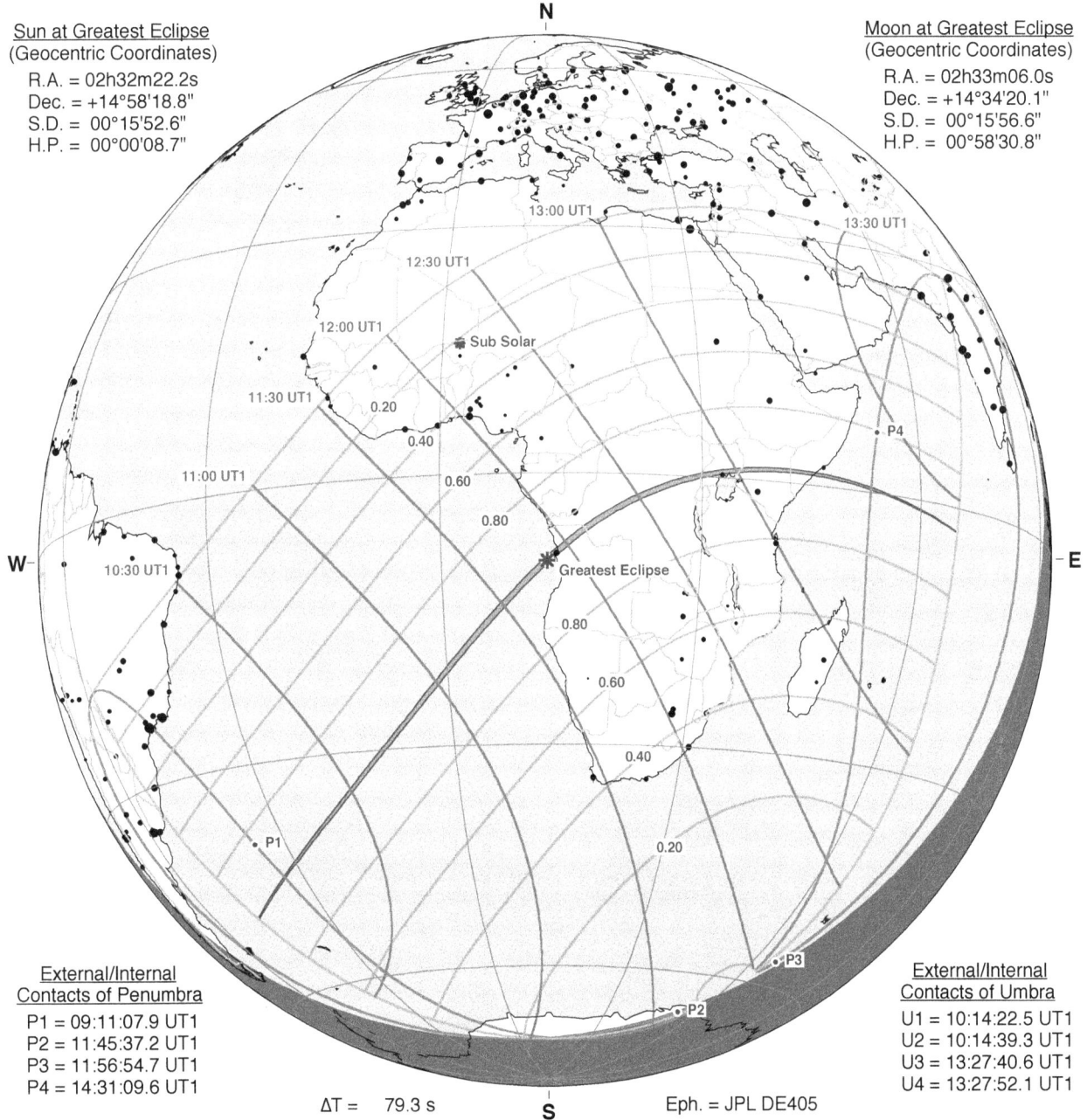

External/Internal
Contacts of Penumbra

P1 = 09:11:07.9 UT1
P2 = 11:45:37.2 UT1
P3 = 11:56:54.7 UT1
P4 = 14:31:09.6 UT1

ΔT = 79.3 s

Eph. = JPL DE405

External/Internal
Contacts of Umbra

U1 = 10:14:22.5 UT1
U2 = 10:14:39.3 UT1
U3 = 13:27:40.6 UT1
U4 = 13:27:52.1 UT1

Circumstances at Greatest Eclipse: 11:51:01.5 UT1

Lat. = 09°37.2'S   Sun Alt. = 63.2°
Long. = 012°09.7'E   Sun Azm. = 336.7°
Path Width = 71.7 km   Duration = 01m50.6s

Circumstances at Greatest Duration: 11:52:09.5 UT1

Lat. = 09°22.9'S   Sun Alt. = 63.2°
Long. = 012°26.2'E   Sun Azm. = 335.4°
Path Width = 71.7 km   Duration = 01m50.6s

Annular Solar Eclipse of 2041 Oct 25

Greatest Eclipse = 01:36:21.7 TD (= 01:35:02.2 UT1)

Eclipse Magnitude = 0.9467   Saros Series = 134
Gamma = 0.4133   Saros Member = 45 of 71

Sun at Greatest Eclipse
(Geocentric Coordinates)
R.A. = 13h59m22.0s
Dec. = -12°10'20.1"
S.D. = 00°16'04.9"
H.P. = 00°00'08.8"

Moon at Greatest Eclipse
(Geocentric Coordinates)
R.A. = 14h00m02.5s
Dec. = -11°49'54.3"
S.D. = 00°15'00.8"
H.P. = 00°55'06.0"

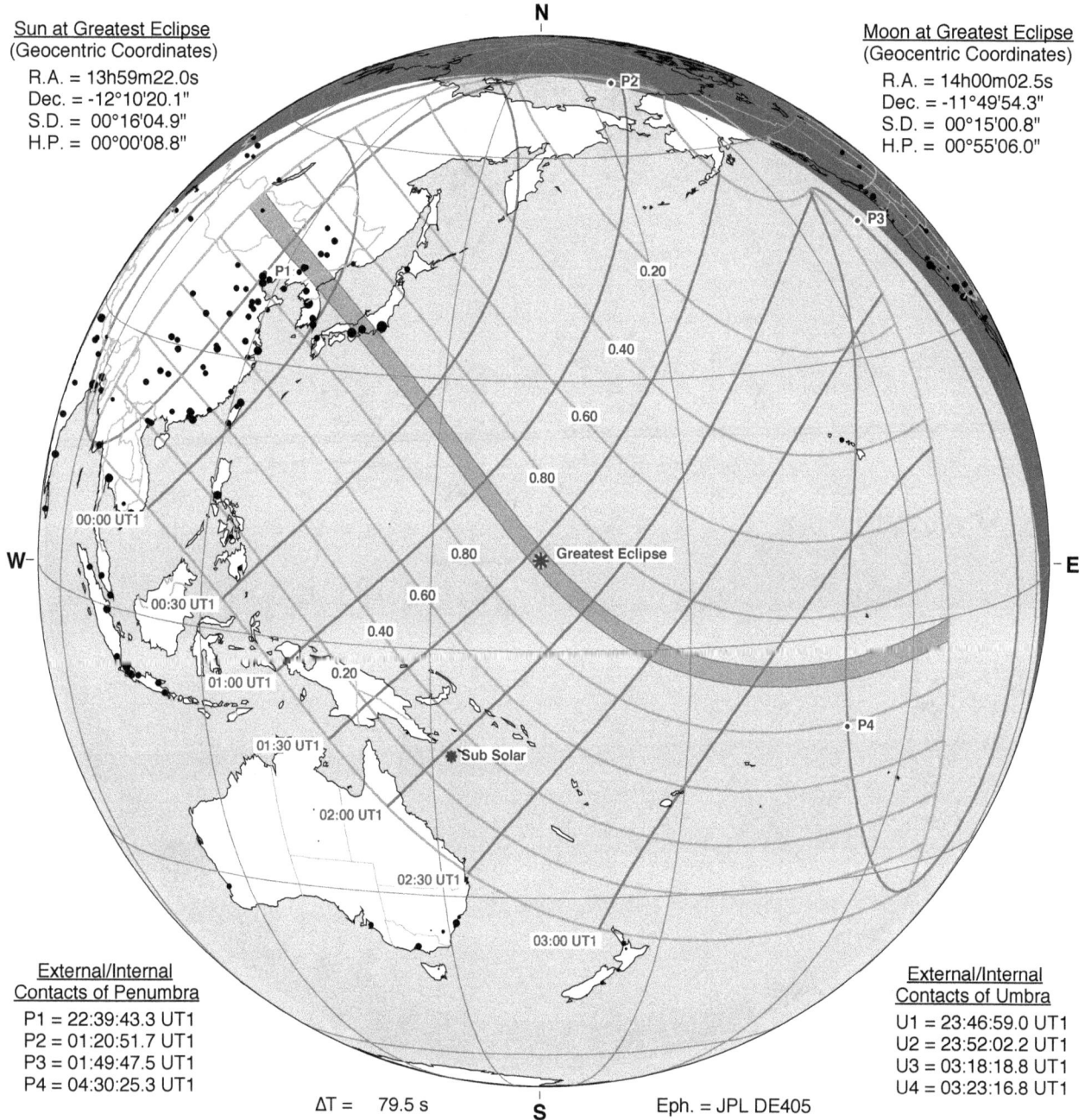

N

P2

P3

0.20

0.40

0.60

0.80

P1

0.80

Greatest Eclipse

00:00 UT1

W —

0.60

— E

0.40

00:30 UT1

0.20

01:00 UT1

P4

01:30 UT1

Sub Solar

02:00 UT1

02:30 UT1

03:00 UT1

External/Internal
Contacts of Penumbra
P1 = 22:39:43.3 UT1
P2 = 01:20:51.7 UT1
P3 = 01:49:47.5 UT1
P4 = 04:30:25.3 UT1

ΔT = 79.5 s   S   Eph. = JPL DE405

External/Internal
Contacts of Umbra
U1 = 23:46:59.0 UT1
U2 = 23:52:02.2 UT1
U3 = 03:18:18.8 UT1
U4 = 03:23:16.8 UT1

Circumstances at Greatest Eclipse: 01:35:02.2 UT1
Lat. = 09°56.0'N   Sun Alt. = 65.5°
Long. = 162°49.8'E   Sun Azm. = 205.6°
Path Width = 213.1 km Duration = 06m07.1s

Circumstances at Greatest Duration: 01:50:35.9 UT1
Lat. = 06°42.8'N   Sun Alt. = 64.2°
Long. = 166°01.4'E   Sun Azm. = 223.0°
Path Width = 218.3 km Duration = 06m08.4s

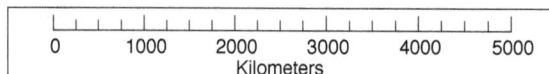

| 0 | 1000 | 2000 | 3000 | 4000 | 5000 |
Kilometers

©2016 F. Espenak
www.EclipseWise.com

Total Solar Eclipse of 2042 Apr 20

Greatest Eclipse = 02:17:30.1 TD (= 02:16:10.4 UT1)

Eclipse Magnitude = 1.0614	Saros Series = 139
Gamma = 0.2956	Saros Member = 31 of 71

Sun at Greatest Eclipse
(Geocentric Coordinates)
R.A. = 01h52m12.4s
Dec. = +11°31'19.4"
S.D. = 00°15'55.3"
H.P. = 00°00'08.8"

Moon at Greatest Eclipse
(Geocentric Coordinates)
R.A. = 01h51m39.9s
Dec. = +11°47'27.9"
S.D. = 00°16'37.6"
H.P. = 01°01'01.4"

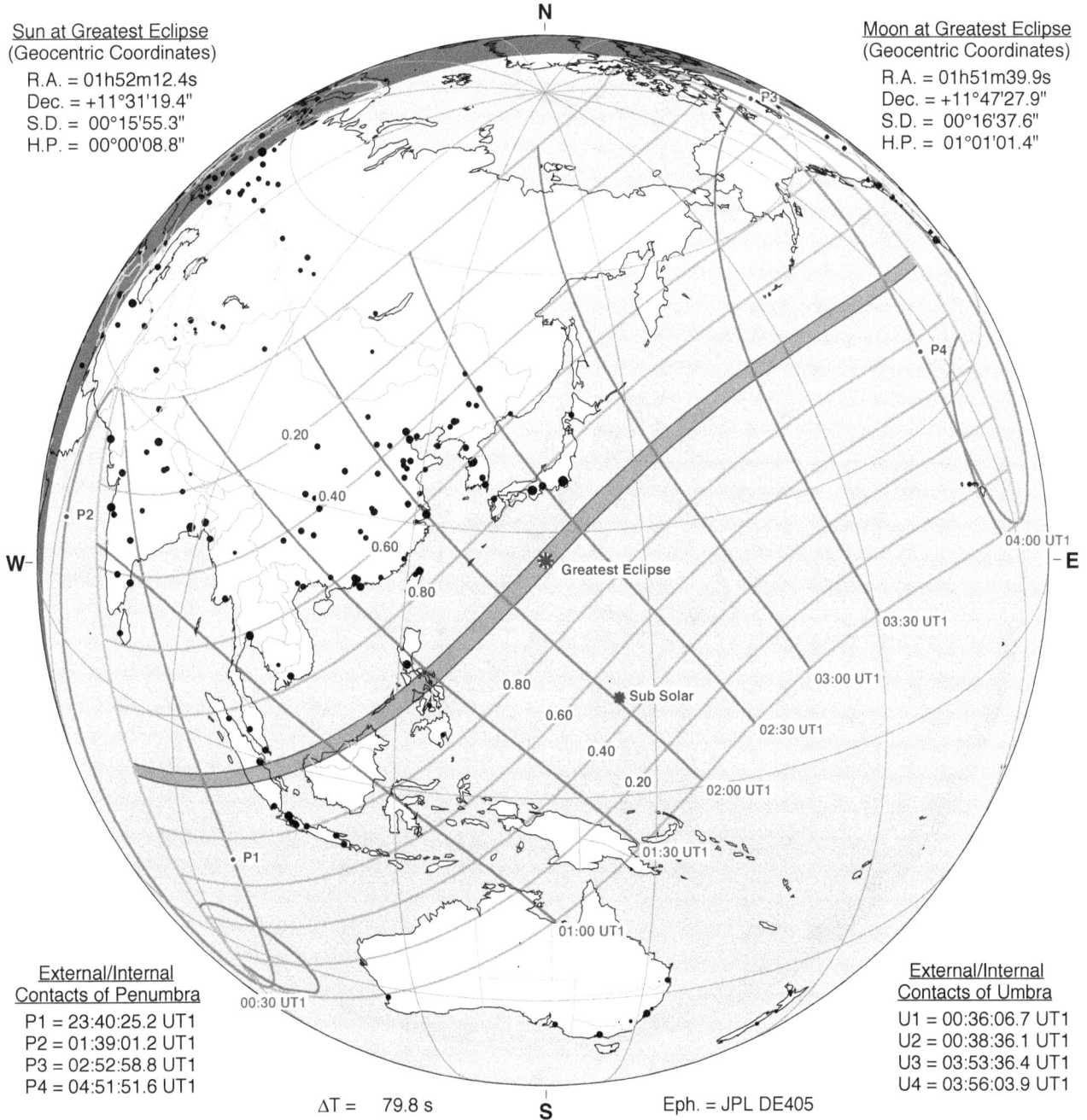

External/Internal
Contacts of Penumbra
P1 = 23:40:25.2 UT1
P2 = 01:39:01.2 UT1
P3 = 02:52:58.8 UT1
P4 = 04:51:51.6 UT1

External/Internal
Contacts of Umbra
U1 = 00:36:06.7 UT1
U2 = 00:38:36.1 UT1
U3 = 03:53:36.4 UT1
U4 = 03:56:03.9 UT1

ΔT = 79.8 s

Eph. = JPL DE405

Circumstances at Greatest Eclipse: 02:16:10.4 UT1

Lat. = 26°57.2'N	Sun Alt. = 72.7°
Long. = 137°17.0'E	Sun Azm. = 151.2°
Path Width = 210.4 km	Duration = 04m51.0s

Circumstances at Greatest Duration: 02:20:00.4 UT1

Lat. = 27°59.1'N	Sun Alt. = 72.5°
Long. = 138°25.2'E	Sun Azm. = 159.0°
Path Width = 209.7 km	Duration = 04m51.2s

Kilometers
0 1000 2000 3000 4000 5000

©2016 F. Espenak
www.EclipseWise.com

Annular Solar Eclipse of 2042 Oct 14

Greatest Eclipse = 02:00:41.9 TD (= 01:59:21.8 UT1)

Eclipse Magnitude = 0.9301   Saros Series = 144
Gamma = -0.3030   Saros Member = 18 of 70

Sun at Greatest Eclipse
(Geocentric Coordinates)
R.A. = 13h17m05.8s
Dec. = -08°08'35.1"
S.D. = 00°16'01.9"
H.P. = 00°00'08.8"

Moon at Greatest Eclipse
(Geocentric Coordinates)
R.A. = 13h16m35.0s
Dec. = -08°23'00.1"
S.D. = 00°14'41.9"
H.P. = 00°53'56.6"

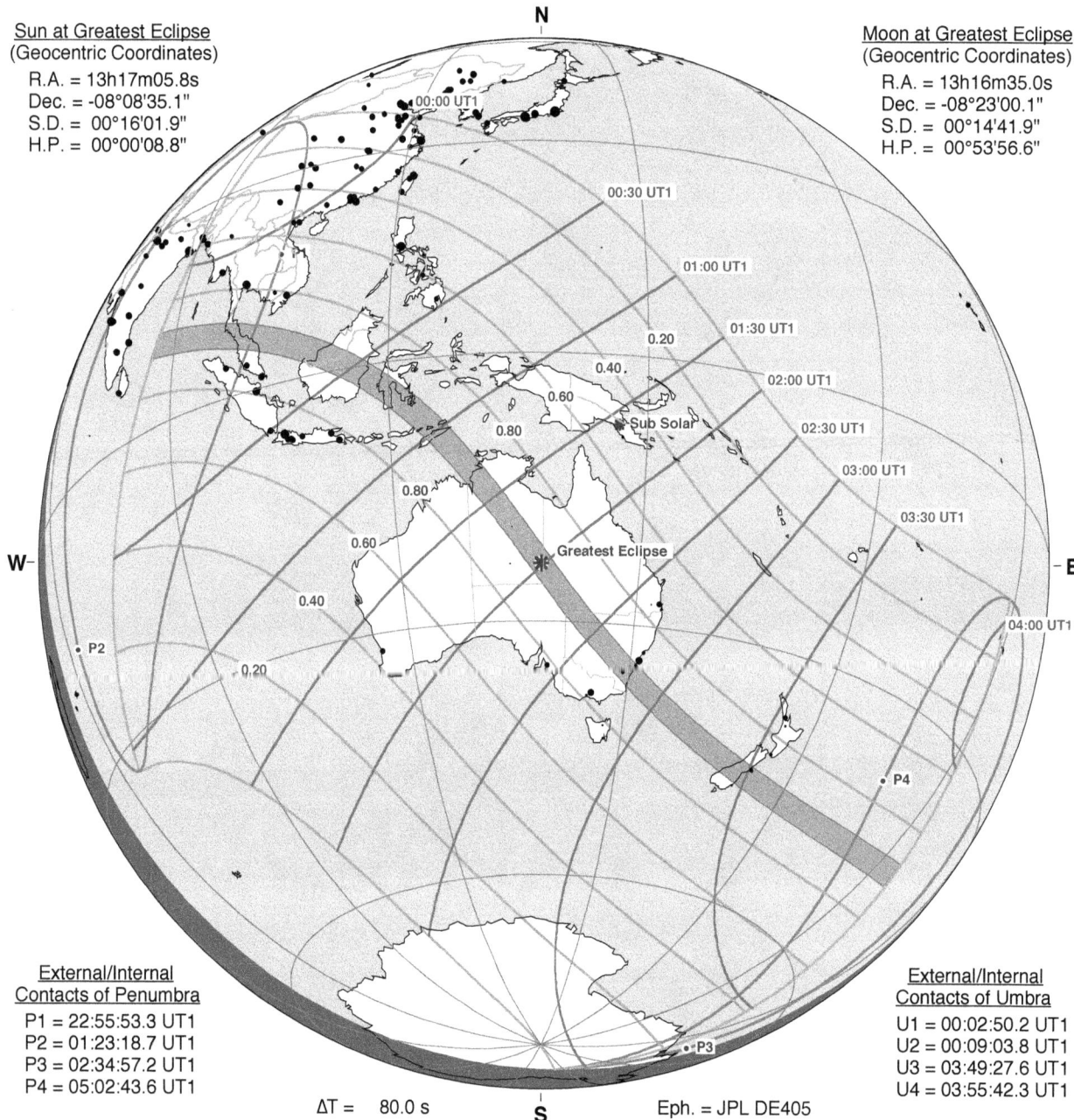

N

00:00 UT1
00:30 UT1
01:00 UT1
01:30 UT1
0.20
0.40
02:00 UT1
0.60
Sub Solar
0.80
02:30 UT1
03:00 UT1
0.80
03:30 UT1
0.60
Greatest Eclipse
W
0.40
E
04:00 UT1
P2
0.20
P4
P3
S

External/Internal Contacts of Penumbra
P1 = 22:55:53.3 UT1
P2 = 01:23:18.7 UT1
P3 = 02:34:57.2 UT1
P4 = 05:02:43.6 UT1

External/Internal Contacts of Umbra
U1 = 00:02:50.2 UT1
U2 = 00:09:03.8 UT1
U3 = 03:49:27.6 UT1
U4 = 03:55:42.3 UT1

ΔT = 80.0 s   Eph. = JPL DE405

Circumstances at Greatest Eclipse: 01:59:21.8 UT1
Lat. = 23°44.7'S   Sun Alt. = 72.2°
Long. = 137°47.4'E   Sun Azm. = 30.1°
Path Width = 273.3 km   Duration = 07m44.2s

Circumstances at Greatest Duration: 02:13:39.4 UT1
Lat. = 27°13.1'S   Sun Alt. = 70.8°
Long. = 140°49.9'E   Sun Azm. = 6.8°
Path Width = 270.8 km   Duration = 07m45.4s

0   1000   2000   3000   4000   5000
Kilometers

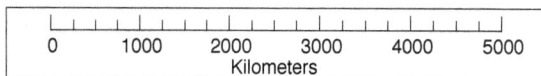

©2016 F. Espenak
www.EclipseWise.com

Total Solar Eclipse of 2043 Apr 09

Greatest Eclipse = 18:57:49.4 TD   (= 18:56:29.1 UT1)

Eclipse Magnitude = 1.0096   Saros Series = 149
Gamma = 1.0031   Saros Member = 22 of 71

Sun at Greatest Eclipse
(Geocentric Coordinates)
R.A. = 01h13m12.2s
Dec. = +07°45'05.1"
S.D. = 00°15'58.1"
H.P. = 00°00'08.8"

Moon at Greatest Eclipse
(Geocentric Coordinates)
R.A. = 01h11m17.3s
Dec. = +08°39'09.1"
S.D. = 00°16'38.0"
H.P. = 01°01'02.7"

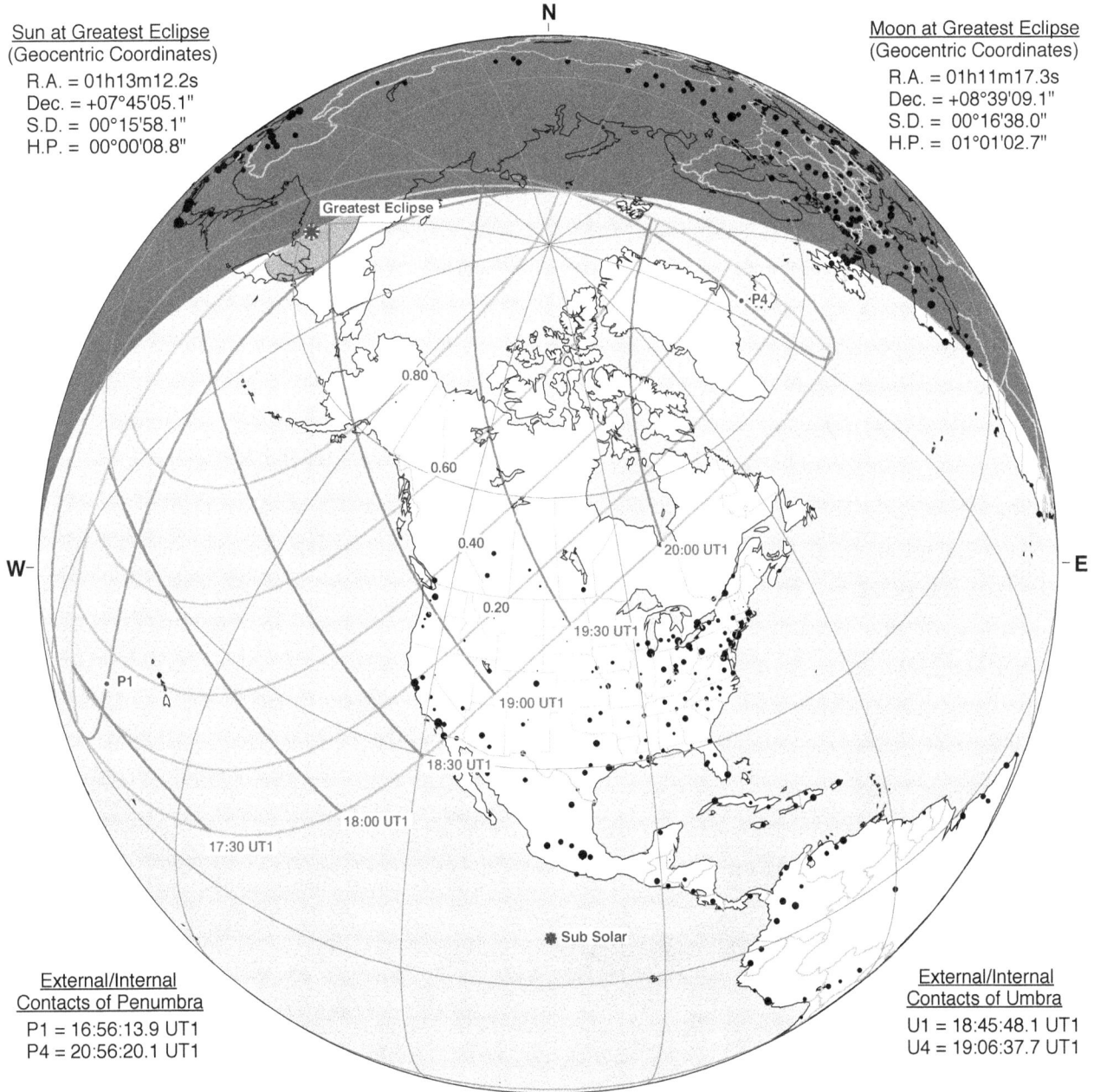

Greatest Eclipse

P4

N
W
E
S

0.80
0.60
0.40
0.20

20:00 UT1
19:30 UT1
19:00 UT1
18:30 UT1
18:00 UT1
17:30 UT1

P1

Sub Solar

External/Internal
Contacts of Penumbra
P1 = 16:56:13.9 UT1
P4 = 20:56:20.1 UT1

External/Internal
Contacts of Umbra
U1 = 18:45:48.1 UT1
U4 = 19:06:37.7 UT1

ΔT = 80.3 s   Eph. = JPL DE405

Circumstances at Greatest Eclipse: 18:56:29.1 UT1
Lat. = 61°18.7'N   Sun Alt. = 0.0°
Long. = 151°56.2'E   Sun Azm. = 73.7°
Path Width = 0.0 km   Duration = 00m00.0s

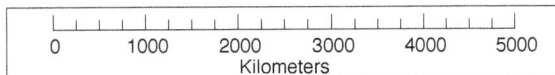

©2016 F. Espenak
www.EclipseWise.com

0   1000   2000   3000   4000   5000
Kilometers

Annular Solar Eclipse of 2043 Oct 03

Greatest Eclipse = 03:01:48.9 TD (= 03:00:28.4 UT1)

Eclipse Magnitude = 0.9497   Saros Series = 154

Gamma = -1.0102   Saros Member = 8 of 71

Sun at Greatest Eclipse
(Geocentric Coordinates)
R.A. = 12h36m02.9s
Dec. = -03°53'04.6"
S.D. = 00°15'58.8"
H.P. = 00°00'08.8"

Moon at Greatest Eclipse
(Geocentric Coordinates)
R.A. = 12h34m15.0s
Dec. = -04°41'56.9"
S.D. = 00°15'05.1"
H.P. = 00°55'21.7"

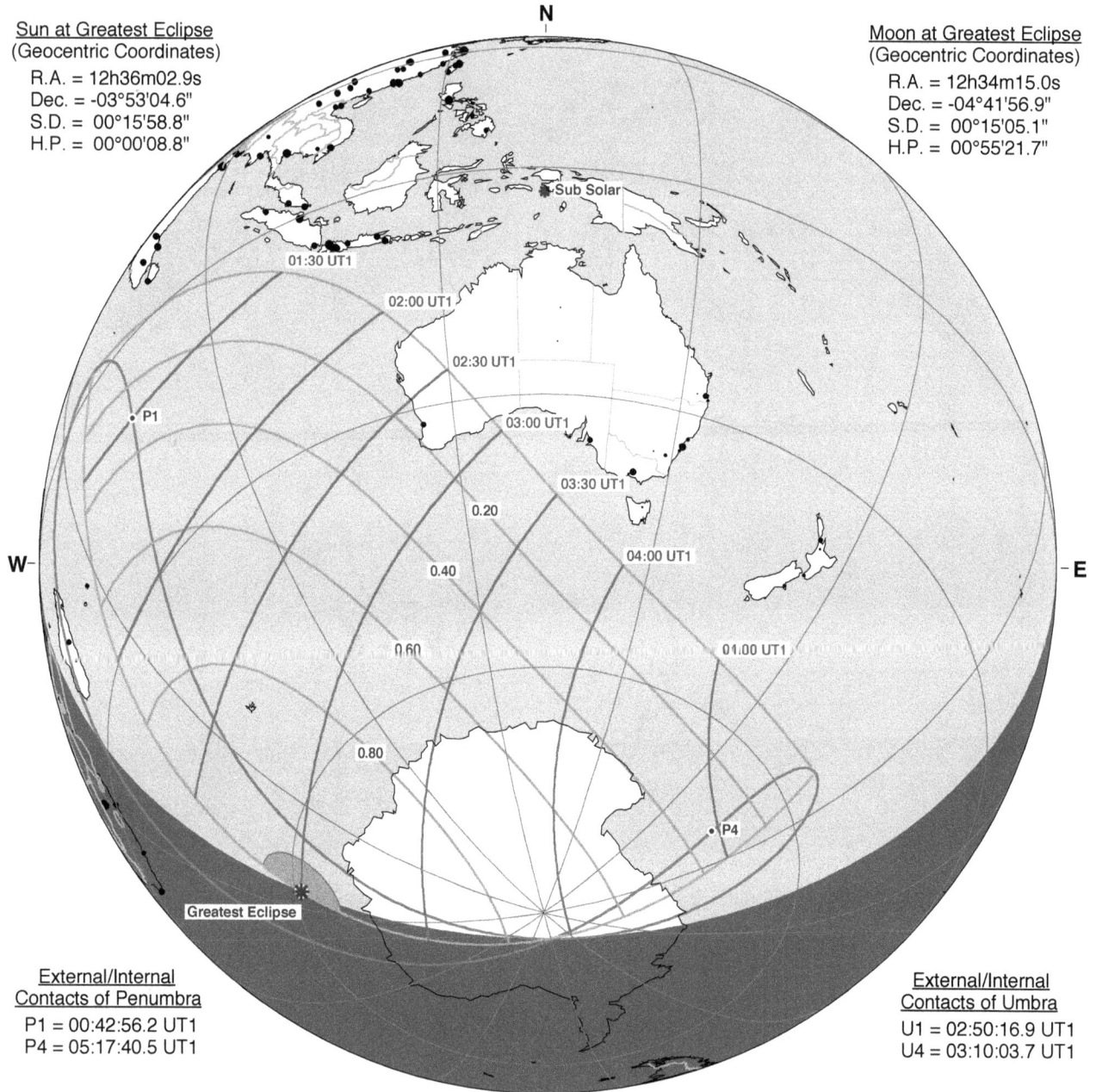

N

Sub Solar

01:30 UT1
02:00 UT1
02:30 UT1
03:00 UT1
03:30 UT1
0.20
04:00 UT1
0.40
05:00 UT1
0.60
0.80

P1

W — — E

P4

Greatest Eclipse

External/Internal Contacts of Penumbra
P1 = 00:42:56.2 UT1
P4 = 05:17:40.5 UT1

External/Internal Contacts of Umbra
U1 = 02:50:16.9 UT1
U4 = 03:10:03.7 UT1

ΔT = 80.5 s   S   Eph. = JPL DE405

Circumstances at Greatest Eclipse: 03:00:28.4 UT1

Lat. = 60°57.6'S	Sun Alt. = 0.0°
Long. = 035°14.0'E	Sun Azm. = 98.0°
Path Width = 0.0 km	Duration = 00m00.0s

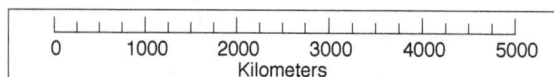

0   1000   2000   3000   4000   5000
Kilometers

Annular Solar Eclipse of 2044 Feb 28

Greatest Eclipse = 20:24:39.5 TD   (= 20:23:18.7 UT1)

Eclipse Magnitude = 0.9600   Saros Series = 121

Gamma = -0.9954   Saros Member = 62 of 71

Sun at Greatest Eclipse
(Geocentric Coordinates)
R.A. = 22h45m44.1s
Dec. = -07°51'30.6"
S.D. = 00°16'08.8"
H.P. = 00°00'08.9"

Moon at Greatest Eclipse
(Geocentric Coordinates)
R.A. = 22h47m30.6s
Dec. = -08°41'25.7"
S.D. = 00°15'29.6"
H.P. = 00°56'51.8"

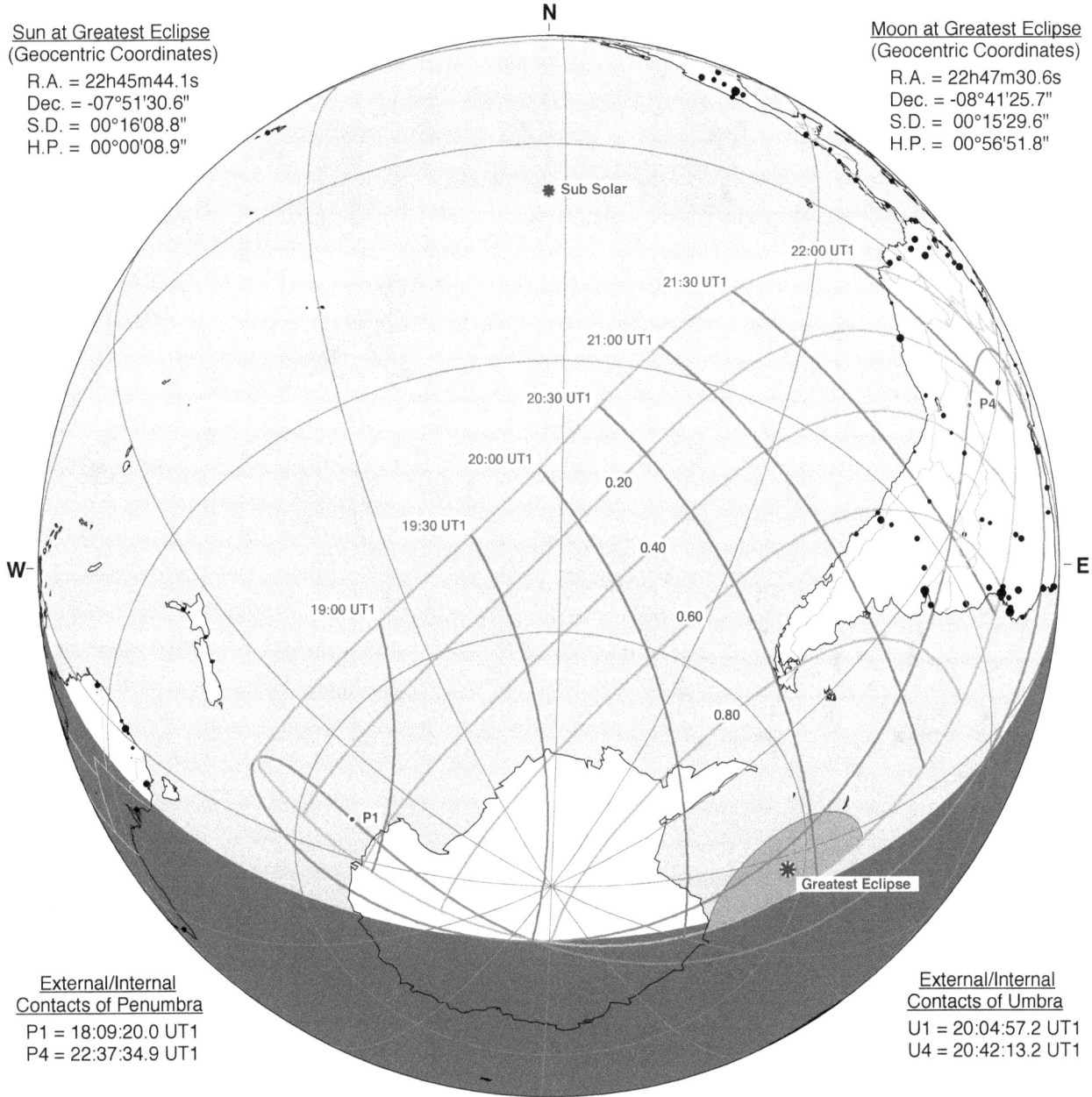

N

Sub Solar

22:00 UT1
21:30 UT1
21:00 UT1
20:30 UT1
20:00 UT1
19:30 UT1
19:00 UT1

P4

0.20
0.40
0.60
0.80

W

E

P1

Greatest Eclipse

S

External/Internal Contacts of Penumbra
P1 = 18:09:20.0 UT1
P4 = 22:37:34.9 UT1

External/Internal Contacts of Umbra
U1 = 20:04:57.2 UT1
U4 = 20:42:13.2 UT1

ΔT =   80.8 s   Eph. = JPL DE405

Circumstances at Greatest Eclipse: 20:23:18.7 UT1

Lat. = 62°10.9'S   Sun Alt. =   3.7°
Long. = 025°38.6'W   Sun Azm. = 260.1°
Path Width =2237.9 km   Duration = 02m27.2s

Circumstances at Greatest Duration: 20:16:25.2 UT1

Lat. = 64°41.7'S   Sun Alt. =   0.0°
Long. = 014°00.7'W   Sun Azm. = 251.4°
Path Width =2582.2 km   Duration = 02m27.4s

0 1000 2000 3000 4000 5000
Kilometers

©2016 F. Espenak
www.EclipseWise.com

Total Solar Eclipse of 2044 Aug 23

Greatest Eclipse = 01:17:01.7 TD (= 01:15:40.6 UT1)

Eclipse Magnitude = 1.0364	Saros Series = 126
Gamma = 0.9613	Saros Member = 49 of 72

Sun at Greatest Eclipse
(Geocentric Coordinates)
R.A. = 10h10m33.4s
Dec. = +11°16'02.2"
S.D. = 00°15'48.9"
H.P. = 00°00'08.7"

Moon at Greatest Eclipse
(Geocentric Coordinates)
R.A. = 10h12m17.2s
Dec. = +12°07'34.4"
S.D. = 00°16'19.6"
H.P. = 00°59'55.1"

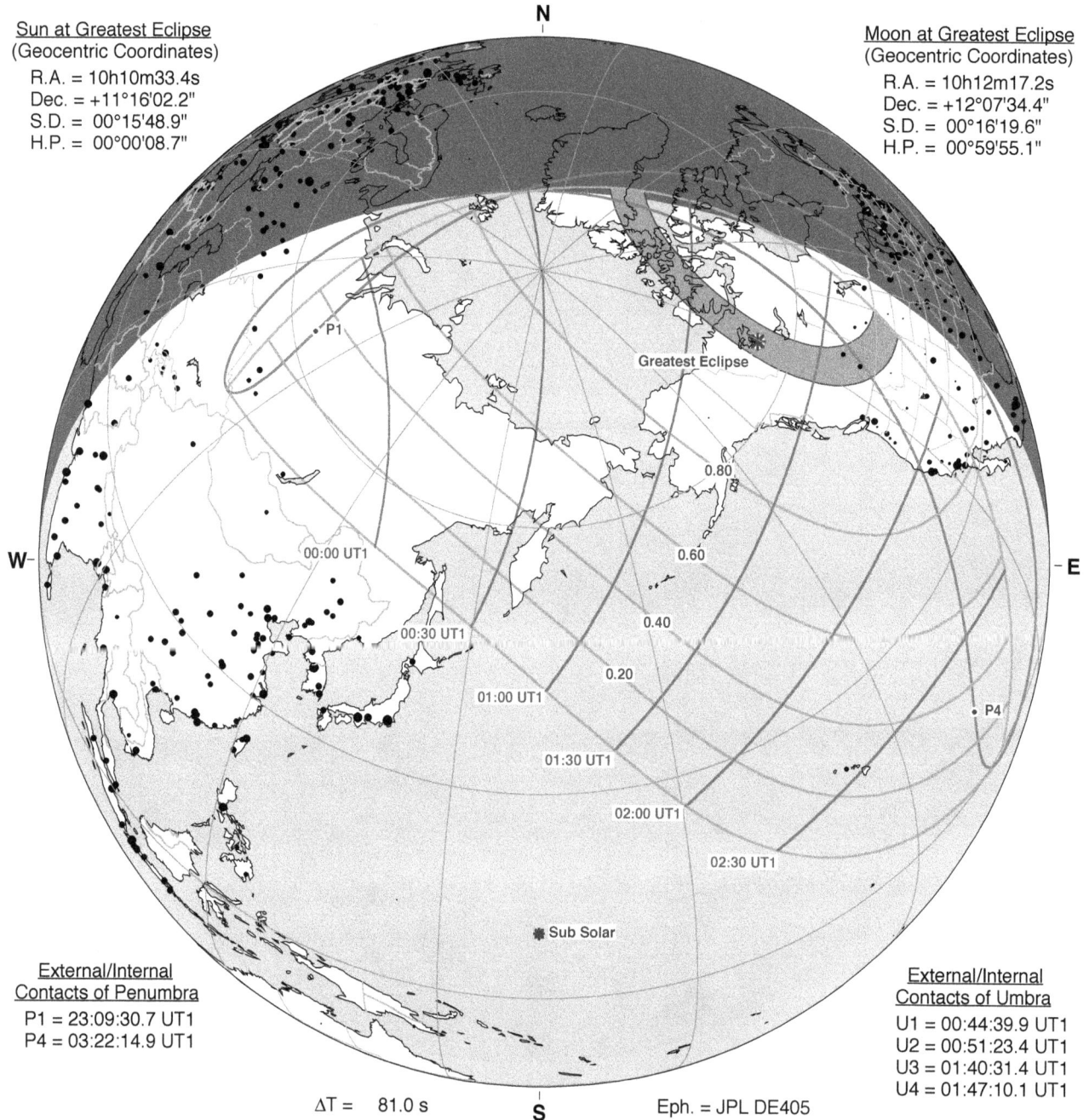

N

P1

Greatest Eclipse

0.80

0.60

0.40

0.20

00:00 UT1

00:30 UT1

01:00 UT1

01:30 UT1

02:00 UT1

02:30 UT1

W

E

P4

Sub Solar

External/Internal Contacts of Penumbra
P1 = 23:09:30.7 UT1
P4 = 03:22:14.9 UT1

External/Internal Contacts of Umbra
U1 = 00:44:39.9 UT1
U2 = 00:51:23.4 UT1
U3 = 01:40:31.4 UT1
U4 = 01:47:10.1 UT1

ΔT = 81.0 s

S

Eph. = JPL DE405

Circumstances at Greatest Eclipse: 01:15:40.6 UT1

Lat. = 64°20.1'N	Sun Alt. = 15.4°
Long. = 120°28.6'W	Sun Azm. = 263.9°
Path Width = 452.7 km	Duration = 02m03.9s

Circumstances at Greatest Duration: 01:15:14.8 UT1

Lat. = 64°34.1'N	Sun Alt. = 15.4°
Long. = 120°25.0'W	Sun Azm. = 263.8°
Path Width = 451.6 km	Duration = 02m03.9s

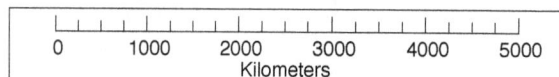

©2016 F. Espenak
www.EclipseWise.com

0 1000 2000 3000 4000 5000
Kilometers

Annular Solar Eclipse of 2045 Feb 16

Greatest Eclipse = 23:56:06.6 TD   (= 23:54:45.3 UT1)

Eclipse Magnitude = 0.9285   Saros Series = 131
Gamma = -0.3125   Saros Member = 52 of 70

Sun at Greatest Eclipse
(Geocentric Coordinates)
R.A. = 22h03m27.1s
Dec. = -11°55'04.8"
S.D. = 00°16'11.2"
H.P. = 00°00'08.9"

Moon at Greatest Eclipse
(Geocentric Coordinates)
R.A. = 22h03m57.6s
Dec. = -12°10'17.7"
S.D. = 00°14'48.9"
H.P. = 00°54'22.2"

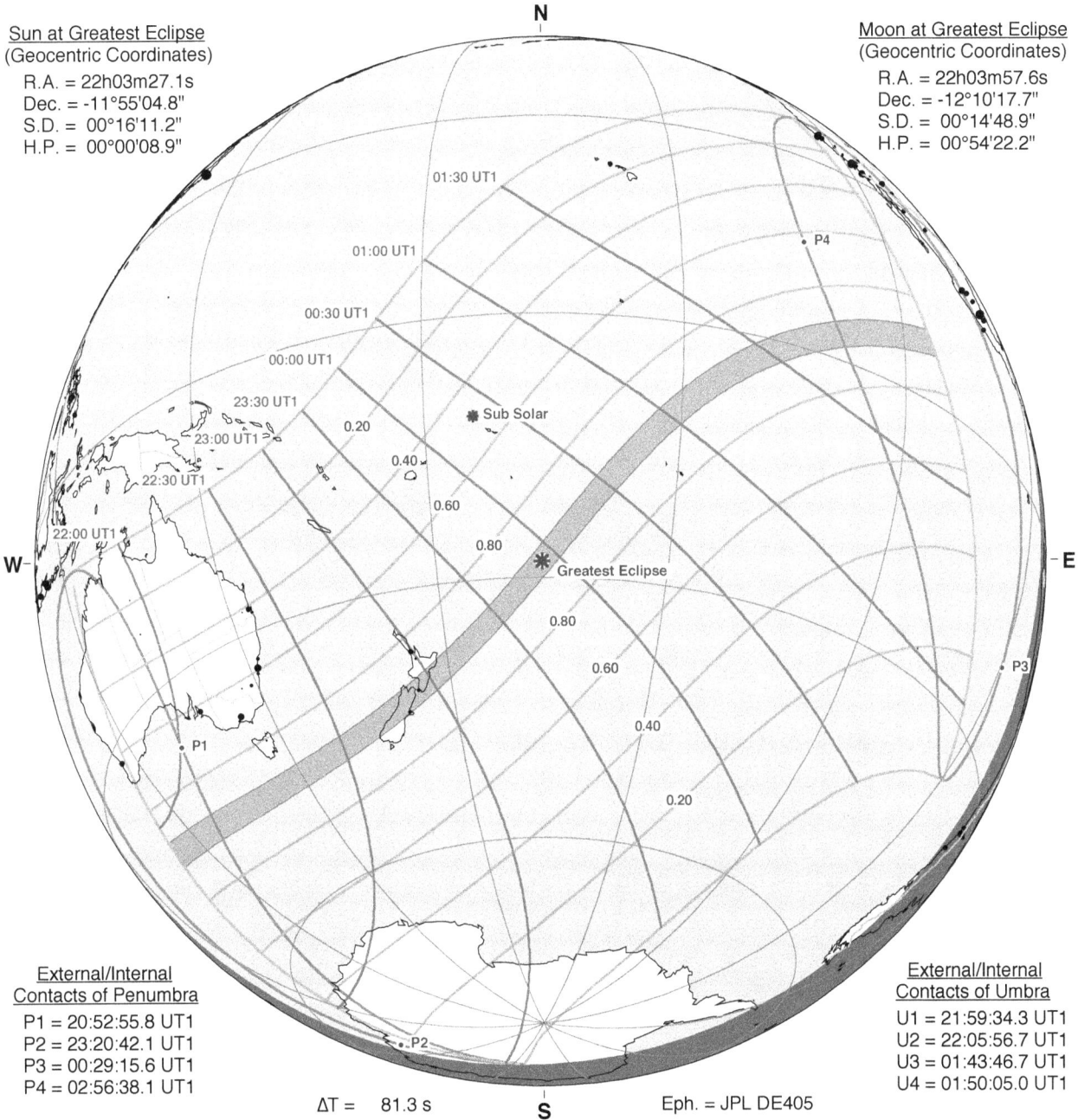

N

01:30 UT1
01:00 UT1
00:30 UT1
00:00 UT1
23:30 UT1
23:00 UT1
22:30 UT1
22:00 UT1

0.20
0.40
0.60
0.80

Sub Solar

Greatest Eclipse

0.80
0.60
0.40
0.20

P4
P3
P1
P2

W
E
S

External/Internal
Contacts of Penumbra
P1 = 20:52:55.8 UT1
P2 = 23:20:42.1 UT1
P3 = 00:29:15.6 UT1
P4 = 02:56:38.1 UT1

External/Internal
Contacts of Umbra
U1 = 21:59:34.3 UT1
U2 = 22:05:56.7 UT1
U3 = 01:43:46.7 UT1
U4 = 01:50:05.0 UT1

ΔT = 81.3 s   Eph. = JPL DE405

Circumstances at Greatest Eclipse: 23:54:45.3 UT1
Lat. = 28°15.4'S   Sun Alt. = 71.6°
Long. = 166°13.8'W   Sun Azm. = 331.0°
Path Width = 281.2 km Duration = 07m46.5s

Circumstances at Greatest Duration: 23:37:15.9 UT1
Lat. = 32°20.8'S   Sun Alt. = 69.6°
Long. = 170°31.5'W   Sun Azm. = 359.1°
Path Width = 278.7 km Duration = 07m48.7s

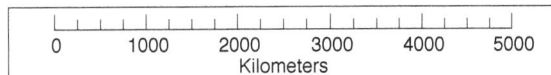

©2016 F. Espenak
www.EclipseWise.com

0   1000   2000   3000   4000   5000
Kilometers

119

Total Solar Eclipse of 2045 Aug 12

Greatest Eclipse = 17:42:39.1 TD   (= 17:41:17.5 UT1)

Eclipse Magnitude = 1.0774   Saros Series = 136
Gamma = 0.2116   Saros Member = 39 of 71

Sun at Greatest Eclipse
(Geocentric Coordinates)
R.A. = 09h31m17.7s
Dec. = +14°40'40.5"
S.D. = 00°15'47.0"
H.P. = 00°00'08.7"

Moon at Greatest Eclipse
(Geocentric Coordinates)
R.A. = 09h31m39.7s
Dec. = +14°52'29.9"
S.D. = 00°16'43.3"
H.P. = 01°01'22.3"

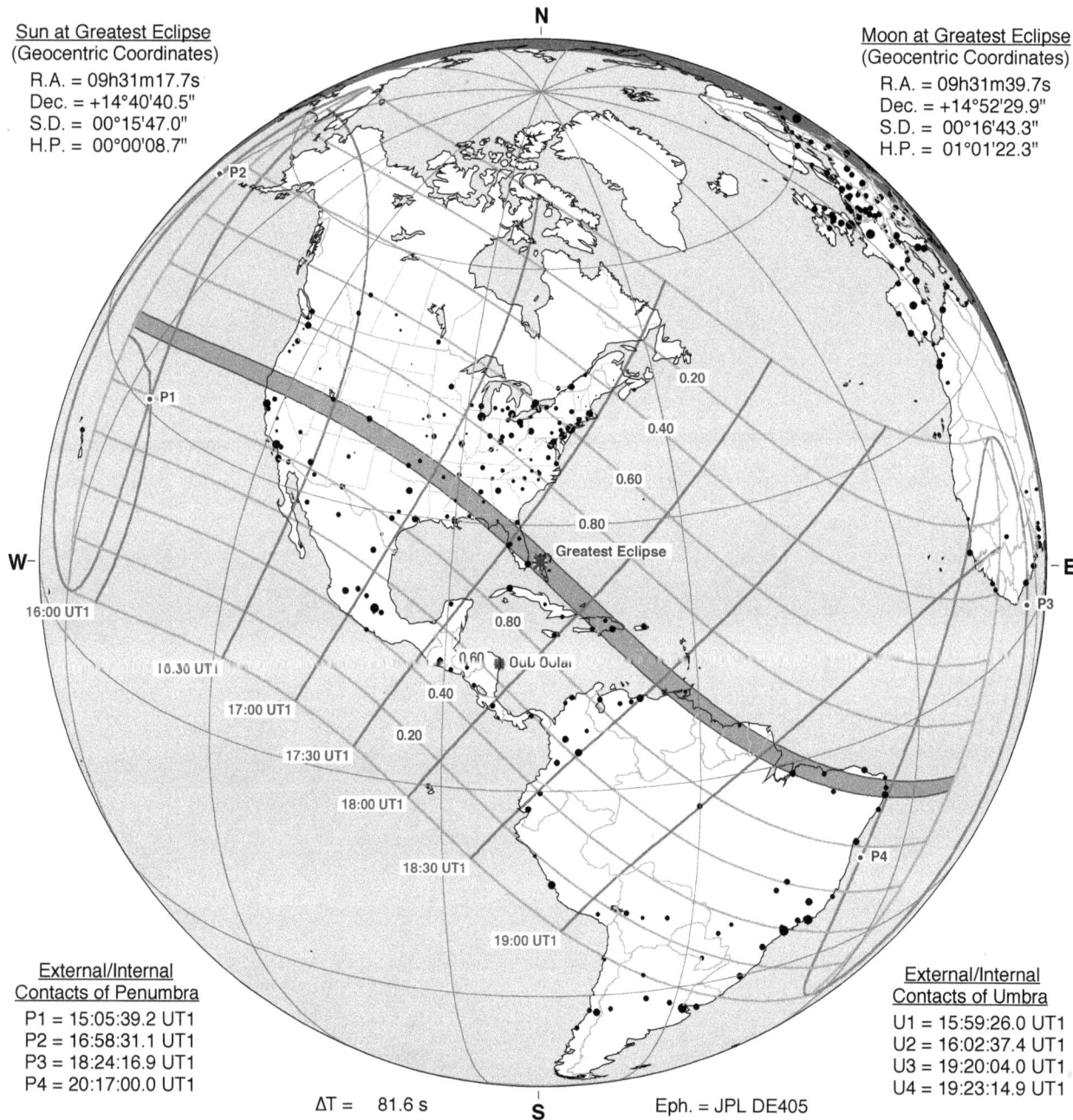

N

0.20
0.40
0.60
0.80
Greatest Eclipse

W —

— E

16:00 UT1

P1

P2

P3

0.80
0.60   Sub Solar
0.40
0.20

16:30 UT1
17:00 UT1
17:30 UT1
18:00 UT1
18:30 UT1
19:00 UT1

P4

External/Internal
Contacts of Penumbra
P1 = 15:05:39.2 UT1
P2 = 16:58:31.1 UT1
P3 = 18:24:16.9 UT1
P4 = 20:17:00.0 UT1

External/Internal
Contacts of Umbra
U1 = 15:59:26.0 UT1
U2 = 16:02:37.4 UT1
U3 = 19:20:04.0 UT1
U4 = 19:23:14.9 UT1

ΔT = 81.6 s

S

Eph. = JPL DE405

Circumstances at Greatest Eclipse: 17:41:17.5 UT1
Lat. = 25°54.6'N   Sun Alt. = 77.6°
Long. = 078°34.1'W   Sun Azm. = 205.7°
Path Width = 255.6 km   Duration = 06m05.7s

Circumstances at Greatest Duration: 17:35:29.1 UT1
Lat. = 27°19.3'N   Sun Alt. = 77.2°
Long. = 080°22.9'W   Sun Azm. = 189.9°
Path Width = 254.8 km   Duration = 06m06.2s

0   1000   2000   3000   4000   5000
Kilometers

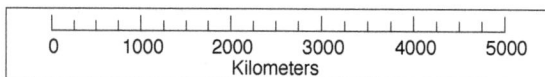

©2016 F. Espenak
www.EclipseWise.com

Annular Solar Eclipse of 2046 Feb 05

Greatest Eclipse = 23:06:26.2 TD   (= 23:05:04.3 UT1)

Eclipse Magnitude = 0.9232   Saros Series = 141
Gamma = 0.3765   Saros Member = 25 of 70

Sun at Greatest Eclipse
(Geocentric Coordinates)
R.A. = 21h19m00.8s
Dec. = -15°38'42.4"
S.D. = 00°16'13.2"
H.P. = 00°00'08.9"

Moon at Greatest Eclipse
(Geocentric Coordinates)
R.A. = 21h18m27.2s
Dec. = -15°20'02.1"
S.D. = 00°14'46.0"
H.P. = 00°54'11.7"

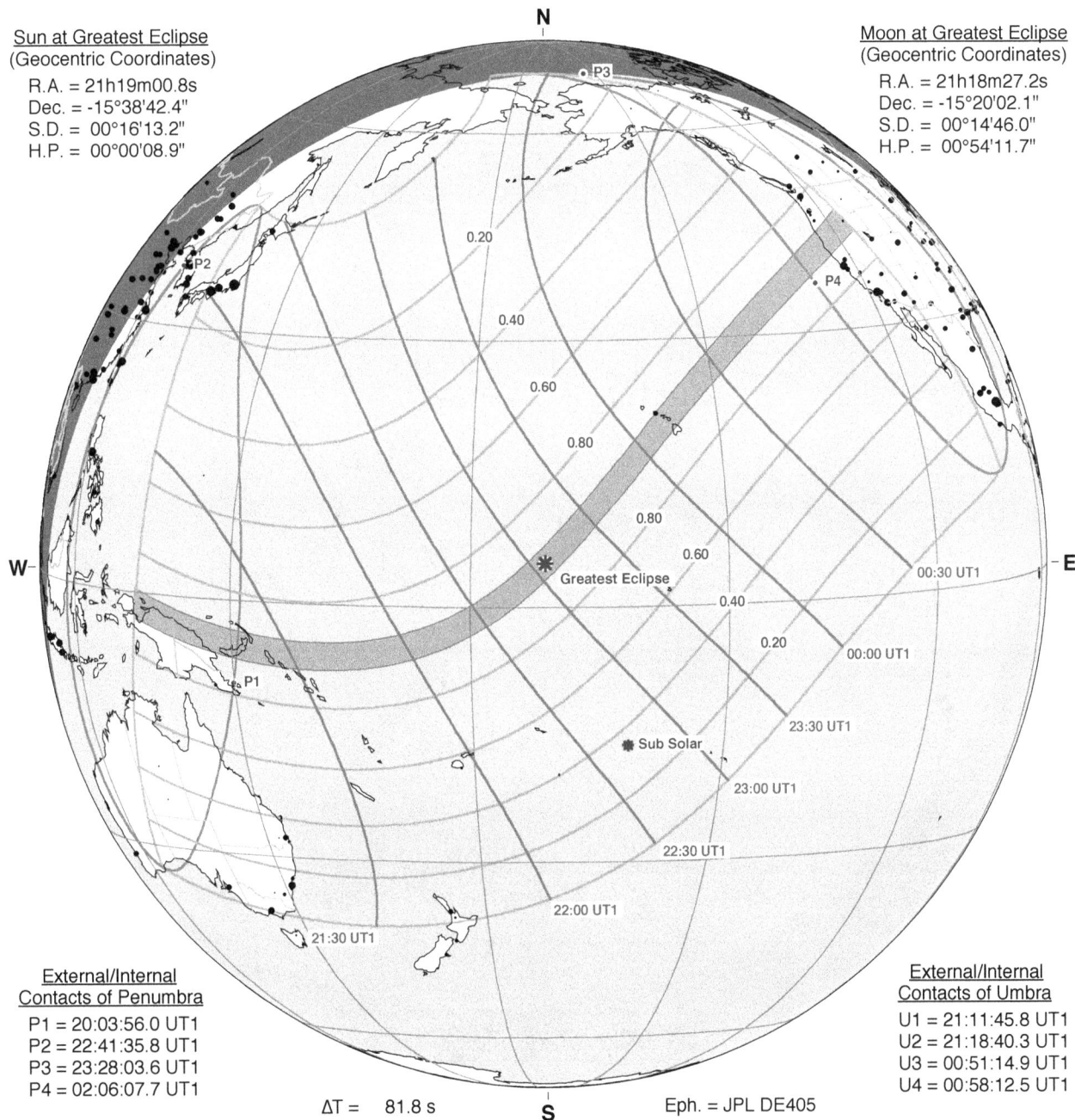

N

P3

0.20
0.40
0.60
0.80
0.80
0.60
0.40
0.20

P2
P4

W — — E

Greatest Eclipse

00:30 UT1
00:00 UT1
23:30 UT1
23:00 UT1
22:30 UT1
22:00 UT1
21:30 UT1

Sub Solar

P1

External/Internal
Contacts of Penumbra
P1 = 20:03:56.0 UT1
P2 = 22:41:35.8 UT1
P3 = 23:28:03.6 UT1
P4 = 02:06:07.7 UT1

External/Internal
Contacts of Umbra
U1 = 21:11:45.8 UT1
U2 = 21:18:40.3 UT1
U3 = 00:51:14.9 UT1
U4 = 00:58:12.5 UT1

ΔT = 81.8 s   S   Eph. = JPL DE405

Circumstances at Greatest Eclipse: 23:05:04.3 UT1
Lat. = 04°46.9'N   Sun Alt. = 67.9°
Long. = 171°25.6'W   Sun Azm. = 157.4°
Path Width = 310.1 km   Duration = 09m42.4s

Circumstances at Greatest Duration: 22:49:00.6 UT1
Lat. = 01°56.1'N   Sun Alt. = 66.4°
Long. = 174°40.8'W   Sun Azm. = 138.7°
Path Width = 316.7 km   Duration = 09m45.6s

0 1000 2000 3000 4000 5000
Kilometers

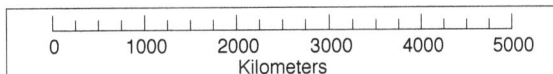

©2016 F. Espenak
www.EclipseWise.com

Total Solar Eclipse of 2046 Aug 02

Greatest Eclipse = 10:21:13.4 TD   (= 10:19:51.3 UT1)

Eclipse Magnitude = 1.0531	Saros Series = 146
Gamma = -0.5350	Saros Member = 29 of 76

Sun at Greatest Eclipse
(Geocentric Coordinates)
R.A. = 08h51m04.7s
Dec. = +17°39'03.1"
S.D. = 00°15'45.5"
H.P. = 00°00'08.7"

Moon at Greatest Eclipse
(Geocentric Coordinates)
R.A. = 08h50m16.1s
Dec. = +17°09'10.7"
S.D. = 00°16'21.8"
H.P. = 01°00'03.4"

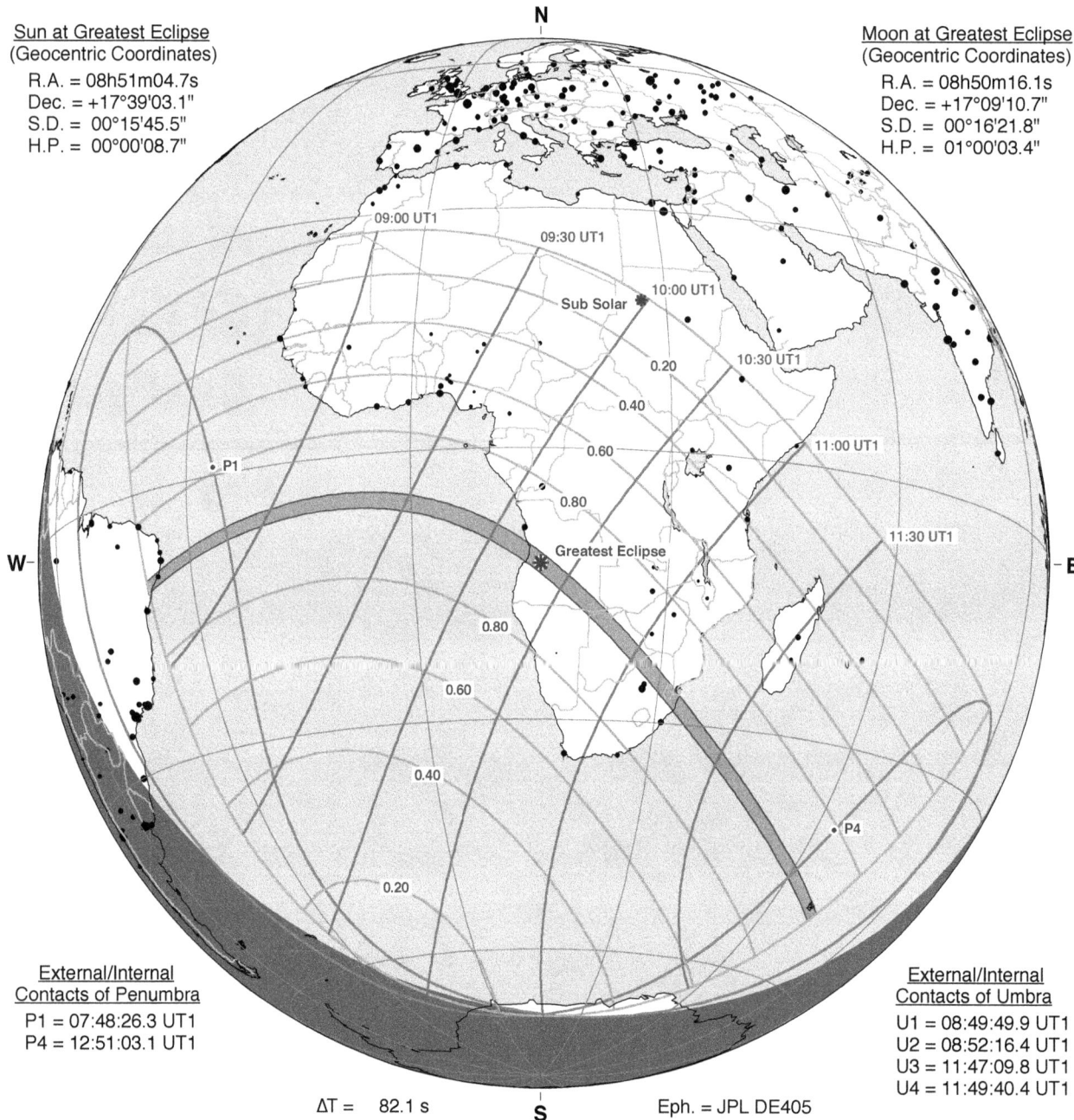

09:00 UT1
09:30 UT1
10:00 UT1
Sub Solar
0.20
0.40
10:30 UT1
0.60
11:00 UT1
0.80
Greatest Eclipse
11:30 UT1
0.80
0.60
0.40
0.20
P1
P4
N
S
W
E

External/Internal Contacts of Penumbra
P1 = 07:48:26.3 UT1
P4 = 12:51:03.1 UT1

External/Internal Contacts of Umbra
U1 = 08:49:49.9 UT1
U2 = 08:52:16.4 UT1
U3 = 11:47:09.8 UT1
U4 = 11:49:40.4 UT1

ΔT =   82.1 s   Eph. = JPL DE405

Circumstances at Greatest Eclipse: 10:19:51.3 UT1

Lat. = 12°43.8'S	Sun Alt. = 57.6°
Long. = 015°08.8'E	Sun Azm. = 20.7°
Path Width = 206.0 km	Duration = 04m51.4s

Circumstances at Greatest Duration: 10:16:03.2 UT1

Lat. = 12°00.3'S	Sun Alt. = 57.5°
Long. = 014°06.4'E	Sun Azm. = 24.4°
Path Width = 207.2 km	Duration = 04m51.6s

Scale: 0 — 1000 — 2000 — 3000 — 4000 — 5000 Kilometers

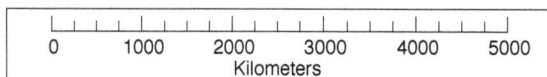

©2016 F. Espenak
www.EclipseWise.com

Partial Solar Eclipse of 2047 Jan 26

Greatest Eclipse = 01:33:17.8 TD   (= 01:31:55.4 UT1)

Eclipse Magnitude = 0.8908   Saros Series = 151
Gamma = 1.0450   Saros Member = 16 of 72

Sun at Greatest Eclipse
(Geocentric Coordinates)
R.A. = 20h33m28.4s
Dec. = -18°46'10.9"
S.D. = 00°16'14.7"
H.P. = 00°00'08.9"

Moon at Greatest Eclipse
(Geocentric Coordinates)
R.A. = 20h32m04.0s
Dec. = -17°50'50.8"
S.D. = 00°15'23.2"
H.P. = 00°56'28.0"

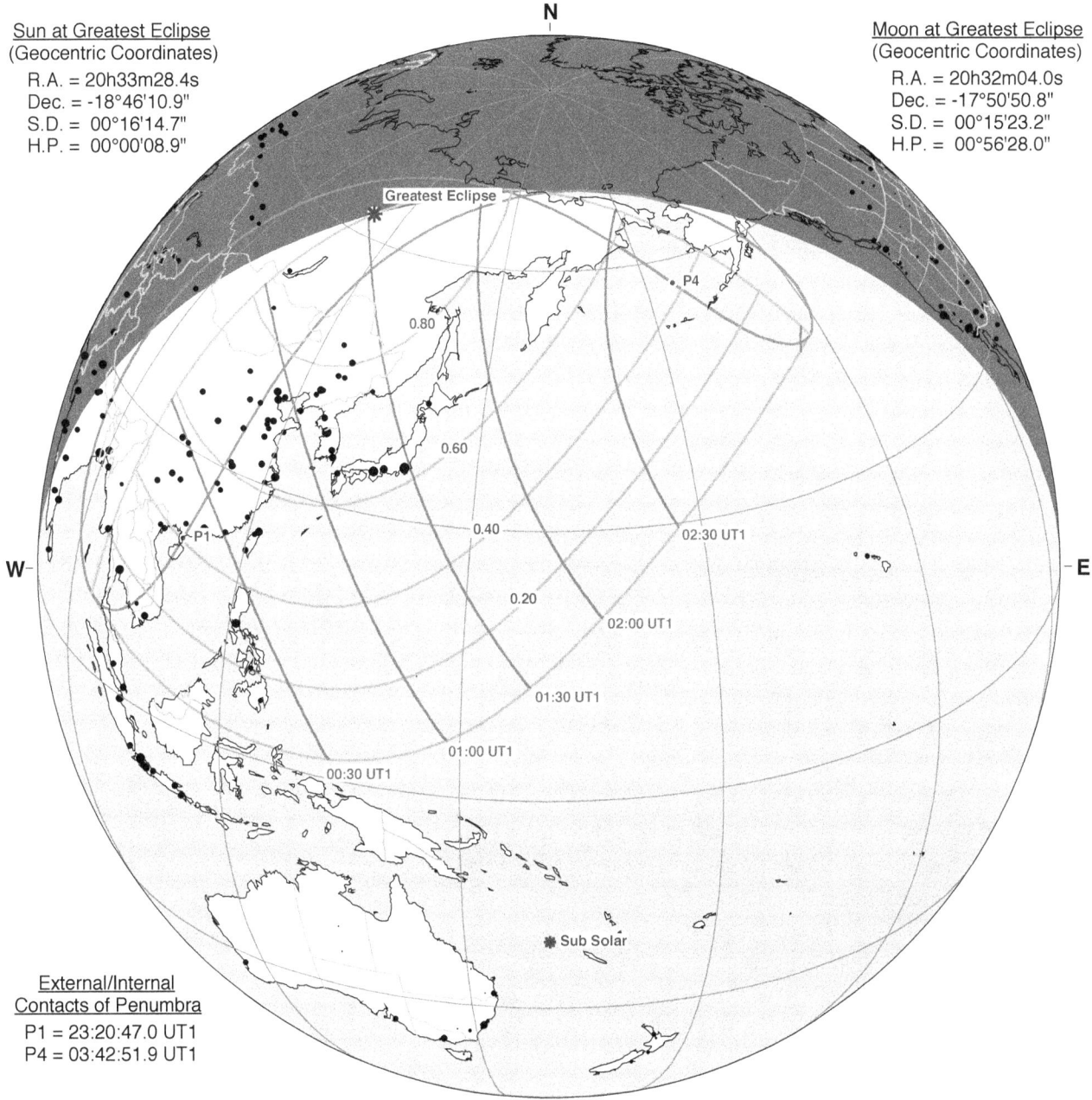

N

Greatest Eclipse

P4

0.80

0.60

0.40   02:30 UT1

0.20   02:00 UT1

01:30 UT1

01:00 UT1

00:30 UT1

P1

W

E

Sub Solar

External/Internal
Contacts of Penumbra
P1 = 23:20:47.0 UT1
P4 = 03:42:51.9 UT1

ΔT =   82.4 s   S   Eph. = JPL DE405

Circumstances at Greatest Eclipse: 01:31:55.4 UT1
Lat. = 62°52.0'N   Sun Alt. =   0.0°
Long. = 111°42.3'E   Sun Azm. = 134.9°

0   1000   2000   3000   4000   5000
Kilometers

©2016 F. Espenak
www.EclipseWise.com

Partial Solar Eclipse of 2047 Jun 23

Greatest Eclipse = 10:52:30.6 TD   (= 10:51:08.0 UT1)

Eclipse Magnitude = 0.3129	Saros Series = 118
Gamma = 1.3766	Saros Member = 70 of 72

N

Sun at Greatest Eclipse
(Geocentric Coordinates)
R.A. = 06h08m27.7s
Dec. = +23°25'10.2"
S.D. = 00°15'44.2"
H.P. = 00°00'08.7"

Moon at Greatest Eclipse
(Geocentric Coordinates)
R.A. = 06h09m05.2s
Dec. = +24°40'56.6"
S.D. = 00°15'07.9"
H.P. = 00°55'32.1"

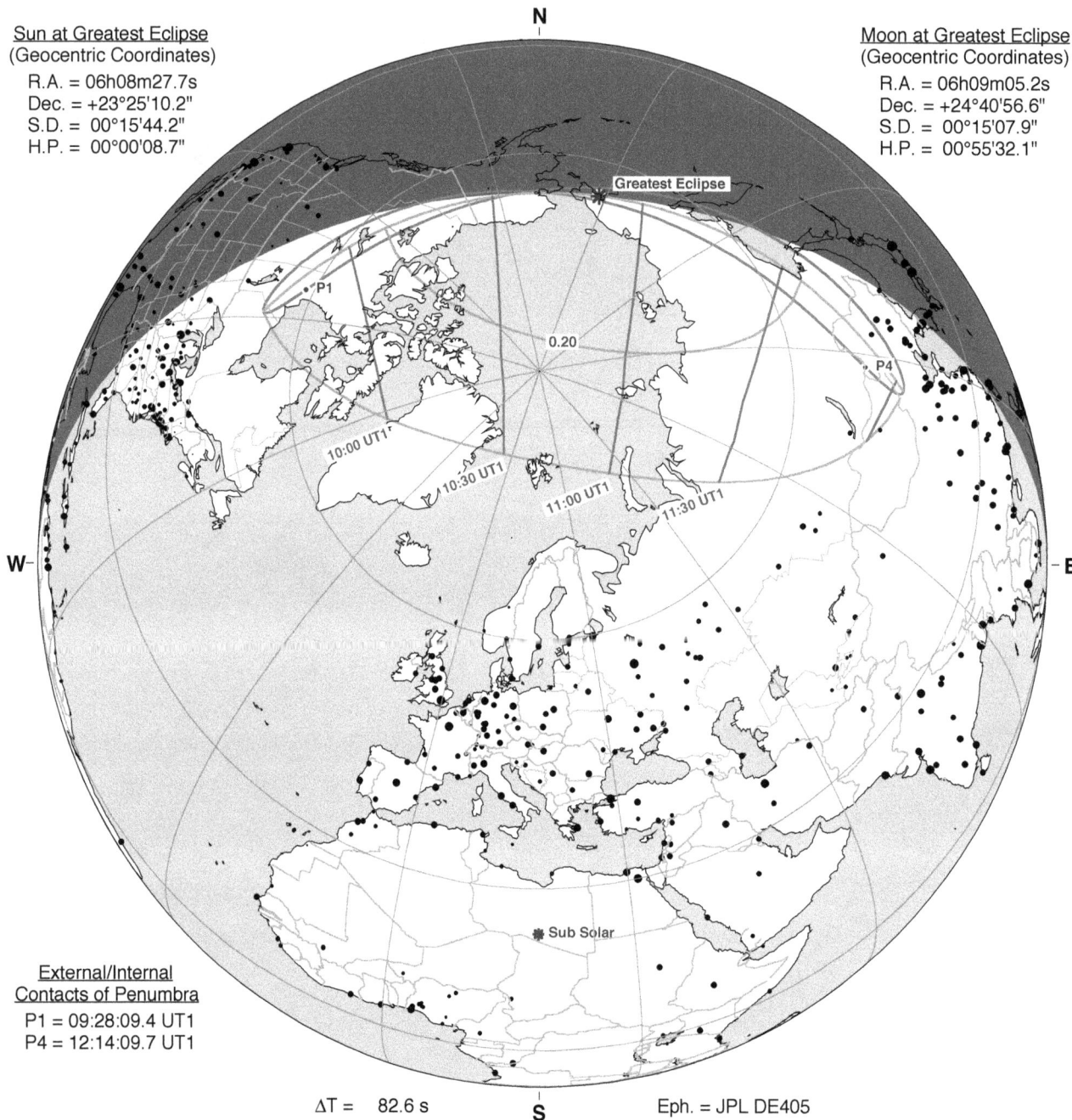

Greatest Eclipse

P1

0.20

P4

10:00 UT1
10:30 UT1
11:00 UT1
11:30 UT1

W

E

Sub Solar

External/Internal
Contacts of Penumbra
P1 = 09:28:09.4 UT1
P4 = 12:14:09.7 UT1

ΔT = 82.6 s

S

Eph. = JPL DE405

Circumstances at Greatest Eclipse: 10:51:08.0 UT1

Lat. = 65°46.0'N	Sun Alt. = 0.0°
Long. = 178°02.4'W	Sun Azm. = 345.5°

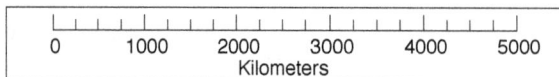

0   1000   2000   3000   4000   5000
Kilometers

©2016 F. Espenak
www.EclipseWise.com

Partial Solar Eclipse of 2047 Jul 22

Greatest Eclipse = 22:36:17.4 TD   (= 22:34:54.7 UT1)

Eclipse Magnitude = 0.3605   Saros Series = 156
Gamma = -1.3477   Saros Member = 3 of 69

Sun at Greatest Eclipse
(Geocentric Coordinates)
R.A. = 08h08m59.7s
Dec. = +20°07'53.9"
S.D. = 00°15'44.5"
H.P. = 00°00'08.7"

Moon at Greatest Eclipse
(Geocentric Coordinates)
R.A. = 08h07m21.2s
Dec. = +18°54'51.1"
S.D. = 00°15'32.1"
H.P. = 00°57'00.9"

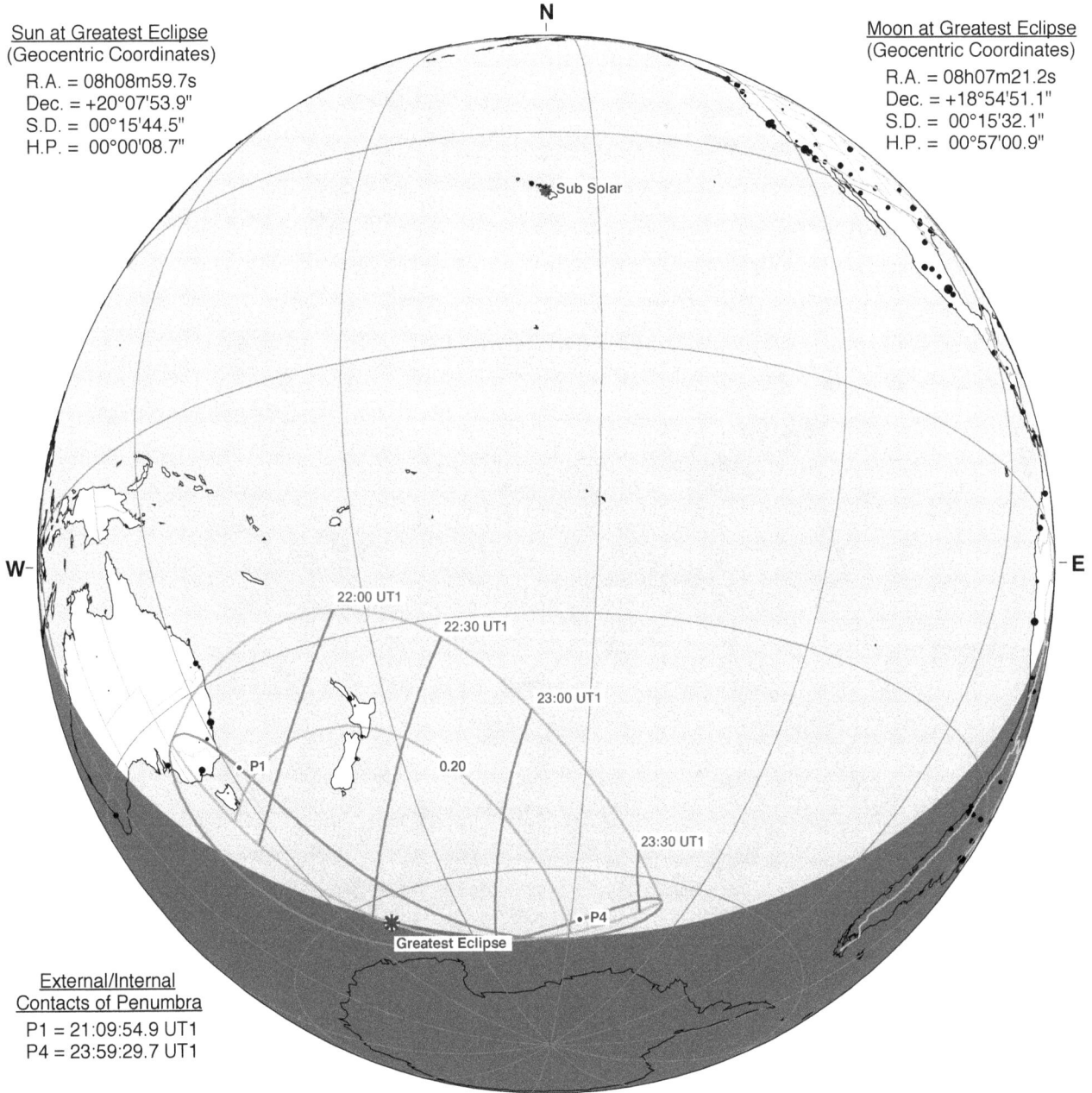

N

Sub Solar

W

E

22:00 UT1
22:30 UT1
23:00 UT1
0.20
23:30 UT1

P1

P4

Greatest Eclipse

External/Internal
Contacts of Penumbra
P1 = 21:09:54.9 UT1
P4 = 23:59:29.7 UT1

ΔT = 82.7 s   S   Eph. = JPL DE405

Circumstances at Greatest Eclipse: 22:34:54.7 UT1
Lat. = 63°26.5'S   Sun Alt. = 0.0°
Long. = 160°08.3'E   Sun Azm. = 39.7°

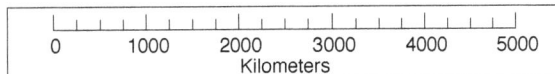

0   1000   2000   3000   4000   5000
Kilometers

©2016 F. Espenak
www.EclipseWise.com

125

Partial Solar Eclipse of 2047 Dec 16

Greatest Eclipse = 23:50:12.3 TD (= 23:48:49.4 UT1)

Eclipse Magnitude = 0.8817   Saros Series = 123
Gamma = -1.0661   Saros Member = 55 of 70

Sun at Greatest Eclipse
(Geocentric Coordinates)
R.A. = 17h37m56.6s
Dec. = -23°20'10.9"
S.D. = 00°16'15.0"
H.P. = 00°00'08.9"

Moon at Greatest Eclipse
(Geocentric Coordinates)
R.A. = 17h38m13.1s
Dec. = -24°24'51.1"
S.D. = 00°16'35.9"
H.P. = 01°00'54.9"

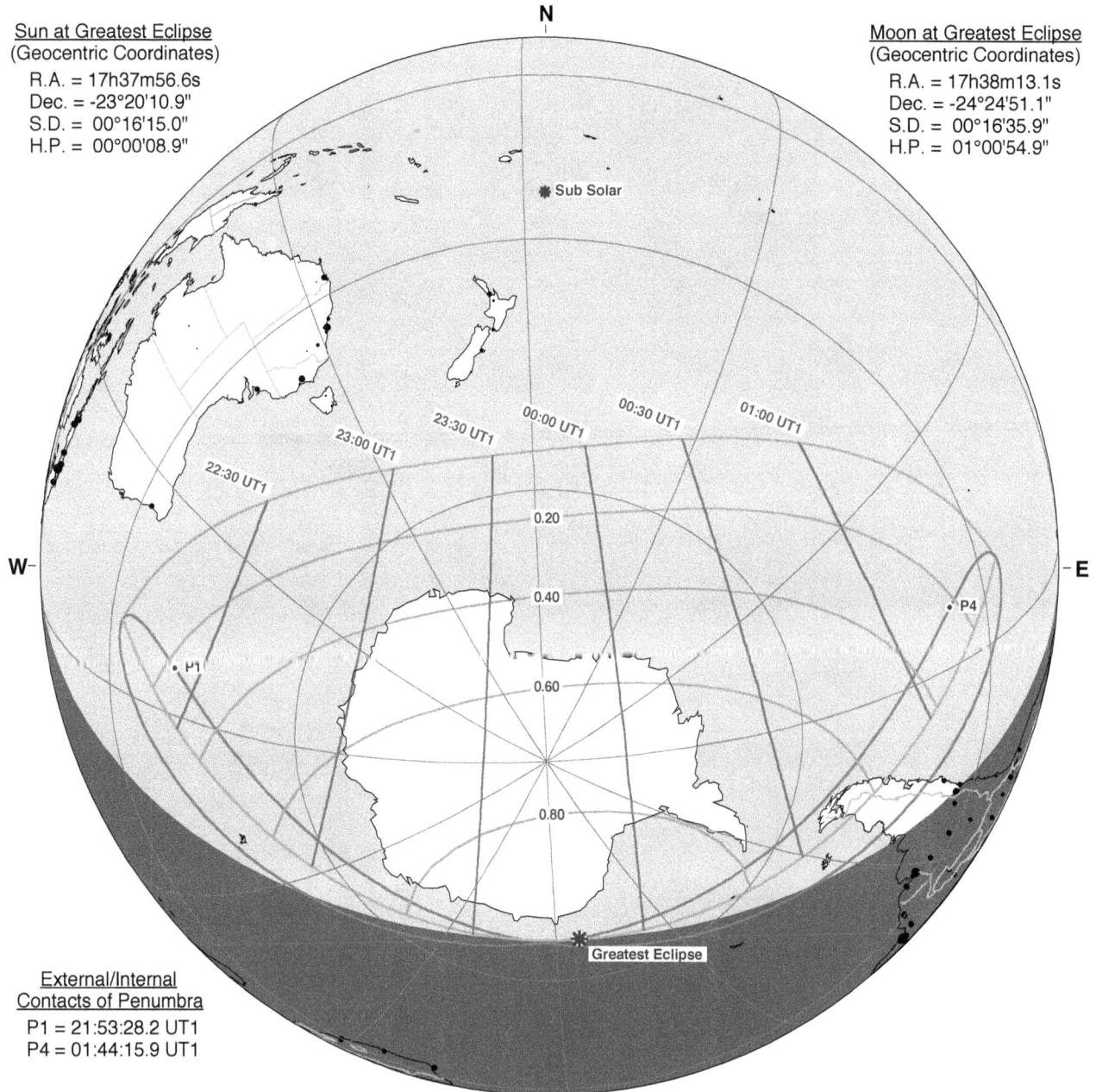

N

Sub Solar

23:00 UT1   23:30 UT1   00:00 UT1   00:30 UT1   01:00 UT1
22:30 UT1

0.20

0.40

W ─   ─ E

P4

P1

0.60

0.80

Greatest Eclipse

**External/Internal
Contacts of Penumbra**
P1 = 21:53:28.2 UT1
P4 = 01:44:15.9 UT1

ΔT = 82.9 s   S   Eph. = JPL DE405

Circumstances at Greatest Eclipse: 23:48:49.4 UT1
Lat. = 66°26.6'S   Sun Alt. = 0.0°
Long. = 006°38.2'W   Sun Azm. = 187.7°

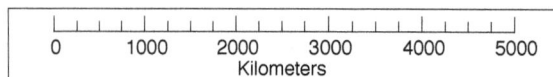

0	1000	2000	3000	4000	5000

Kilometers

©2016 F. Espenak
www.EclipseWise.com

Annular Solar Eclipse of 2048 Jun 11

Greatest Eclipse = 12:58:52.8 TD   (= 12:57:29.6 UT1)

Eclipse Magnitude = 0.9441   Saros Series = 128
Gamma = 0.6468   Saros Member = 60 of 73

Sun at Greatest Eclipse
(Geocentric Coordinates)
R.A. = 05h22m03.9s
Dec. = +23°08'47.0"
S.D. = 00°15'45.1"
H.P. = 00°00'08.7"

Moon at Greatest Eclipse
(Geocentric Coordinates)
R.A. = 05h22m09.1s
Dec. = +23°43'34.6"
S.D. = 00°14'42.3"
H.P. = 00°53'58.0"

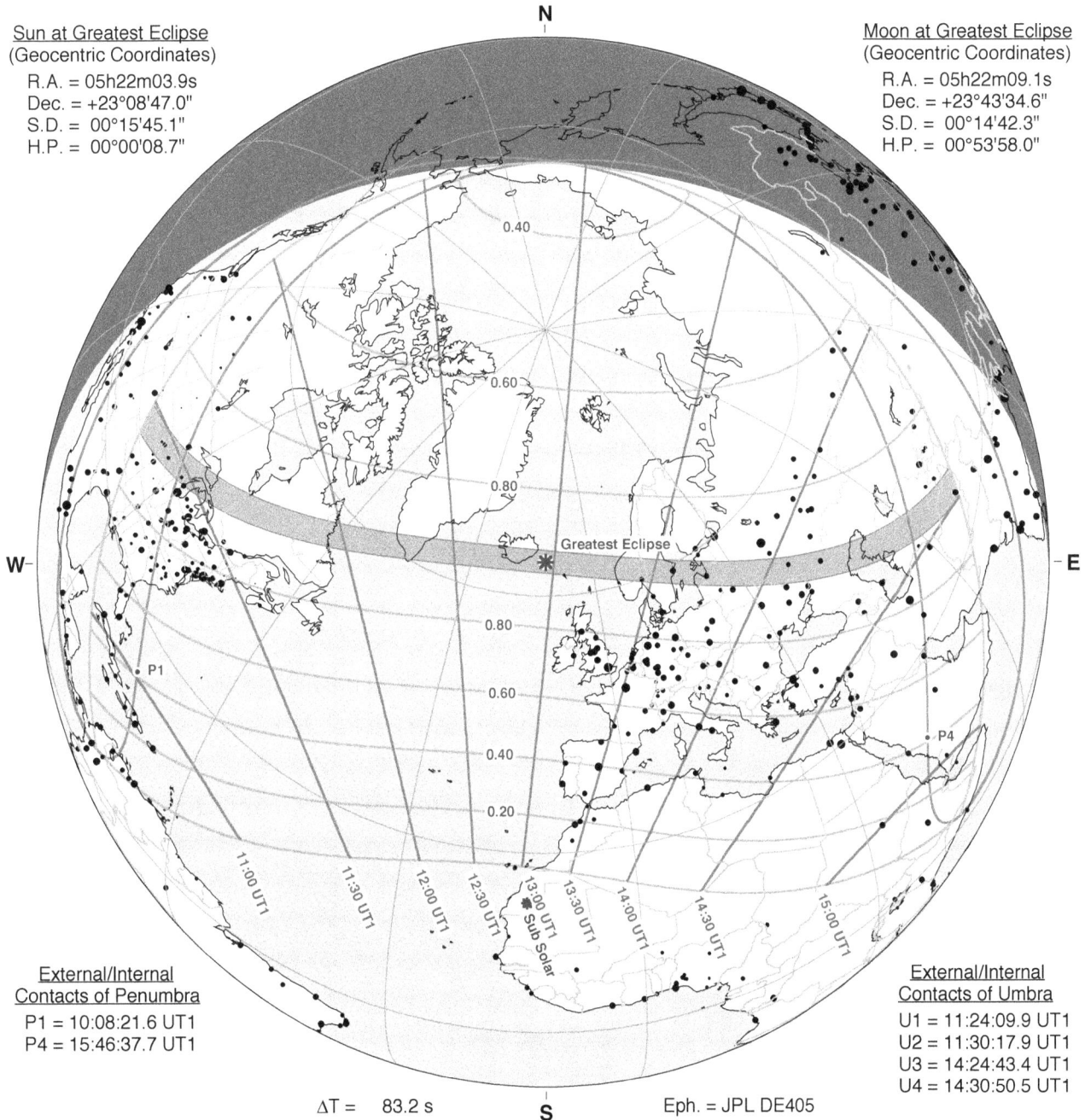

External/Internal
Contacts of Penumbra
P1 = 10:08:21.6 UT1
P4 = 15:46:37.7 UT1

External/Internal
Contacts of Umbra
U1 = 11:24:09.9 UT1
U2 = 11:30:17.9 UT1
U3 = 14:24:43.4 UT1
U4 = 14:30:50.5 UT1

ΔT = 83.2 s   Eph. = JPL DE405

Circumstances at Greatest Eclipse: 12:57:29.6 UT1
Lat. = 63°41.3'N   Sun Alt. = 49.4°
Long. = 011°32.5'W   Sun Azm. = 184.0°
Path Width = 271.5 km   Duration = 04m58.3s

Circumstances at Greatest Duration: 12:56:04.3 UT1
Lat. = 63°43.9'N   Sun Alt. = 49.4°
Long. = 012°43.9'W   Sun Azm. = 181.9°
Path Width = 271.5 km   Duration = 04m58.3s

©2016 F. Espenak
www.EclipseWise.com

Kilometers: 0 1000 2000 3000 4000 5000

Total Solar Eclipse of 2048 Dec 05

Greatest Eclipse = 15:35:26.7 TD   (= 15:34:03.2 UT1)

Eclipse Magnitude = 1.0440	Saros Series = 133
Gamma = -0.3973	Saros Member = 47 of 72

Sun at Greatest Eclipse
(Geocentric Coordinates)
R.A. = 16h51m20.5s
Dec. = -22°29'40.9"
S.D. = 00°16'13.8"
H.P. = 00°00'08.9"

Moon at Greatest Eclipse
(Geocentric Coordinates)
R.A. = 16h51m18.6s
Dec. = -22°53'56.4"
S.D. = 00°16'40.9"
H.P. = 01°01'13.3"

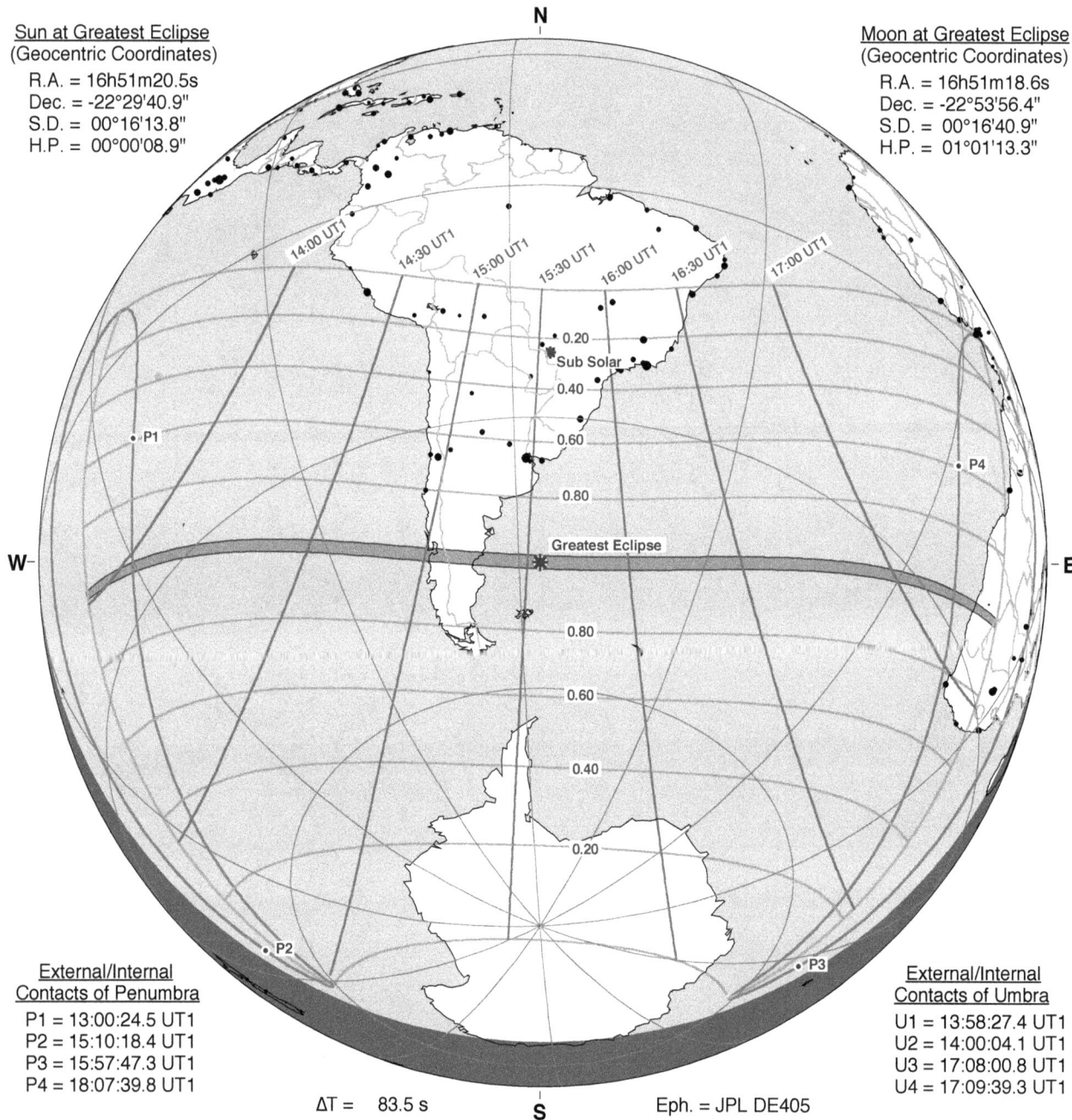

N

14:00 UT1   14:30 UT1   15:00 UT1   15:30 UT1   16:00 UT1   16:30 UT1   17:00 UT1

P1

0.20
Sub Solar
0.40
0.60
0.80

Greatest Eclipse

P4

W — — E

0.80
0.60
0.40
0.20

P2

P3

S

External/Internal Contacts of Penumbra
P1 = 13:00:24.5 UT1
P2 = 15:10:18.4 UT1
P3 = 15:57:47.3 UT1
P4 = 18:07:39.8 UT1

ΔT = 83.5 s

Eph. = JPL DE405

External/Internal Contacts of Umbra
U1 = 13:58:27.4 UT1
U2 = 14:00:04.1 UT1
U3 = 17:08:00.8 UT1
U4 = 17:09:39.3 UT1

Circumstances at Greatest Eclipse: 15:34:03.2 UT1

Lat. = 46°07.9'S	Sun Alt. = 66.4°
Long. = 056°24.4'W	Sun Azm. = 1.4°
Path Width = 160.3 km	Duration = 03m27.6s

Circumstances at Greatest Duration: 15:34:47.8 UT1

Lat. = 46°08.5'S	Sun Alt. = 66.4°
Long. = 056°00.0'W	Sun Azm. = 0.0°
Path Width = 160.3 km	Duration = 03m27.6s

0   1000   2000   3000   4000   5000
Kilometers

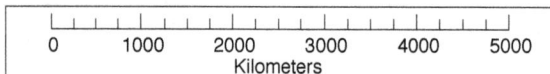

Annular Solar Eclipse of 2049 May 31

Greatest Eclipse = 13:59:58.8 TD (= 13:58:35.0 UT1)

Eclipse Magnitude = 0.9631

Gamma = -0.1187

Saros Series = 138

Saros Member = 33 of 70

Sun at Greatest Eclipse
(Geocentric Coordinates)
R.A. = 04h35m51.4s
Dec. = +22°01'26.4"
S.D. = 00°15'46.5"
H.P. = 00°00'08.7"

Moon at Greatest Eclipse
(Geocentric Coordinates)
R.A. = 04h35m52.6s
Dec. = +21°54'56.7"
S.D. = 00°14'57.8"
H.P. = 00°54'55.1"

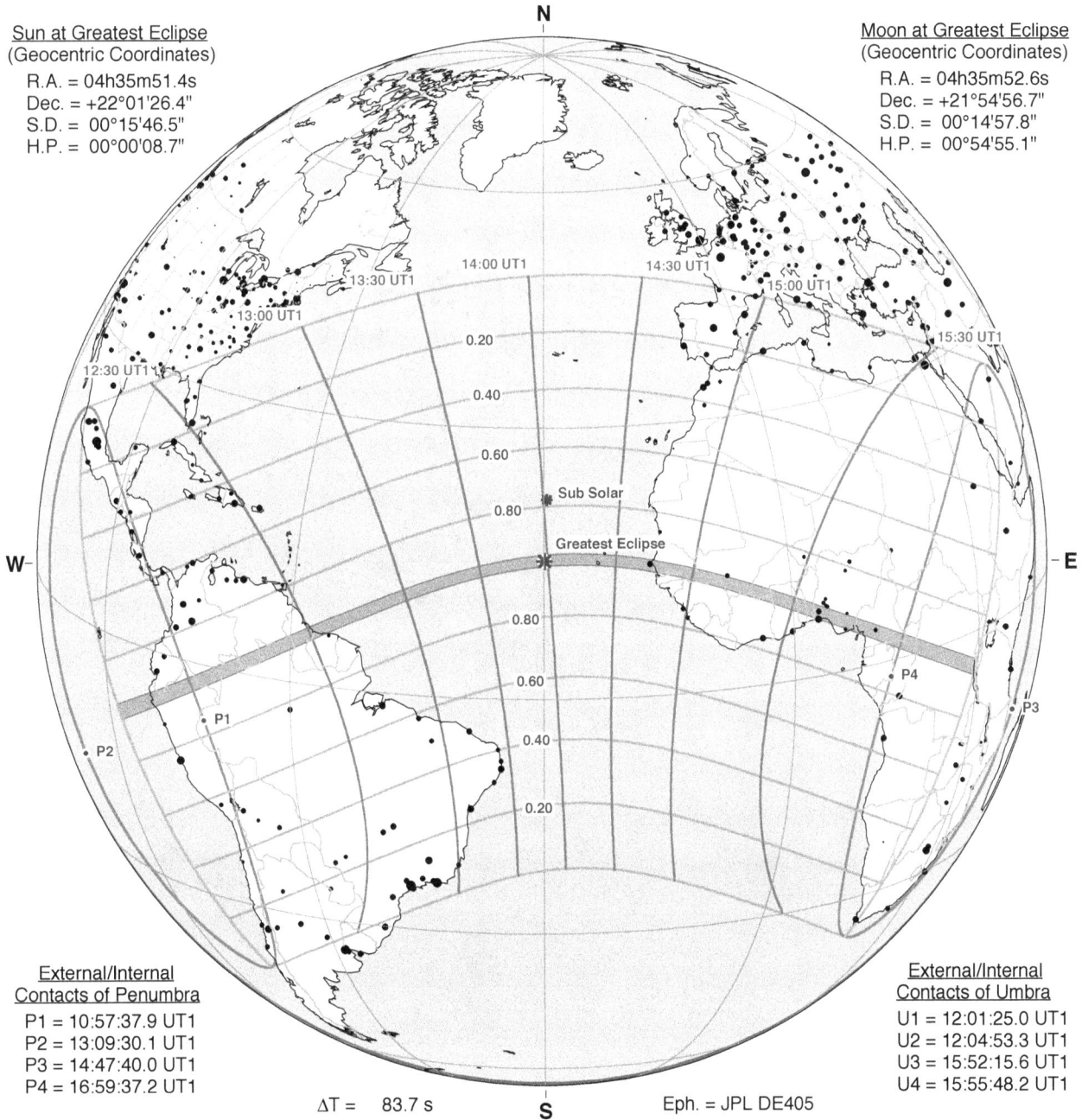

External/Internal Contacts of Penumbra
P1 = 10:57:37.9 UT1
P2 = 13:09:30.1 UT1
P3 = 14:47:40.0 UT1
P4 = 16:59:37.2 UT1

ΔT = 83.7 s

Eph. = JPL DE405

External/Internal Contacts of Umbra
U1 = 12:01:25.0 UT1
U2 = 12:04:53.3 UT1
U3 = 15:52:15.6 UT1
U4 = 15:55:48.2 UT1

Circumstances at Greatest Eclipse: 13:58:35.0 UT1

Lat. = 15°18.6'N
Long. = 029°53.3'W
Path Width = 134.4 km
Sun Alt. = 83.3°
Sun Azm. = 357.6°
Duration = 04m45.3s

Circumstances at Greatest Duration: 14:04:03.2 UT1

Lat. = 15°23.8'N
Long. = 028°27.6'W
Path Width = 134.6 km
Sun Alt. = 82.8°
Sun Azm. = 336.6°
Duration = 04m45.5s

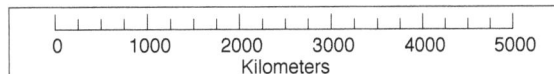

©2016 F. Espenak
www.EclipseWise.com

129

Hybrid Solar Eclipse of 2049 Nov 25

Greatest Eclipse = 05:33:47.9 TD (= 05:32:23.9 UT1)

Eclipse Magnitude = 1.0057   Saros Series = 143
Gamma = 0.2943   Saros Member = 25 of 72

Sun at Greatest Eclipse
(Geocentric Coordinates)
R.A. = 16h05m24.9s
Dec. = -20°49'25.8"
S.D. = 00°16'12.0"
H.P. = 00°00'08.9"

Moon at Greatest Eclipse
(Geocentric Coordinates)
R.A. = 16h05m31.7s
Dec. = -20°32'13.6"
S.D. = 00°16'02.3"
H.P. = 00°58'51.9"

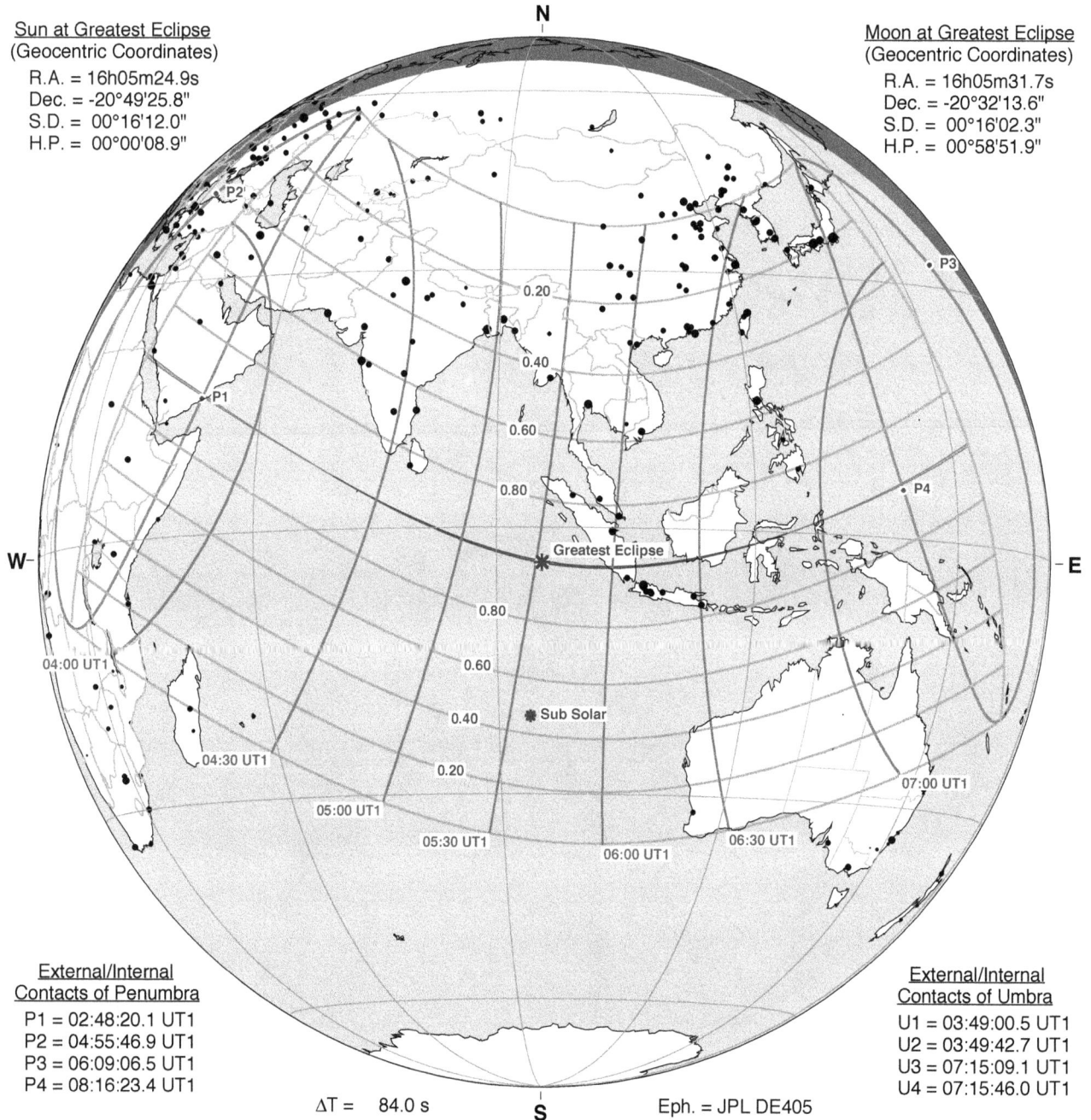

External/Internal Contacts of Penumbra
P1 = 02:48:20.1 UT1
P2 = 04:55:46.9 UT1
P3 = 06:09:06.5 UT1
P4 = 08:16:23.4 UT1

External/Internal Contacts of Umbra
U1 = 03:49:00.5 UT1
U2 = 03:49:42.7 UT1
U3 = 07:15:09.1 UT1
U4 = 07:15:46.0 UT1

ΔT = 84.0 s   Eph. = JPL DE405

Circumstances at Greatest Eclipse: 05:32:23.9 UT1
Lat. = 03°48.2'S   Sun Alt. = 72.9°
Long. = 095°12.7'E   Sun Azm. = 185.0°
Path Width = 20.7 km   Duration = 00m37.5s

Circumstances at Greatest Duration: 03:49:21.6 UT1
Lat. = 21°02.0'N   Sun Alt. = 0.0°
Long. = 037°48.4'E   Sun Azm. = 112.4°
Path Width = 40.0 km   Duration = 00m41.8s

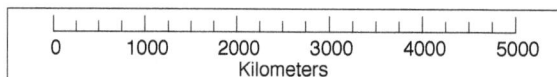

©2016 F. Espenak
www.EclipseWise.com

Hybrid Solar Eclipse of 2050 May 20

Greatest Eclipse = 20:42:50.2 TD   (= 20:41:25.9 UT1)

Eclipse Magnitude = 1.0038   Saros Series = 148
Gamma = -0.8688   Saros Member = 23 of 75

Sun at Greatest Eclipse
(Geocentric Coordinates)

R.A. = 03h51m25.4s
Dec. = +20°09'01.9"
S.D. = 00°15'48.3"
H.P. = 00°00'08.7"

Moon at Greatest Eclipse
(Geocentric Coordinates)

R.A. = 03h51m49.6s
Dec. = +19°19'17.1"
S.D. = 00°15'44.7"
H.P. = 00°57'47.0"

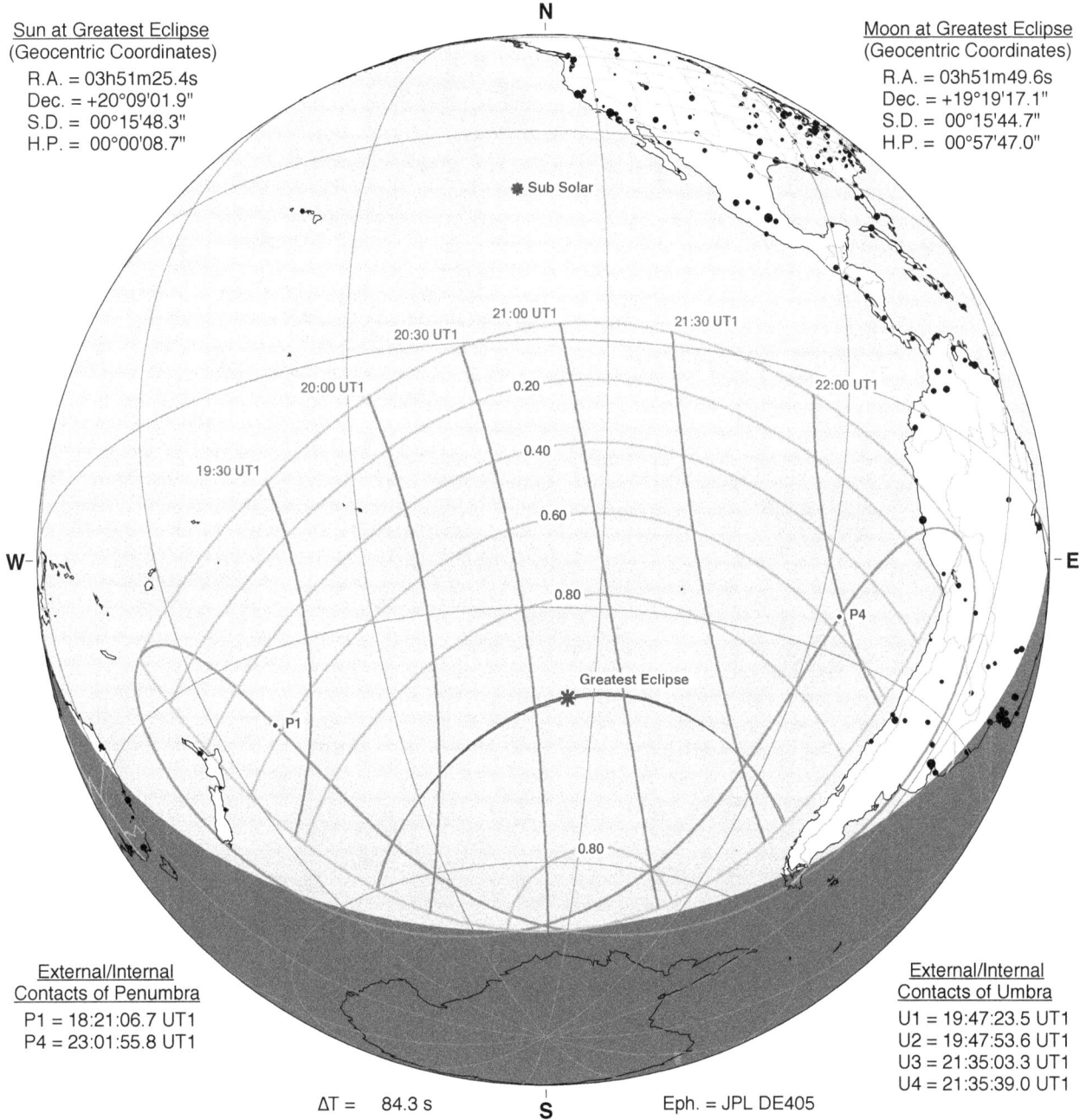

N

* Sub Solar

21:00 UT1
20:30 UT1   21:30 UT1
20:00 UT1
0.20
22:00 UT1
19:30 UT1
0.40

W   0.60   E

0.80

P4

Greatest Eclipse

P1

0.80

External/Internal Contacts of Penumbra

P1 = 18:21:06.7 UT1
P4 = 23:01:55.8 UT1

External/Internal Contacts of Umbra

U1 = 19:47:23.5 UT1
U2 = 19:47:53.6 UT1
U3 = 21:35:03.3 UT1
U4 = 21:35:39.0 UT1

ΔT = 84.3 s   S   Eph. = JPL DE405

Circumstances at Greatest Eclipse: 20:41:25.9 UT1

Lat. = 40°05.5'S   Sun Alt. = 29.4°
Long. = 123°45.1'W   Sun Azm. = 352.0°
Path Width = 26.7 km   Duration = 00m21.3s

Circumstances at Greatest Duration: 20:39:25.9 UT1

Lat. = 40°14.9'S   Sun Alt. = 29.3°
Long. = 124°36.1'W   Sun Azm. = 353.4°
Path Width = 26.6 km   Duration = 00m21.4s

©2016 F. Espenak
www.EclipseWise.com

	0	1000	2000	3000	4000	5000

Kilometers

Partial Solar Eclipse of 2050 Nov 14

Greatest Eclipse = 13:30:52.8 TD   (= 13:29:28.2 UT1)

Eclipse Magnitude = 0.8874	Saros Series = 153
Gamma = 1.0447	Saros Member = 11 of 70

N

Sun at Greatest Eclipse
(Geocentric Coordinates)
R.A. = 15h19m50.5s
Dec. = -18°21'19.2"
S.D. = 00°16'09.8"
H.P. = 00°00'08.9"

Moon at Greatest Eclipse
(Geocentric Coordinates)
R.A. = 15h20m29.5s
Dec. = -17°24'01.9"
S.D. = 00°15'10.6"
H.P. = 00°55'41.9"

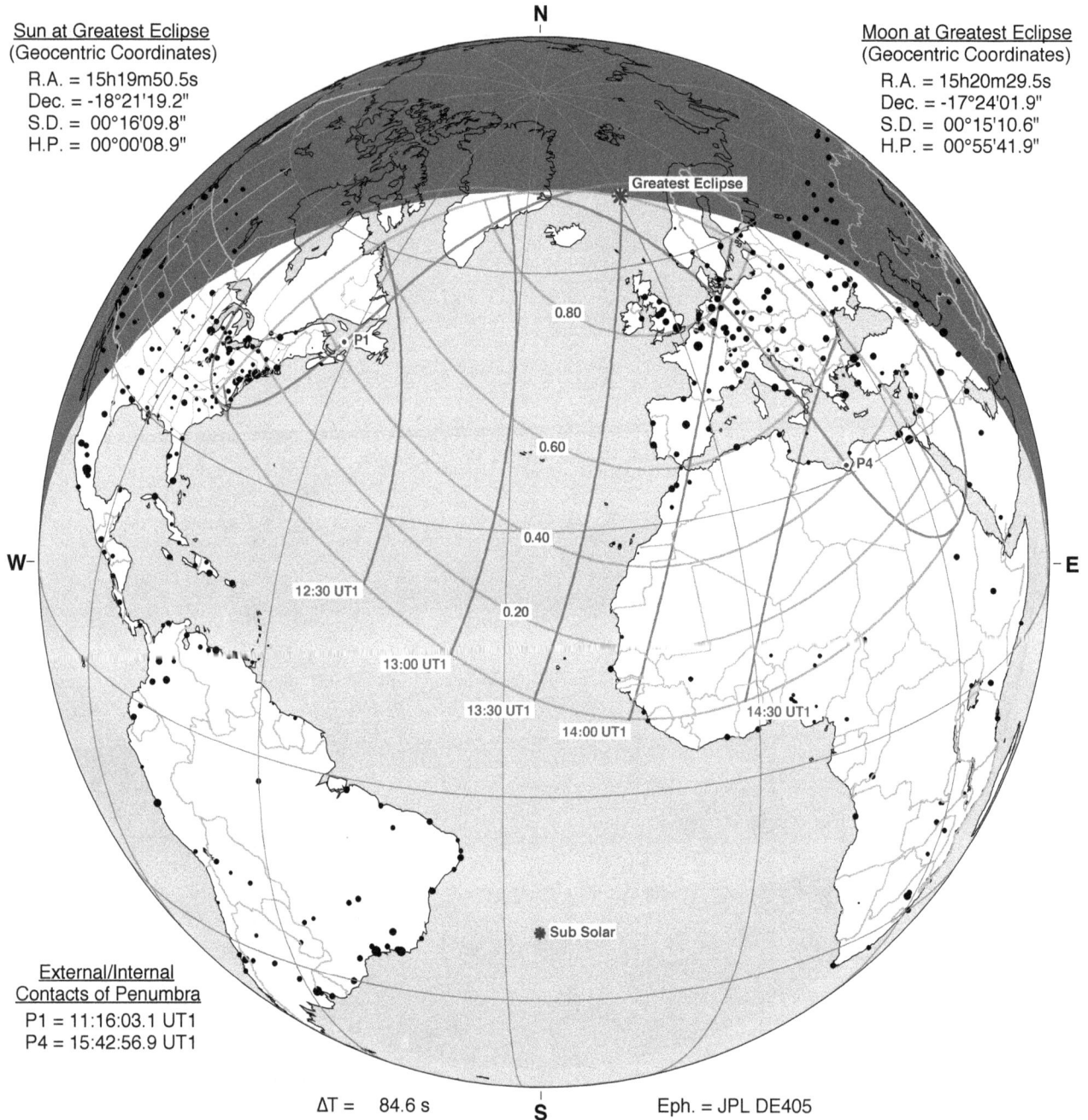

Greatest Eclipse

0.80

0.60

0.40

0.20

P1

P4

W

E

12:30 UT1

13:00 UT1

13:30 UT1

14:00 UT1

14:30 UT1

Sub Solar

External/Internal
Contacts of Penumbra
P1 = 11:16:03.1 UT1
P4 = 15:42:56.9 UT1

ΔT = 84.6 s

S

Eph. = JPL DE405

Circumstances at Greatest Eclipse: 13:29:28.2 UT1

Lat. = 69°31.7'N	Sun Alt. = 0.0°
Long. = 000°59.0'E	Sun Azm. = 205.8°

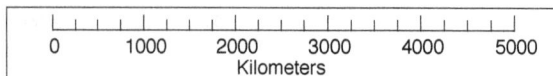

0   1000   2000   3000   4000   5000
Kilometers

Partial Solar Eclipse of 2051 Apr 11

Greatest Eclipse = 02:10:38.6 TD   (= 02:09:13.8 UT1)

Eclipse Magnitude = 0.9849
Gamma = 1.0169

Saros Series = 120
Saros Member = 63 of 71

Sun at Greatest Eclipse
(Geocentric Coordinates)
R.A. = 01h18m13.3s
Dec. = +08°15'12.8"
S.D. = 00°15'57.8"
H.P. = 00°00'08.8"

Moon at Greatest Eclipse
(Geocentric Coordinates)
R.A. = 01h17m01.7s
Dec. = +09°14'52.8"
S.D. = 00°16'42.8"
H.P. = 01°01'20.2"

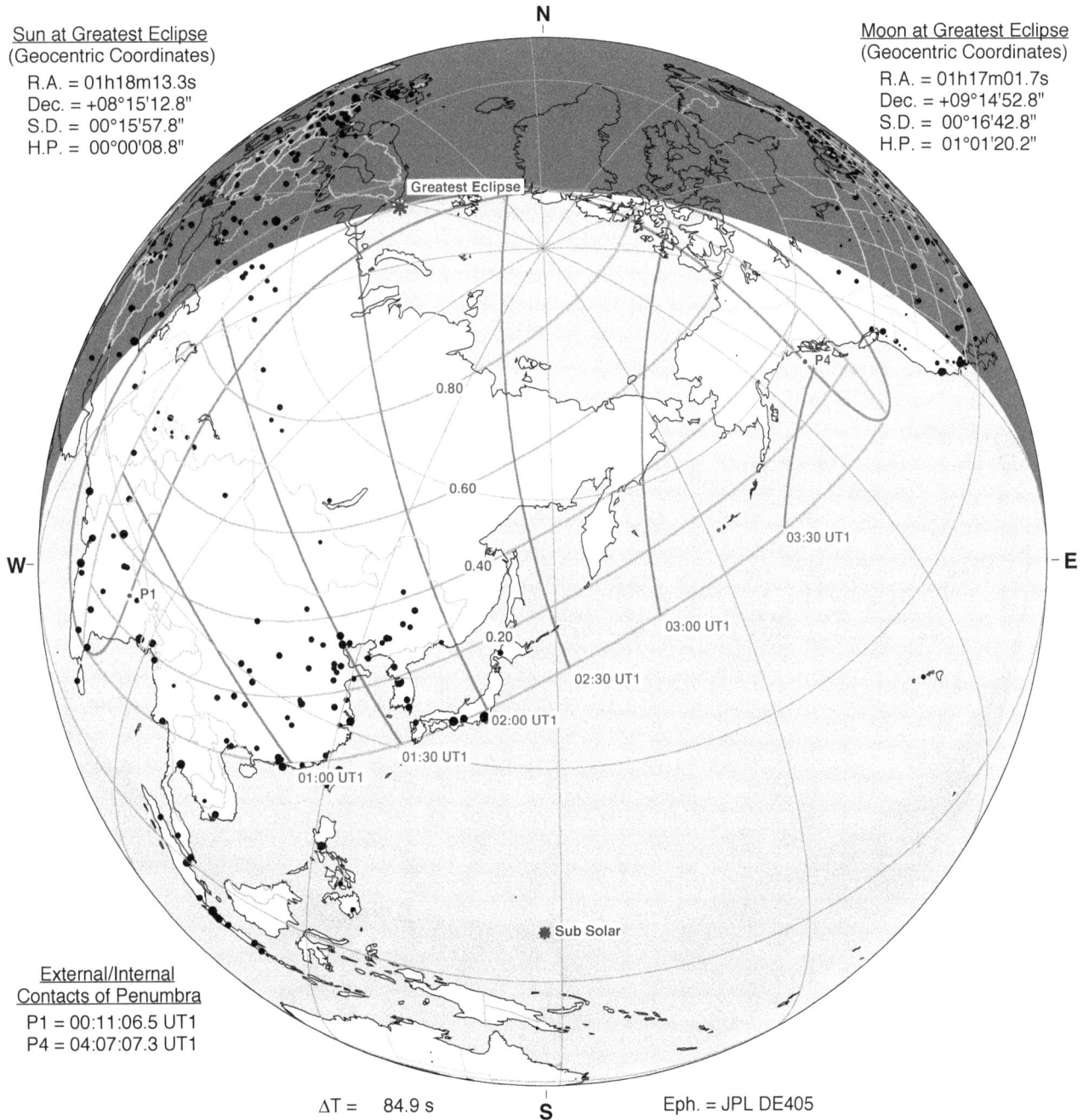

Greatest Eclipse

0.80
0.60
0.40
0.20

P4
03:30 UT1
03:00 UT1
02:30 UT1
02:00 UT1
01:30 UT1
01:00 UT1
P1

Sub Solar

N
W
E
S

External/Internal Contacts of Penumbra
P1 = 00:11:06.5 UT1
P4 = 04:07:07.3 UT1

ΔT = 84.9 s

Eph. = JPL DE405

Circumstances at Greatest Eclipse: 02:09:13.8 UT1
Lat. = 71°37.6'N
Long. = 032°08.7'E
Sun Alt. = 0.0°
Sun Azm. = 62.9°

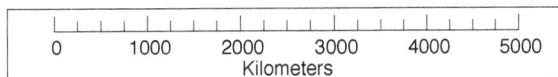

0 1000 2000 3000 4000 5000
Kilometers

©2016 F. Espenak
www.EclipseWise.com

133

Partial Solar Eclipse of 2051 Oct 04

Greatest Eclipse = 21:02:14.5 TD   (= 21:00:49.4 UT1)

Eclipse Magnitude = 0.6024

Gamma = -1.2094

Saros Series = 125

Saros Member = 56 of 73

Sun at Greatest Eclipse
(Geocentric Coordinates)
R.A. = 12h42m39.3s
Dec. = -04°35'05.4"
S.D. = 00°15'59.2"
H.P. = 00°00'08.8"

Moon at Greatest Eclipse
(Geocentric Coordinates)
R.A. = 12h41m20.9s
Dec. = -05°37'21.2"
S.D. = 00°14'44.4"
H.P. = 00°54'05.8"

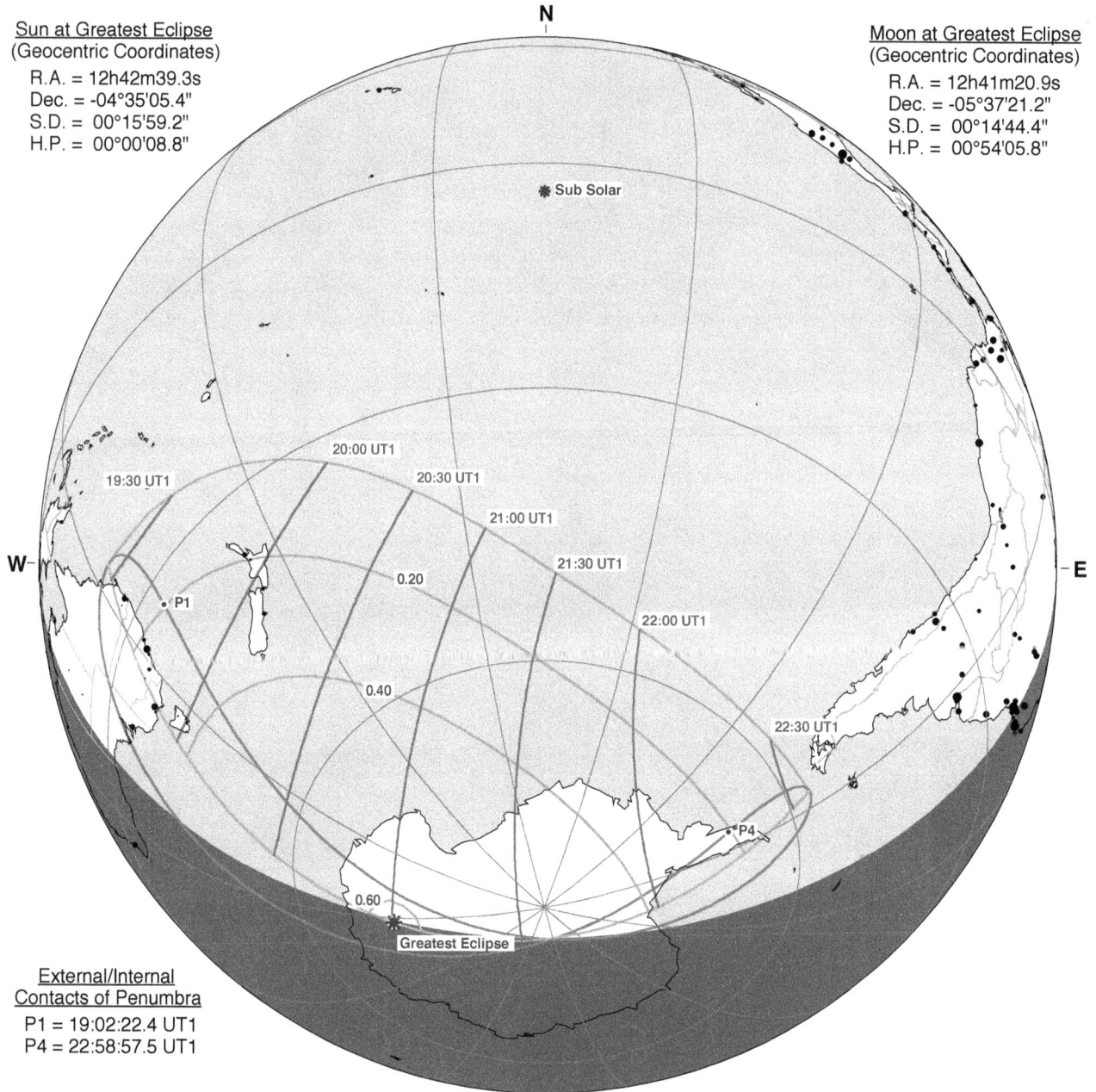

N

* Sub Solar

20:00 UT1

19:30 UT1

20:30 UT1

21:00 UT1

21:30 UT1

0.20

22:00 UT1

W

E

P1

0.40

22:30 UT1

P4

0.60

* Greatest Eclipse

External/Internal
Contacts of Penumbra
P1 = 19:02:22.4 UT1
P4 = 22:58:57.5 UT1

ΔT = 85.1 s

S

Eph. = JPL DE405

Circumstances at Greatest Eclipse: 21:00:49.4 UT1
Lat. = 72°02.7'S   Sun Alt. =   0.0°
Long. = 117°41.2'E   Sun Azm. = 105.0°

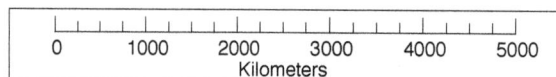

0	1000	2000	3000	4000	5000

Kilometers

©2016 F. Espenak
www.EclipseWise.com

Total Solar Eclipse of 2052 Mar 30

Greatest Eclipse = 18:31:52.9 TD   (= 18:30:27.4 UT1)

Eclipse Magnitude = 1.0466   Saros Series = 130
Gamma = 0.3239   Saros Member = 54 of 73

Sun at Greatest Eclipse
(Geocentric Coordinates)
R.A. = 00h39m33.8s
Dec. = +04°15'25.9"
S.D. = 00°16'00.7"
H.P. = 00°00'08.8"

Moon at Greatest Eclipse
(Geocentric Coordinates)
R.A. = 00h39m10.3s
Dec. = +04°34'05.5"
S.D. = 00°16'29.6"
H.P. = 01°00'31.8"

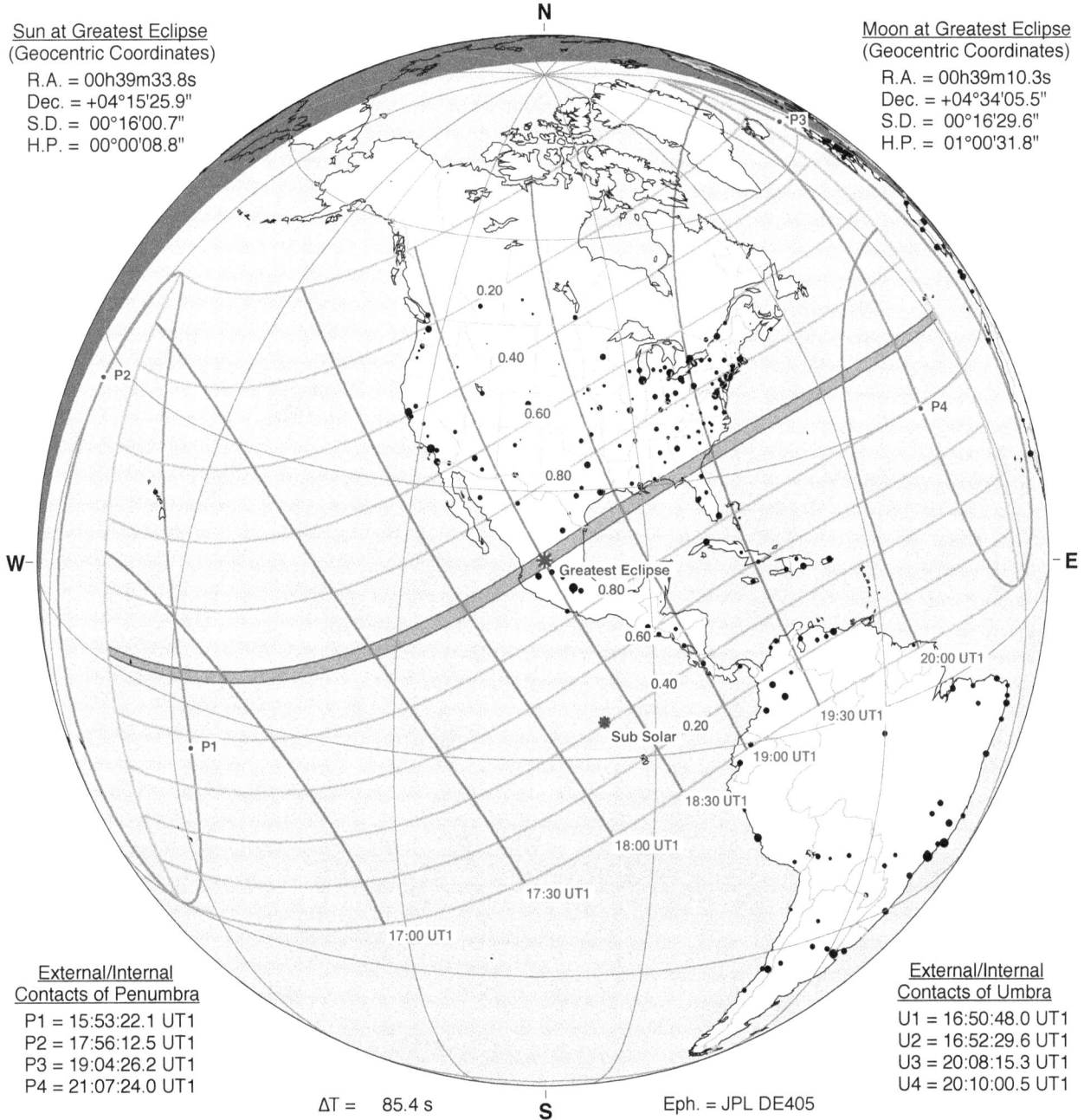

N

P3

0.20
0.40
0.60
0.80

P2

P4

W —

Greatest Eclipse
0.80

— E

0.60

0.40

20:00 UT1

0.20

19:30 UT1

Sub Solar

19:00 UT1

P1

18:30 UT1

18:00 UT1

17:30 UT1

17:00 UT1

External/Internal
Contacts of Penumbra
P1 = 15:53:22.1 UT1
P2 = 17:56:12.5 UT1
P3 = 19:04:26.2 UT1
P4 = 21:07:24.0 UT1

External/Internal
Contacts of Umbra
U1 = 16:50:48.0 UT1
U2 = 16:52:29.6 UT1
U3 = 20:08:15.3 UT1
U4 = 20:10:00.5 UT1

ΔT = 85.4 s

S

Eph. = JPL DE405

Circumstances at Greatest Eclipse: 18:30:27.4 UT1
Lat. = 22°22.8'N
Long. = 102°34.9'W
Path Width = 163.7 km
Sun Alt. = 71.0°
Sun Azm. = 161.3°
Duration = 04m08.2s

Circumstances at Greatest Duration: 18:32:33.3 UT1
Lat. = 22°45.2'N
Long. = 101°55.6'W
Path Width = 163.5 km
Sun Alt. = 70.9°
Sun Azm. = 165.1°
Duration = 04m08.2s

0   1000   2000   3000   4000   5000
Kilometers

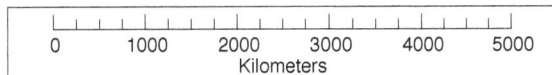

©2016 F. Espenak
www.EclipseWise.com

Annular Solar Eclipse of 2052 Sep 22

Greatest Eclipse = 23:39:09.7 TD   (= 23:37:44.0 UT1)

Eclipse Magnitude = 0.9734   Saros Series = 135
Gamma = -0.4480   Saros Member = 41 of 71

Sun at Greatest Eclipse
(Geocentric Coordinates)
R.A. = 12h02m27.0s
Dec. = -00°15'55.5"
S.D. = 00°15'56.2"
H.P. = 00°00'08.8"

Moon at Greatest Eclipse
(Geocentric Coordinates)
R.A. = 12h01m56.4s
Dec. = -00°39'49.3"
S.D. = 00°15'17.9"
H.P. = 00°56'08.8"

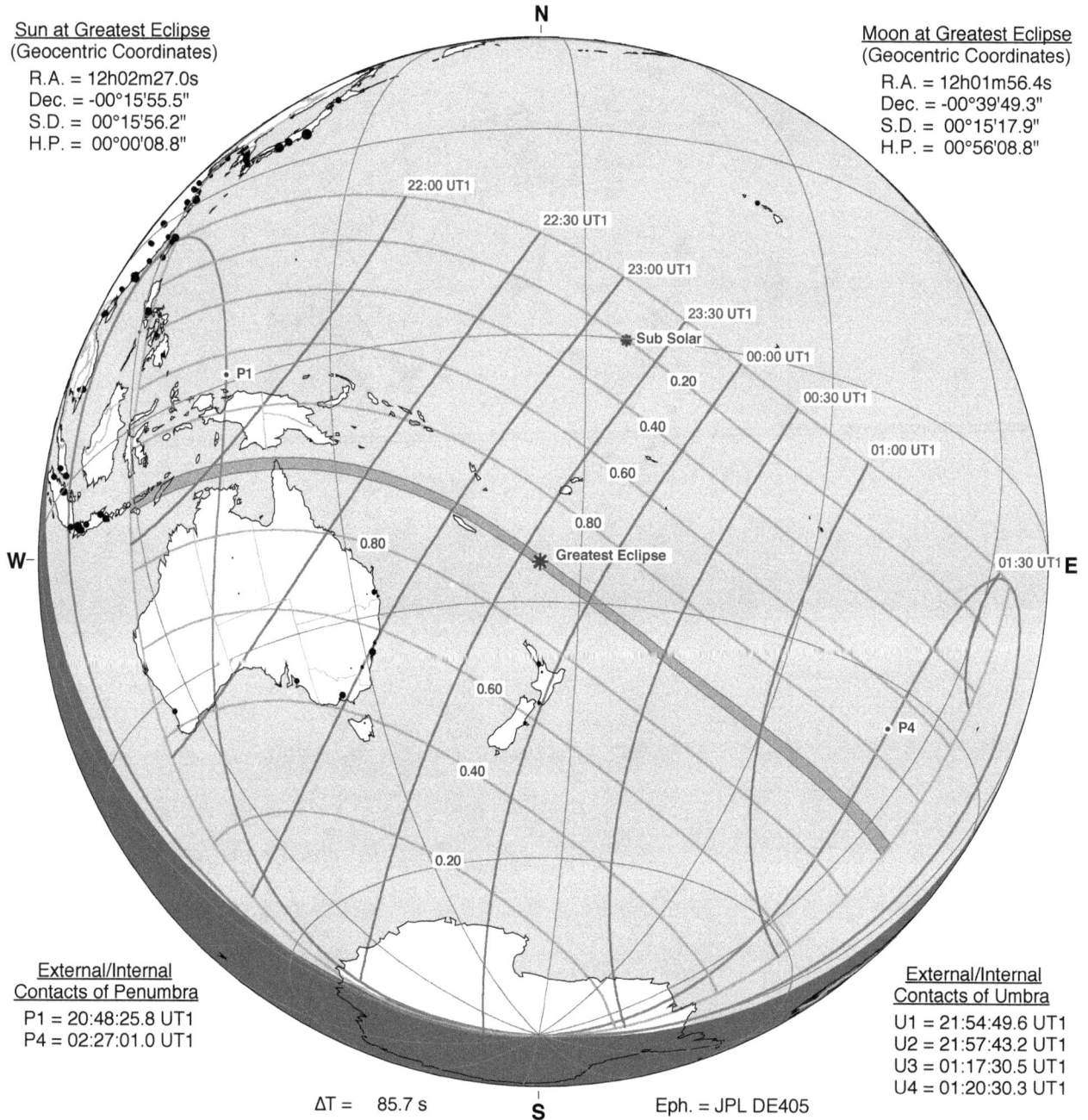

N

22:00 UT1
22:30 UT1
23:00 UT1
23:30 UT1
Sub Solar
00:00 UT1
0.20
00:30 UT1
0.40
01:00 UT1
0.60
0.80
Greatest Eclipse
0.80
01:30 UT1

W — — E

P1
P4

0.60
0.40
0.20

S

External/Internal Contacts of Penumbra
P1 = 20:48:25.8 UT1
P4 = 02:27:01.0 UT1

External/Internal Contacts of Umbra
U1 = 21:54:49.6 UT1
U2 = 21:57:43.2 UT1
U3 = 01:17:30.5 UT1
U4 = 01:20:30.3 UT1

ΔT = 85.7 s   Eph. = JPL DE405

Circumstances at Greatest Eclipse: 23:37:44.0 UT1
Lat. = 25°41.3'S   Sun Alt. = 63.2°
Long. = 174°56.0'E   Sun Azm. = 19.6°
Path Width = 106.0 km   Duration = 02m50.9s

Circumstances at Greatest Duration: 00:04:03.2 UT1
Lat. = 30°17.6'S   Sun Alt. = 59.5°
Long. = 177°20.4'W   Sun Azm. = 348.9°
Path Width = 106.1 km   Duration = 02m51.5s

0   1000   2000   3000   4000   5000
Kilometers

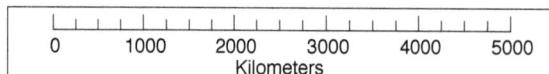

©2016 F. Espenak
www.EclipseWise.com

Annular Solar Eclipse of 2053 Mar 20

Greatest Eclipse = 07:08:19.4 TD   (= 07:06:53.4 UT1)

Eclipse Magnitude = 0.9919	Saros Series = 140
Gamma = -0.4089	Saros Member = 31 of 71

Sun at Greatest Eclipse
(Geocentric Coordinates)
R.A. = 00h00m30.3s
Dec. = +00°03'17.2"
S.D. = 00°16'03.6"
H.P. = 00°00'08.8"

Moon at Greatest Eclipse
(Geocentric Coordinates)
R.A. = 00h00m59.0s
Dec. = -00°19'05.6"
S.D. = 00°15'41.9"
H.P. = 00°57'37.0"

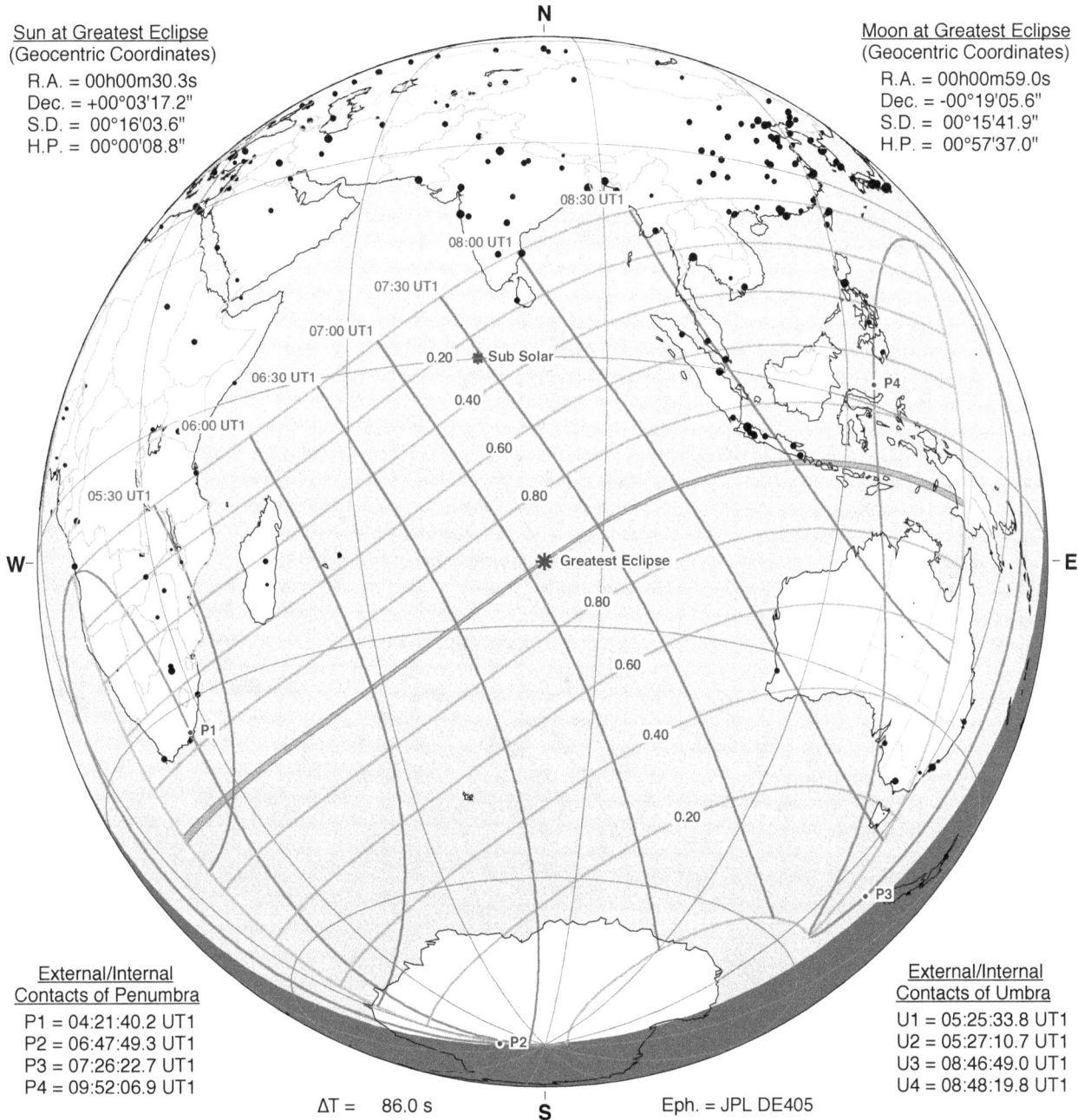

N

08:30 UT1
08:00 UT1
07:30 UT1
07:00 UT1
06:30 UT1
0.20 Sub Solar
06:00 UT1
0.40
05:30 UT1
0.60
0.80

W

* Greatest Eclipse
0.80
0.60
0.40
0.20

E

P1
P4
P2
P3

External/Internal
Contacts of Penumbra
P1 = 04:21:40.2 UT1
P2 = 06:47:49.3 UT1
P3 = 07:26:22.7 UT1
P4 = 09:52:06.9 UT1

External/Internal
Contacts of Umbra
U1 = 05:25:33.8 UT1
U2 = 05:27:10.7 UT1
U3 = 08:46:49.0 UT1
U4 = 08:48:19.8 UT1

ΔT = 86.0 s

S

Eph. = JPL DE405

Circumstances at Greatest Eclipse: 07:06:53.4 UT1

Lat. = 23°00.9'S	Sun Alt. = 65.7°
Long. = 082°54.1'E	Sun Azm. = 340.8°
Path Width = 31.1 km	Duration = 00m49.5s

Circumstances at Greatest Duration: 05:26:22.2 UT1

Lat. = 42°07.6'S	Sun Alt. = 0.0°
Long. = 010°16.4'E	Sun Azm. = 90.0°
Path Width = 93.4 km	Duration = 01m28.3s

0 1000 2000 3000 4000 5000
Kilometers

©2016 F. Espenak
www.EclipseWise.com

Total Solar Eclipse of 2053 Sep 12

Greatest Eclipse = 09:34:08.9 TD   (= 09:32:42.6 UT1)

Eclipse Magnitude = 1.0328	Saros Series = 145
Gamma = 0.3140	Saros Member = 24 of 77

Sun at Greatest Eclipse
(Geocentric Coordinates)
R.A. = 11h23m36.1s
Dec. = +03°55'14.2"
S.D. = 00°15'53.4"
H.P. = 00°00'08.7"

Moon at Greatest Eclipse
(Geocentric Coordinates)
R.A. = 11h23m58.5s
Dec. = +04°12'57.2"
S.D. = 00°16'09.4"
H.P. = 00°59'17.8"

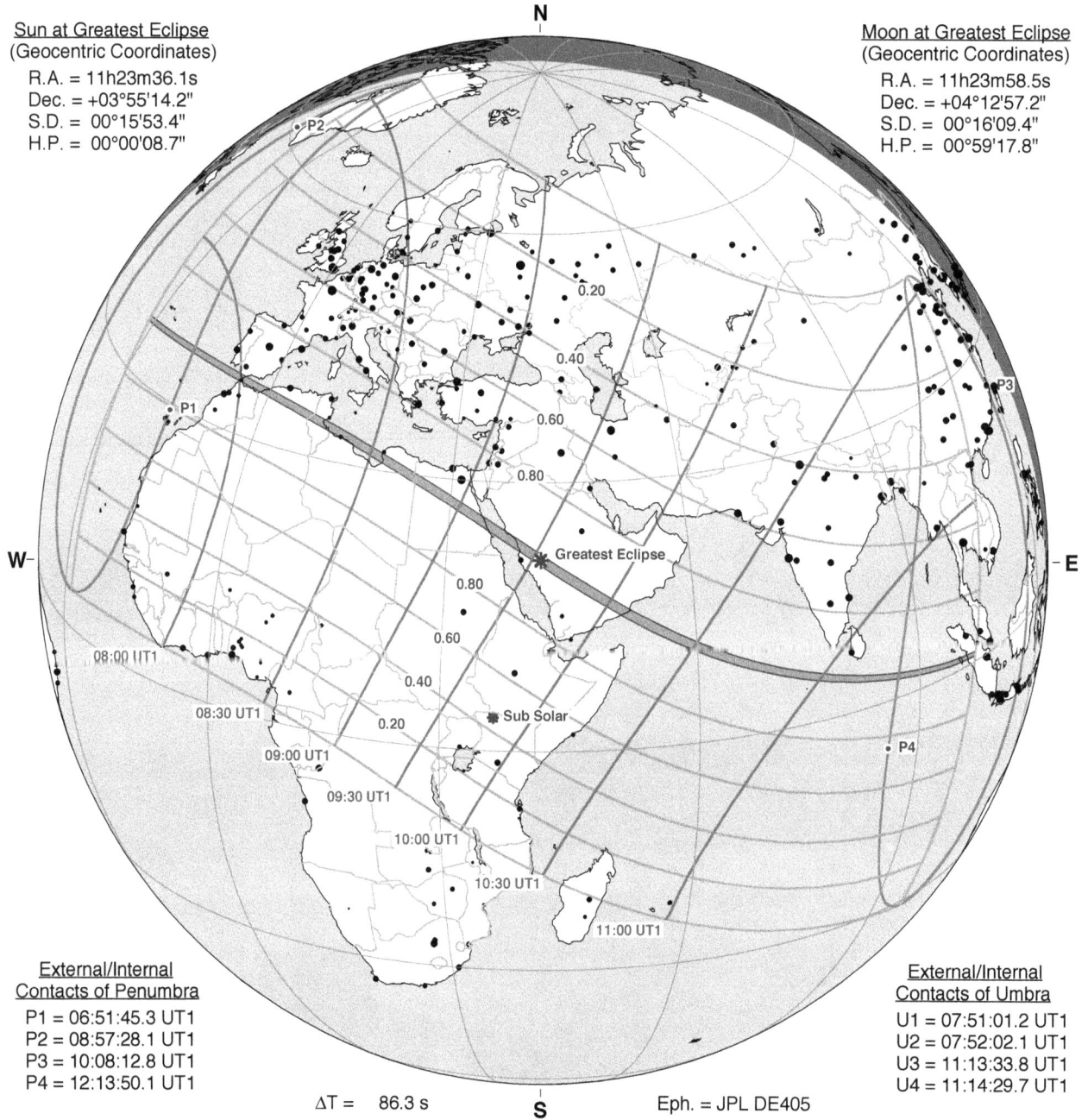

External/Internal
Contacts of Penumbra
P1 = 06:51:45.3 UT1
P2 = 08:57:28.1 UT1
P3 = 10:08:12.8 UT1
P4 = 12:13:50.1 UT1

External/Internal
Contacts of Umbra
U1 = 07:51:01.2 UT1
U2 = 07:52:02.1 UT1
U3 = 11:13:33.8 UT1
U4 = 11:14:29.7 UT1

ΔT = 86.3 s

Eph. = JPL DE405

Circumstances at Greatest Eclipse: 09:32:42.6 UT1

Lat. = 21°28.1'N	Sun Alt. = 71.6°
Long. = 041°39.8'E	Sun Azm. = 198.6°
Path Width = 116.4 km	Duration = 03m04.1s

Circumstances at Greatest Duration: 09:29:56.2 UT1

Lat. = 21°56.9'N	Sun Alt. = 71.5°
Long. = 040°50.4'E	Sun Azm. = 193.6°
Path Width = 116.3 km	Duration = 03m04.2s

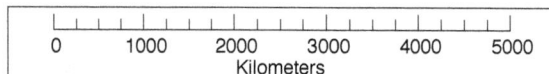

©2016 F. Espenak
www.EclipseWise.com

Partial Solar Eclipse of 2054 Mar 09

Greatest Eclipse = 12:33:40.5 TD (= 12:32:13.9 UT1)

Eclipse Magnitude = 0.6678   Saros Series = 150
Gamma = -1.1711   Saros Member = 19 of 71

Sun at Greatest Eclipse
(Geocentric Coordinates)
R.A. = 23h20m07.5s
Dec. = -04°17'25.4"
S.D. = 00°16'06.6"
H.P. = 00°00'08.9"

Moon at Greatest Eclipse
(Geocentric Coordinates)
R.A. = 23h21m24.6s
Dec. = -05°18'27.6"
S.D. = 00°14'55.7"
H.P. = 00°54'47.2"

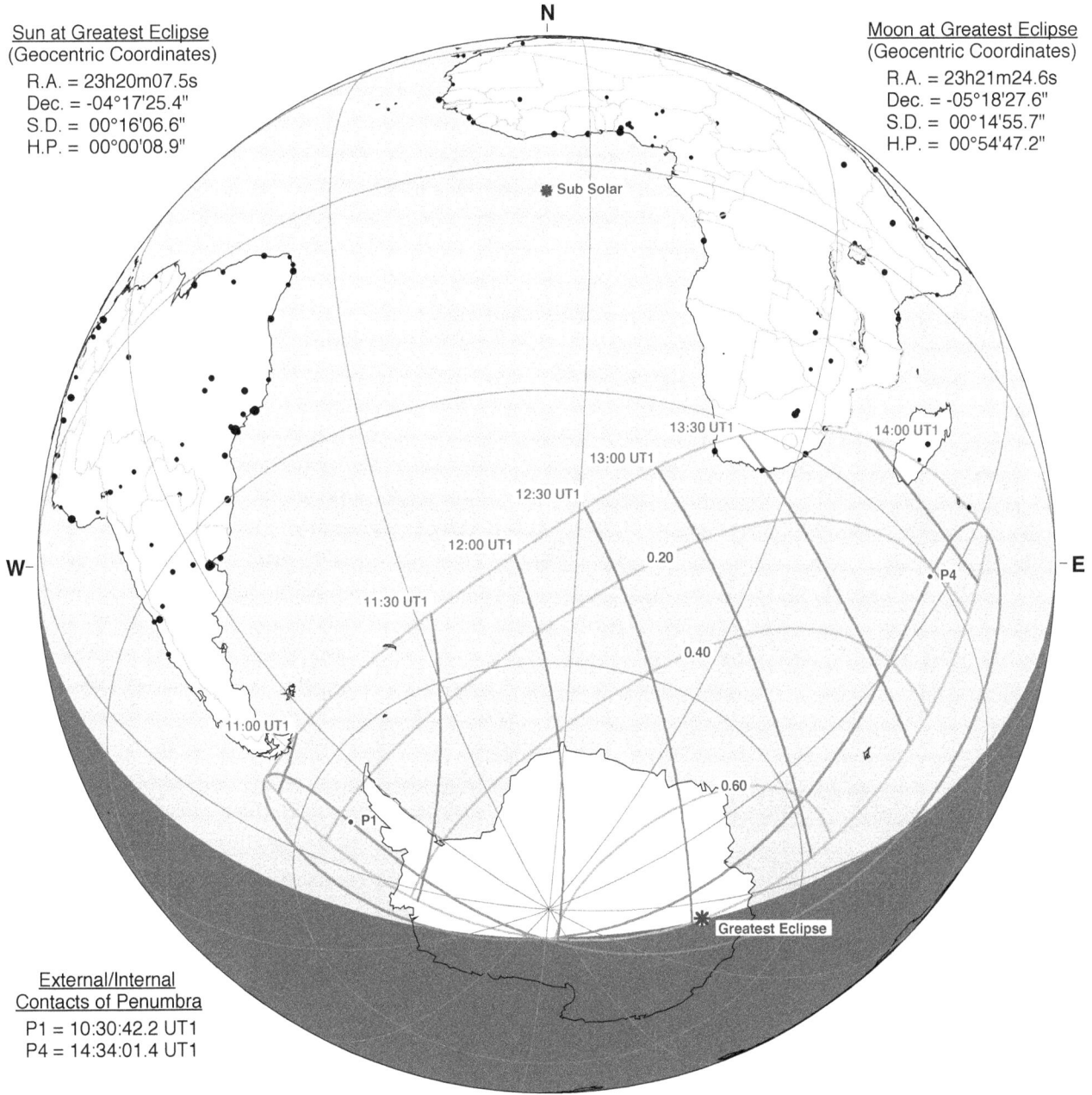

N

* Sub Solar

13:30 UT1   14:00 UT1
13:00 UT1
12:30 UT1
12:00 UT1   0.20
11:30 UT1   P4

W   E

0.40

11:00 UT1
0.60
P1

* Greatest Eclipse

External/Internal
Contacts of Penumbra
P1 = 10:30:42.2 UT1
P4 = 14:34:01.4 UT1

ΔT = 86.6 s   S   Eph. = JPL DE405

Circumstances at Greatest Eclipse: 12:32:13.9 UT1
Lat. = 72°02.6'S   Sun Alt. = 0.0°
Long. = 097°51.8'E   Sun Azm. = 256.0°

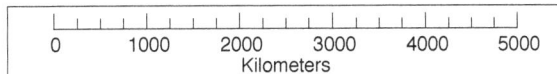

```
0    1000   2000   3000   4000   5000
            Kilometers
```

©2016 F. Espenak
www.EclipseWise.com

Partial Solar Eclipse of 2054 Aug 03

Greatest Eclipse = 18:04:02.1 TD   (= 18:02:35.2 UT1)

Eclipse Magnitude = 0.0656	Saros Series = 117
Gamma = -1.4941	Saros Member = 71 of 71

Sun at Greatest Eclipse
(Geocentric Coordinates)
R.A. = 08h56m24.5s
Dec. = +17°17'09.2"
S.D. = 00°15'45.7"
H.P. = 00°00'08.7"

Moon at Greatest Eclipse
(Geocentric Coordinates)
R.A. = 08h55m14.2s
Dec. = +15°47'22.5"
S.D. = 00°16'41.7"
H.P. = 01°01'16.5"

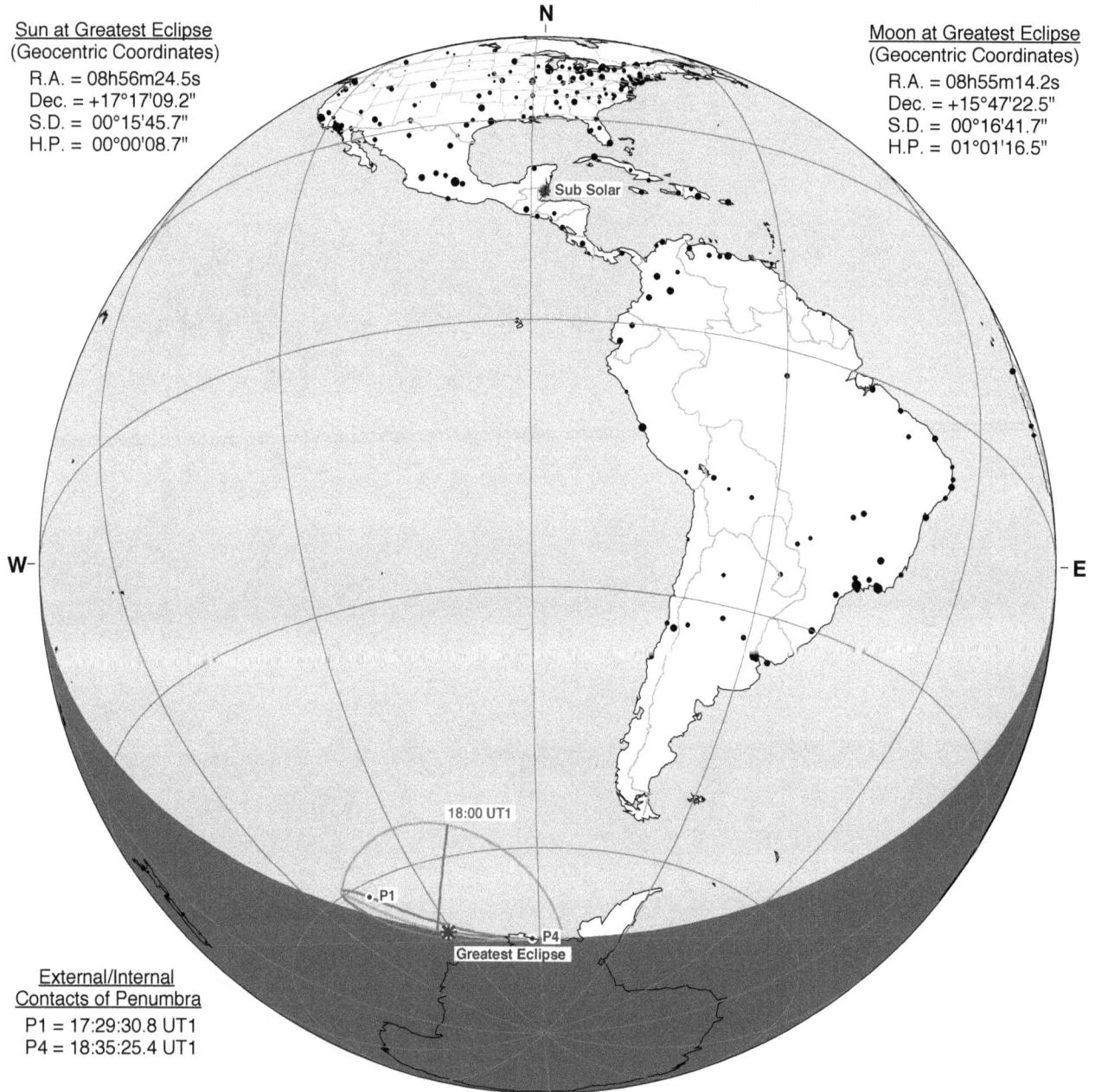

N

Sub Solar

W

E

18:00 UT1

P1

P4

Greatest Eclipse

External/Internal
Contacts of Penumbra
P1 = 17:29:30.8 UT1
P4 = 18:35:25.4 UT1

ΔT = 86.9 s

S

Eph. = JPL DE405

Circumstances at Greatest Eclipse: 18:02:35.2 UT1

Lat. = 69°46.4'S	Sun Alt. = 0.0°
Long. = 121°24.7'W	Sun Azm. = 30.7°

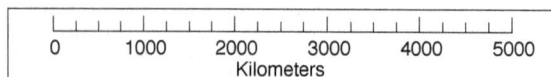

0   1000   2000   3000   4000   5000
Kilometers

Partial Solar Eclipse of 2054 Sep 02

Greatest Eclipse = 01:09:33.7 TD   (= 01:08:06.7 UT1)

Eclipse Magnitude = 0.9793   Saros Series = 155
Gamma = 1.0215   Saros Member = 8 of 71

N

Sun at Greatest Eclipse
(Geocentric Coordinates)
R.A. = 10h45m28.2s
Dec. = +07°52'58.6"
S.D. = 00°15'50.9"
H.P. = 00°00'08.7"

Moon at Greatest Eclipse
(Geocentric Coordinates)
R.A. = 10h46m40.4s
Dec. = +08°52'49.8"
S.D. = 00°16'42.0"
H.P. = 01°01'17.5"

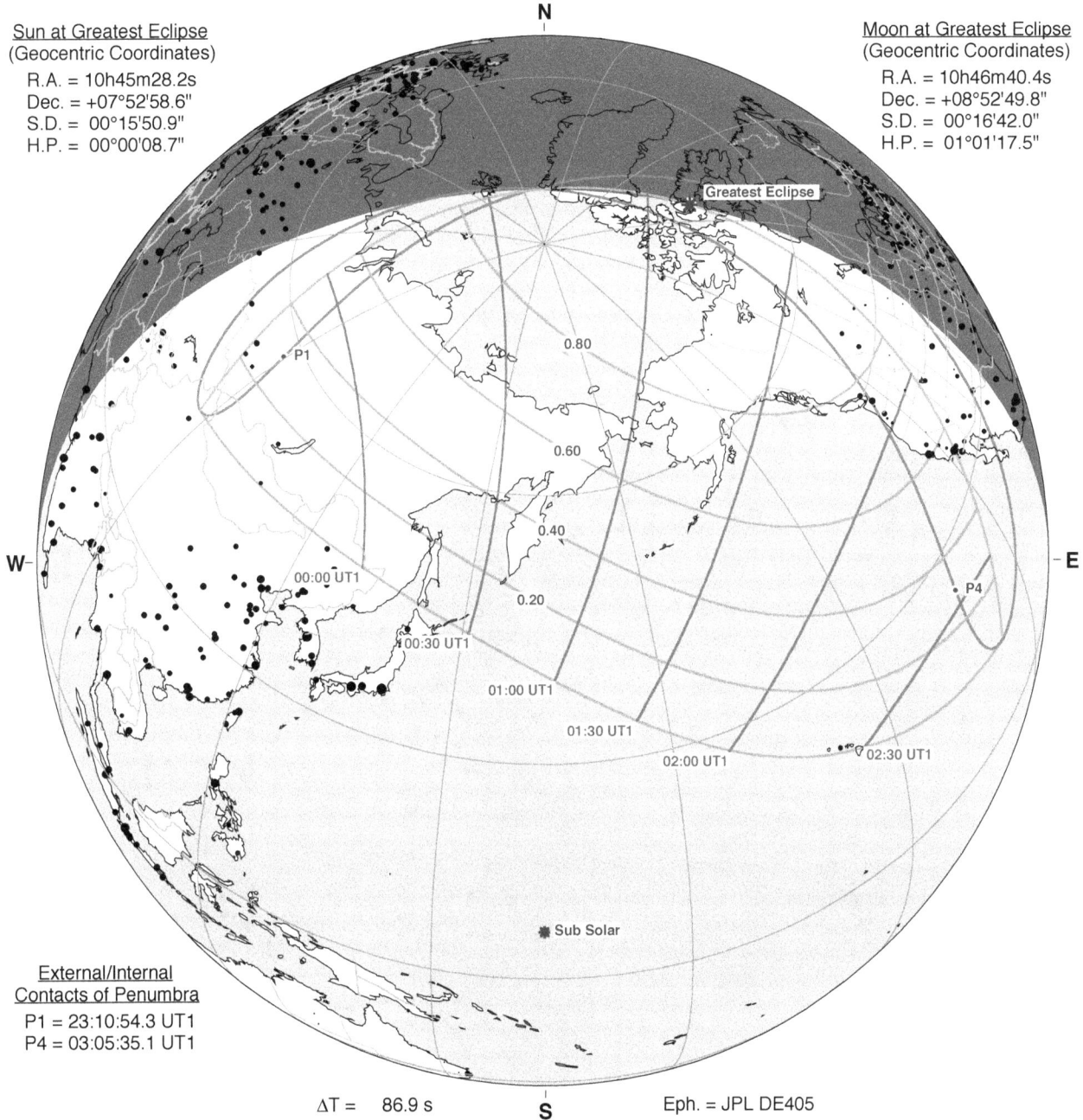

Greatest Eclipse

P1

0.80

0.60

0.40

W

0.20

00:00 UT1

E

00:30 UT1

P4

01:00 UT1

01:30 UT1

02:30 UT1

02:00 UT1

Sub Solar

External/Internal
Contacts of Penumbra
P1 = 23:10:54.3 UT1
P4 = 03:05:35.1 UT1

ΔT =   86.9 s   **S**   Eph. = JPL DE405

Circumstances at Greatest Eclipse: 01:08:06.7 UT1
Lat. = 71°40.9'N   Sun Alt. =   0.0°
Long. = 082°24.8'W   Sun Azm. = 295.9°

| 0 | 1000 | 2000 | 3000 | 4000 | 5000 |
Kilometers

©2016 F. Espenak
www.EclipseWise.com

Partial Solar Eclipse of 2055 Jan 27

Greatest Eclipse = 17:54:05.3 TD (= 17:52:38.1 UT1)

Eclipse Magnitude = 0.6932	Saros Series = 122
Gamma = 1.1550	Saros Member = 60 of 70

N

Sun at Greatest Eclipse
(Geocentric Coordinates)
R.A. = 20h40m41.0s
Dec. = -18°19'18.9"
S.D. = 00°16'14.5"
H.P. = 00°00'08.9"

Moon at Greatest Eclipse
(Geocentric Coordinates)
R.A. = 20h39m58.6s
Dec. = -17°17'11.5"
S.D. = 00°14'53.3"
H.P. = 00°54'38.4"

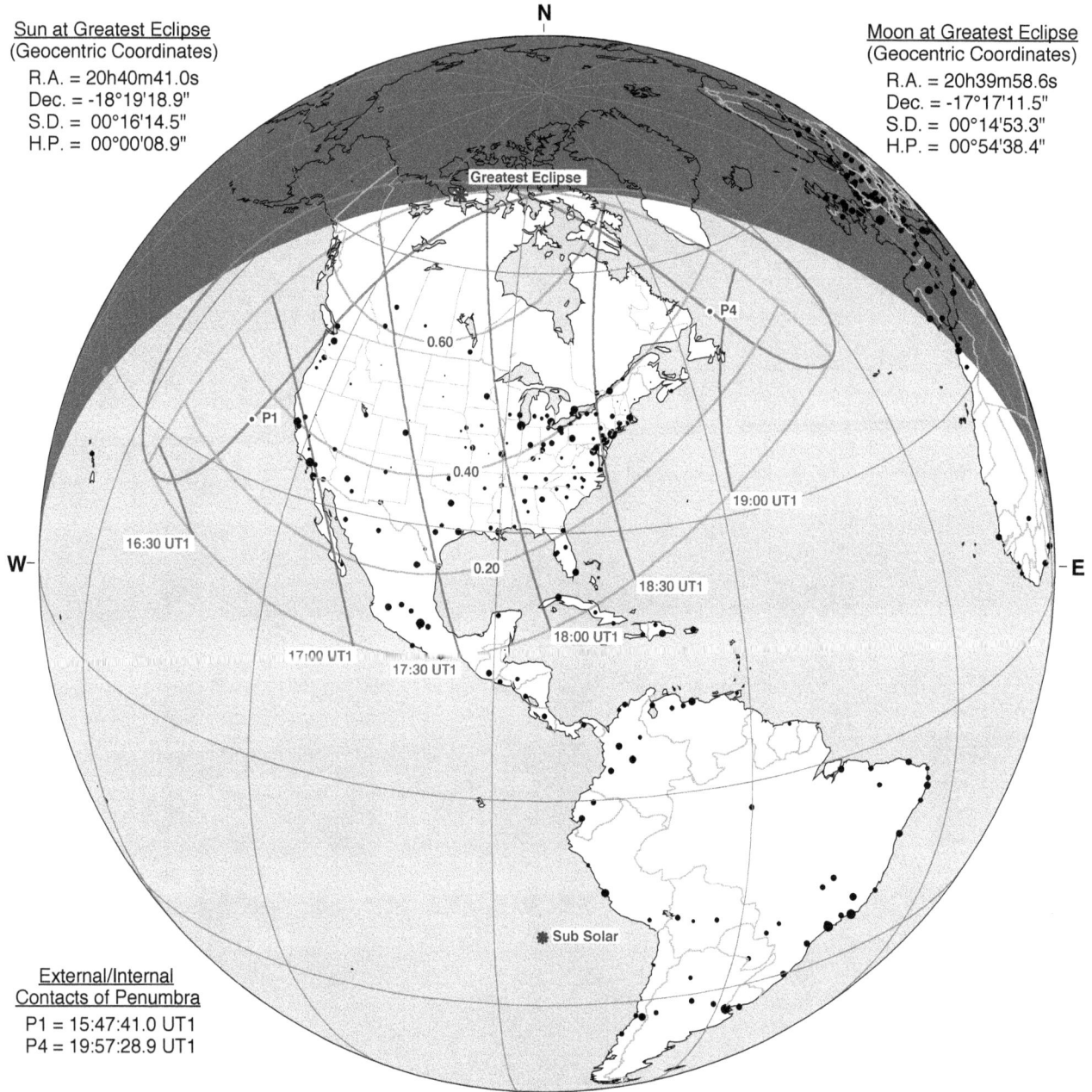

Greatest Eclipse

P4

0.60

P1

0.40

19:00 UT1

16:30 UT1

0.20

18:30 UT1

W

18:00 UT1

E

17:00 UT1

17:30 UT1

Sub Solar

External/Internal
Contacts of Penumbra
P1 = 15:47:41.0 UT1
P4 = 19:57:28.9 UT1

ΔT = 87.2 s

S

Eph. = JPL DE405

Circumstances at Greatest Eclipse: 17:52:38.1 UT1

Lat. = 69°32.9'N	Sun Alt. = 0.0°
Long. = 112°19.0'W	Sun Azm. = 154.1°

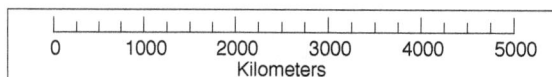

0	1000	2000	3000	4000	5000

Kilometers

©2016 F. Espenak
www.EclipseWise.com

Total Solar Eclipse of 2055 Jul 24

Greatest Eclipse = 09:57:50.3 TD   (= 09:56:22.7 UT1)

Eclipse Magnitude = 1.0359	Saros Series = 127
Gamma = -0.8012	Saros Member = 60 of 82

Sun at Greatest Eclipse
(Geocentric Coordinates)
R.A. = 08h15m04.2s
Dec. = +19°48'43.3"
S.D. = 00°15'44.6"
H.P. = 00°00'08.7"

Moon at Greatest Eclipse
(Geocentric Coordinates)
R.A. = 08h14m39.2s
Dec. = +19°01'42.7"
S.D. = 00°16'09.1"
H.P. = 00°59'16.7"

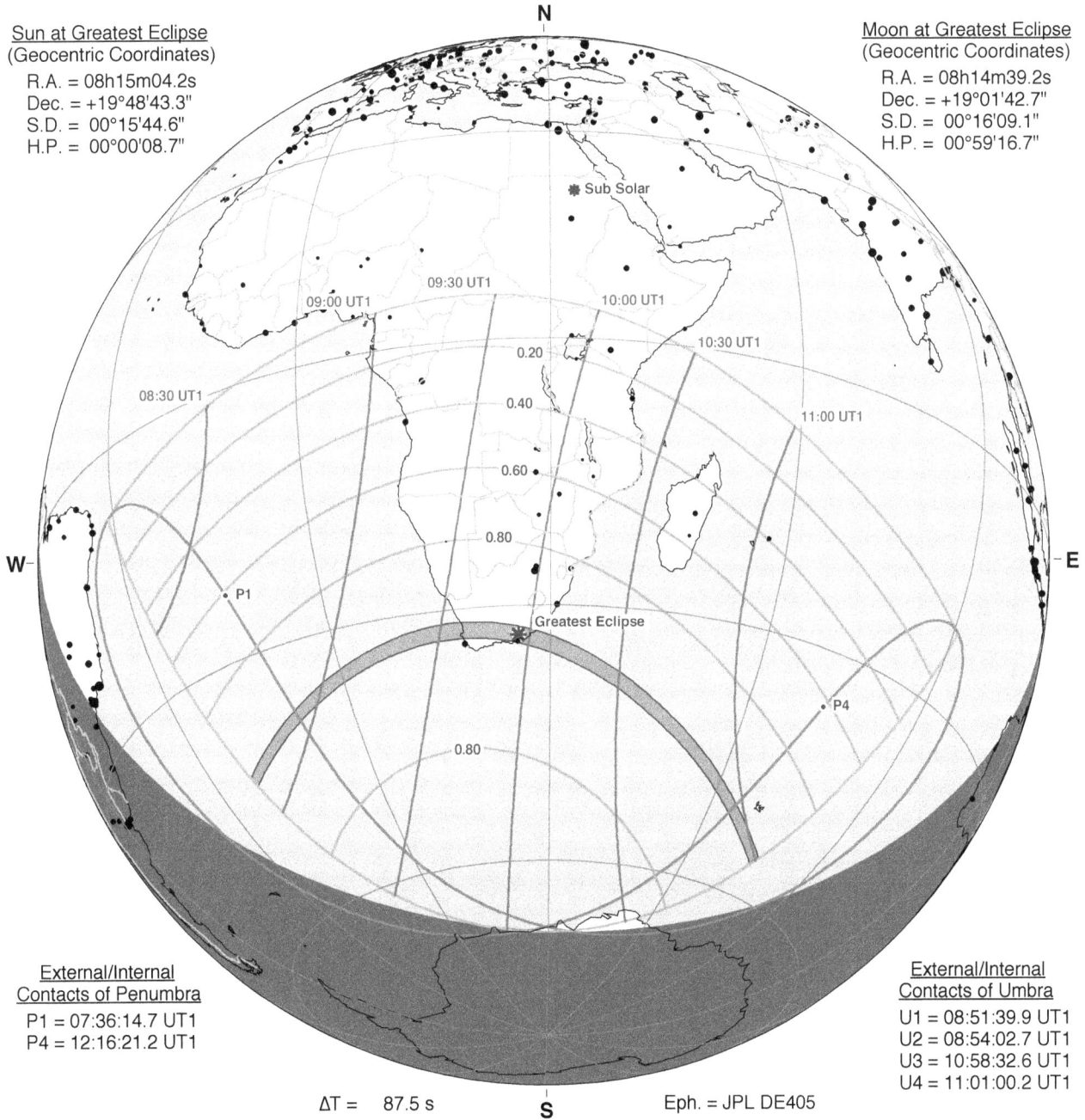

N

Sub Solar

09:30 UT1
09:00 UT1
10:00 UT1
10:30 UT1
08:30 UT1
11:00 UT1

0.20
0.40
0.60
0.80

W

E

P1

0.80

Greatest Eclipse

P4

External/Internal
Contacts of Penumbra
P1 = 07:36:14.7 UT1
P4 = 12:16:21.2 UT1

External/Internal
Contacts of Umbra
U1 = 08:51:39.9 UT1
U2 = 08:54:02.7 UT1
U3 = 10:58:32.6 UT1
U4 = 11:01:00.2 UT1

ΔT =   87.5 s

S

Eph. = JPL DE405

Circumstances at Greatest Eclipse: 09:56:22.7 UT1

Lat. = 33°16.9'S	Sun Alt. = 36.5°
Long. = 025°43.3'E	Sun Azm. = 8.0°
Path Width = 201.8 km	Duration = 03m16.8s

Circumstances at Greatest Duration: 09:56:38.1 UT1

Lat. = 33°18.1'S	Sun Alt. = 36.5°
Long. = 025°49.2'E	Sun Azm. = 7.8°
Path Width = 201.7 km	Duration = 03m16.8s

0   1000   2000   3000   4000   5000
Kilometers

Annular Solar Eclipse of 2056 Jan 16

Greatest Eclipse = 22:16:45.2 TD   (= 22:15:17.3 UT1)

Eclipse Magnitude = 0.9760	Saros Series = 132
Gamma = 0.4199	Saros Member = 48 of 71

Sun at Greatest Eclipse
(Geocentric Coordinates)
R.A. = 19h54m06.4s
Dec. = -20°50'41.3"
S.D. = 00°16'15.5"
H.P. = 00°00'08.9"

Moon at Greatest Eclipse
(Geocentric Coordinates)
R.A. = 19h53m57.0s
Dec. = -20°26'45.0"
S.D. = 00°15'38.4"
H.P. = 00°57'23.8"

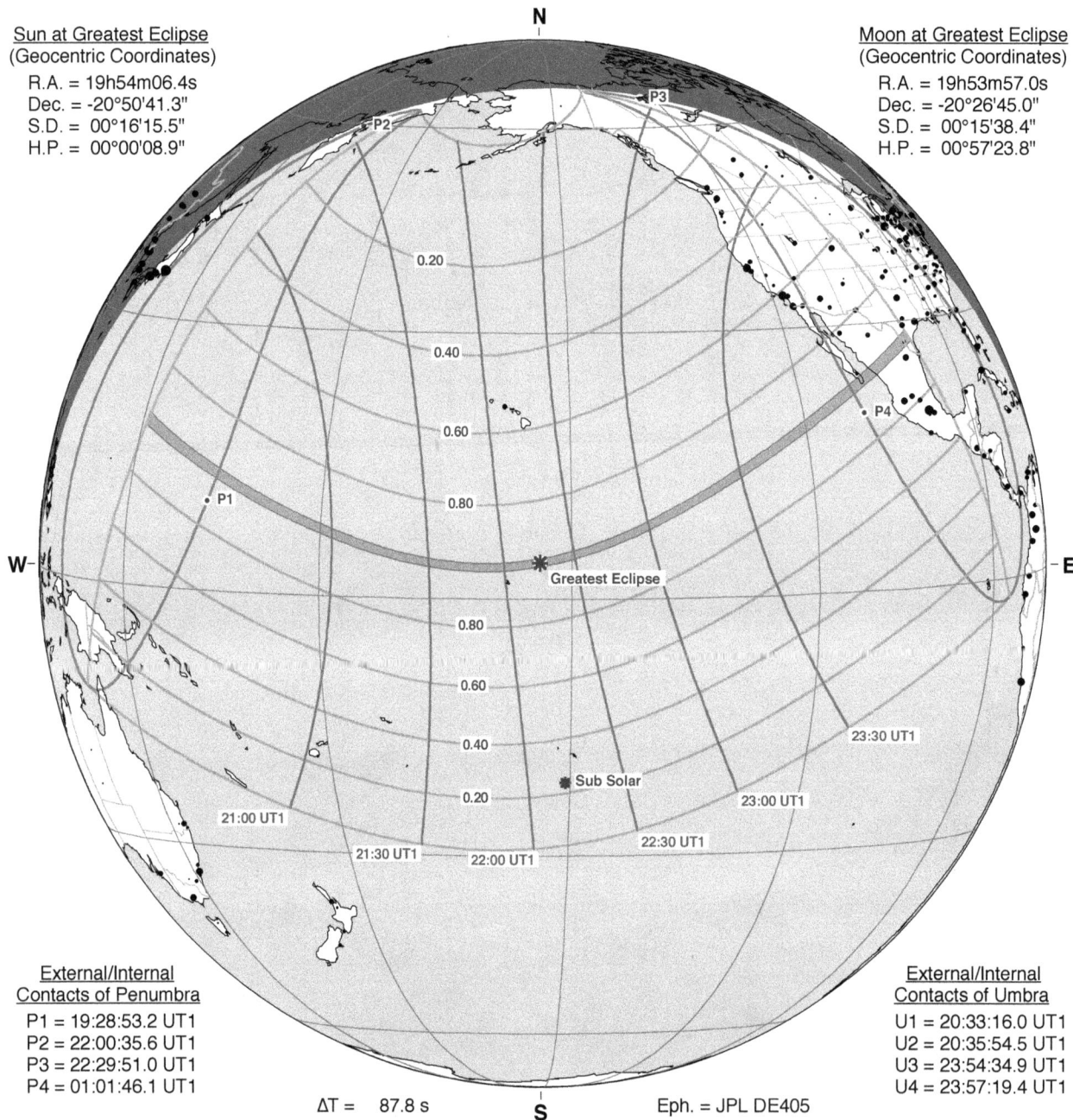

**External/Internal
Contacts of Penumbra**
P1 = 19:28:53.2 UT1
P2 = 22:00:35.6 UT1
P3 = 22:29:51.0 UT1
P4 = 01:01:46.1 UT1

**External/Internal
Contacts of Umbra**
U1 = 20:33:16.0 UT1
U2 = 20:35:54.5 UT1
U3 = 23:54:34.9 UT1
U4 = 23:57:19.4 UT1

ΔT =   87.8 s   Eph. = JPL DE405

Circumstances at Greatest Eclipse: 22:15:17.3 UT1

Lat. = 03°54.9'N	Sun Alt. = 65.1°
Long. = 153°36.2'W	Sun Azm. = 175.1°
Path Width = 94.5 km	Duration = 02m52.4s

Circumstances at Greatest Duration: 22:18:47.3 UT1

Lat. = 04°05.0'N	Sun Alt. = 65.1°
Long. = 152°40.5'W	Sun Azm. = 179.1°
Path Width = 94.5 km	Duration = 02m52.5s

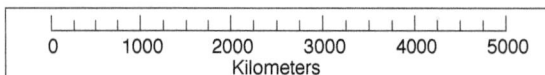

©2016 F. Espenak
www.EclipseWise.com

Annular Solar Eclipse of 2056 Jul 12

Greatest Eclipse = 20:21:59.4 TD (= 20:20:31.2 UT1)

Eclipse Magnitude = 0.9878   Saros Series = 137
Gamma = -0.0426   Saros Member = 38 of 70

Sun at Greatest Eclipse
(Geocentric Coordinates)
R.A. = 07h31m42.7s
Dec. = +21°45'32.5"
S.D. = 00°15'44.0"
H.P. = 00°00'08.7"

Moon at Greatest Eclipse
(Geocentric Coordinates)
R.A. = 07h31m42.1s
Dec. = +21°43'09.5"
S.D. = 00°15'17.9"
H.P. = 00°56'08.9"

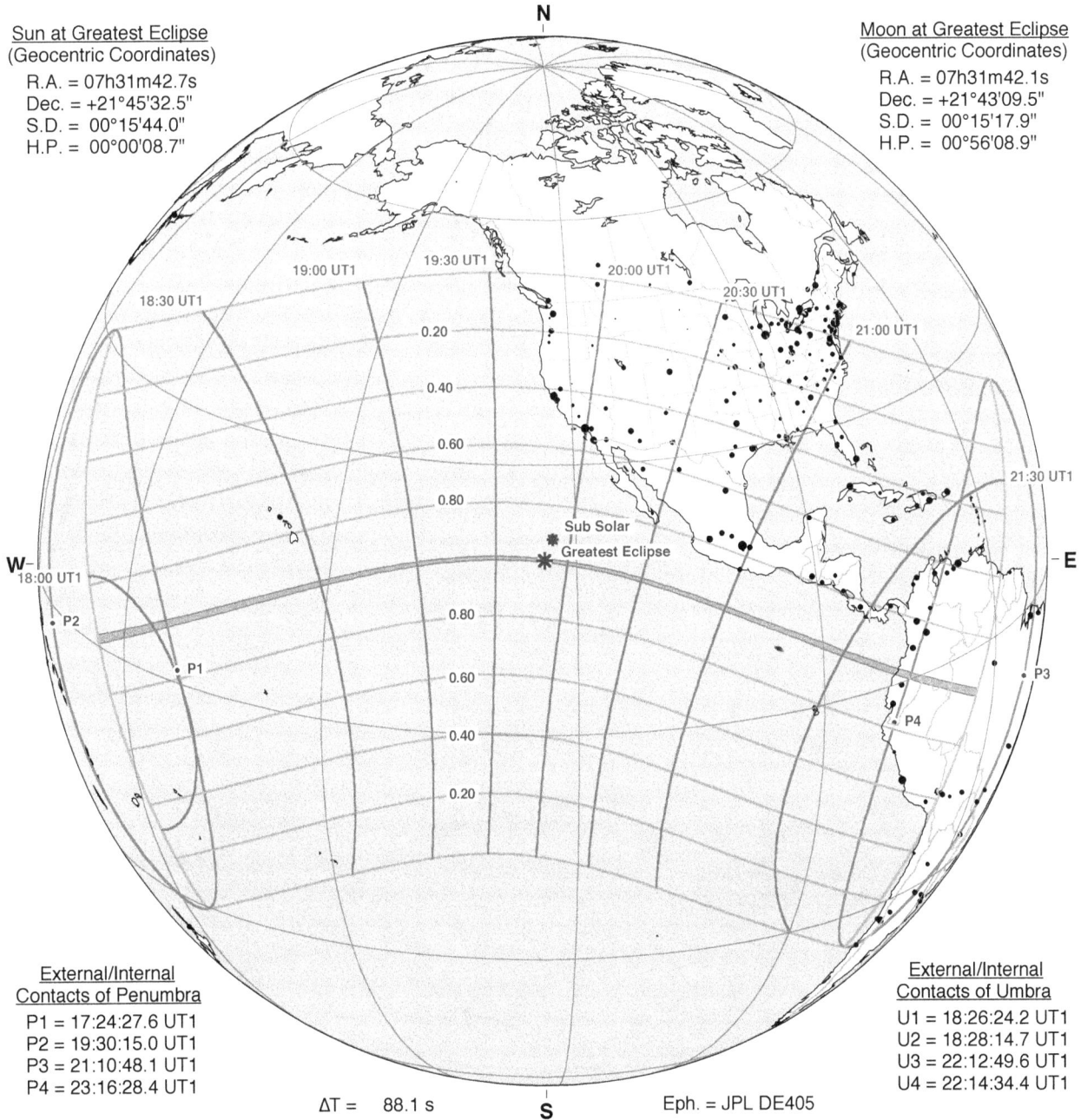

N

19:00 UT1 19:30 UT1 20:00 UT1 20:30 UT1 21:00 UT1
18:30 UT1
0.20
18:00 UT1 0.40
0.60
0.80
21:30 UT1
W Sub Solar
Greatest Eclipse E
18:00 UT1
P2
P1 0.80
0.60
P3
0.40 P4
0.20

S

External/Internal
Contacts of Penumbra
P1 = 17:24:27.6 UT1
P2 = 19:30:15.0 UT1
P3 = 21:10:48.1 UT1
P4 = 23:16:28.4 UT1

External/Internal
Contacts of Umbra
U1 = 18:26:24.2 UT1
U2 = 18:28:14.7 UT1
U3 = 22:12:49.6 UT1
U4 = 22:14:34.4 UT1

ΔT = 88.1 s   Eph. = JPL DE405

Circumstances at Greatest Eclipse: 20:20:31.2 UT1
Lat. = 19°26.4'N   Sun Alt. = 87.7°
Long. = 123°48.5'W   Sun Azm. = 3.3°
Path Width = 43.2 km   Duration = 01m25.9s

Circumstances at Greatest Duration: 18:27:19.5 UT1
Lat. = 00°41.1'N   Sun Alt. = 0.0°
Long. = 174°21.7'E   Sun Azm. = 68.2°
Path Width = 104.6 km   Duration = 01m49.4s

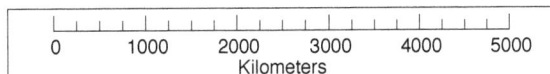

0 1000 2000 3000 4000 5000
Kilometers

©2016 F. Espenak
www.EclipseWise.com

Total Solar Eclipse of 2057 Jan 05

Greatest Eclipse = 09:47:52.2 TD (= 09:46:23.7 UT1)

Eclipse Magnitude = 1.0287	Saros Series = 142
Gamma = -0.2837	Saros Member = 25 of 72

Sun at Greatest Eclipse
(Geocentric Coordinates)
R.A. = 19h07m25.3s
Dec. = -22°31'37.8"
S.D. = 00°16'15.9"
H.P. = 00°00'08.9"

Moon at Greatest Eclipse
(Geocentric Coordinates)
R.A. = 19h07m26.6s
Dec. = -22°48'43.6"
S.D. = 00°16'27.8"
H.P. = 01°00'25.4"

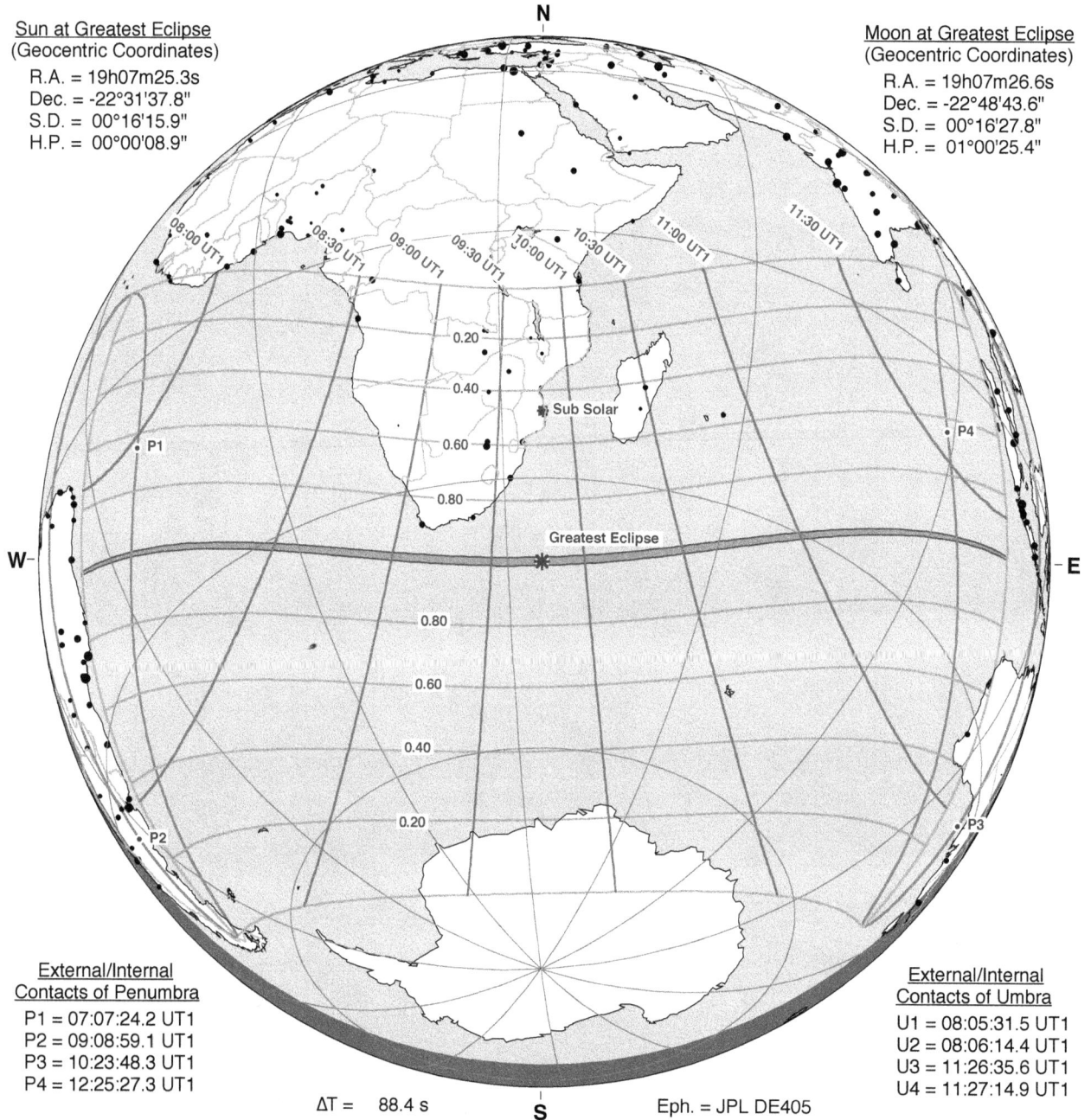

External/Internal Contacts of Penumbra
P1 = 07:07:24.2 UT1
P2 = 09:08:59.1 UT1
P3 = 10:23:48.3 UT1
P4 = 12:25:27.3 UT1

ΔT = 88.4 s

External/Internal Contacts of Umbra
U1 = 08:05:31.5 UT1
U2 = 08:06:14.4 UT1
U3 = 11:26:35.6 UT1
U4 = 11:27:14.9 UT1

Eph. = JPL DE405

Circumstances at Greatest Eclipse: 09:46:23.7 UT1

Lat. = 39°13.0'S	Sun Alt. = 73.3°
Long. = 035°06.8'E	Sun Azm. = 358.9°
Path Width = 101.6 km	Duration = 02m28.7s

Circumstances at Greatest Duration: 09:44:47.7 UT1

Lat. = 39°13.8'S	Sun Alt. = 73.3°
Long. = 034°23.1'E	Sun Azm. = 2.5°
Path Width = 101.6 km	Duration = 02m28.7s

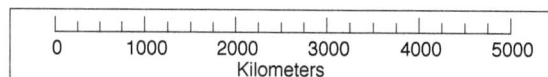

©2016 F. Espenak
www.EclipseWise.com

Annular Solar Eclipse of 2057 Jul 01

Greatest Eclipse = 23:40:15.3 TD (= 23:38:46.6 UT1)

Eclipse Magnitude = 0.9464   Saros Series = 147
Gamma = 0.7455   Saros Member = 25 of 80

N

Sun at Greatest Eclipse
(Geocentric Coordinates)
R.A. = 06h46m13.5s
Dec. = +23°00'23.1"
S.D. = 00°15'43.9"
H.P. = 00°00'08.6"

Moon at Greatest Eclipse
(Geocentric Coordinates)
R.A. = 06h46m10.1s
Dec. = +23°40'36.4"
S.D. = 00°14'44.6"
H.P. = 00°54'06.5"

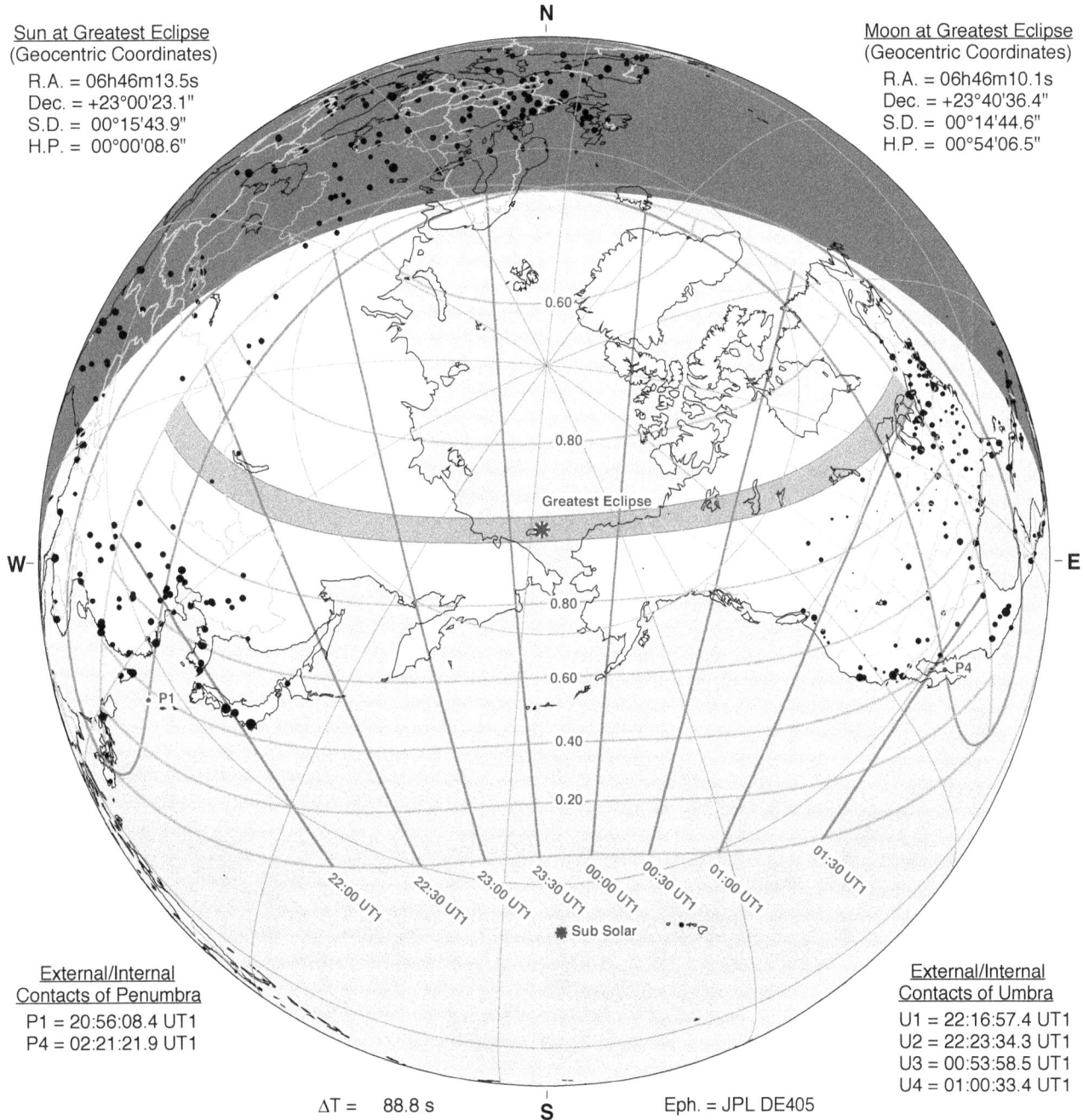

0.60

0.80

Greatest Eclipse

W
E

0.80

0.60

P4

0.40

P1

0.20

01:30 UT1
01:00 UT1
00:30 UT1
00:00 UT1
23:30 UT1
23:00 UT1
22:30 UT1
22:00 UT1

Sub Solar

External/Internal
Contacts of Penumbra
P1 = 20:56:08.4 UT1
P4 = 02:21:21.9 UT1

External/Internal
Contacts of Umbra
U1 = 22:16:57.4 UT1
U2 = 22:23:34.3 UT1
U3 = 00:53:58.5 UT1
U4 = 01:00:33.4 UT1

ΔT = 88.8 s   **S**   Eph. = JPL DE405

Circumstances at Greatest Eclipse: 23:38:46.6 UT1

Lat. = 71°29.6'N	Sun Alt. = 41.5°
Long. = 176°17.0'W	Sun Azm. = 176.8°
Path Width = 298.2 km	Duration = 04m22.5s

Circumstances at Greatest Duration: 23:38:04.2 UT1

Lat. = 71°28.4'N	Sun Alt. = 41.5°
Long. = 177°10.8'W	Sun Azm. = 175.4°
Path Width = 298.2 km	Duration = 04m22.5s

©2016 F. Espenak
www.EclipseWise.com

0   1000   2000   3000   4000   5000
Kilometers

Total Solar Eclipse of 2057 Dec 26

Greatest Eclipse = 01:14:35.2 TD (= 01:13:06.1 UT1)

Eclipse Magnitude = 1.0348	Saros Series = 152
Gamma = -0.9405	Saros Member = 15 of 70

Sun at Greatest Eclipse
(Geocentric Coordinates)
R.A. = 18h20m37.6s
Dec. = -23°20'50.0"
S.D. = 00°16'15.6"
H.P. = 00°00'08.9"

Moon at Greatest Eclipse
(Geocentric Coordinates)
R.A. = 18h20m22.1s
Dec. = -24°18'21.4"
S.D. = 00°16'44.3"
H.P. = 01°01'25.8"

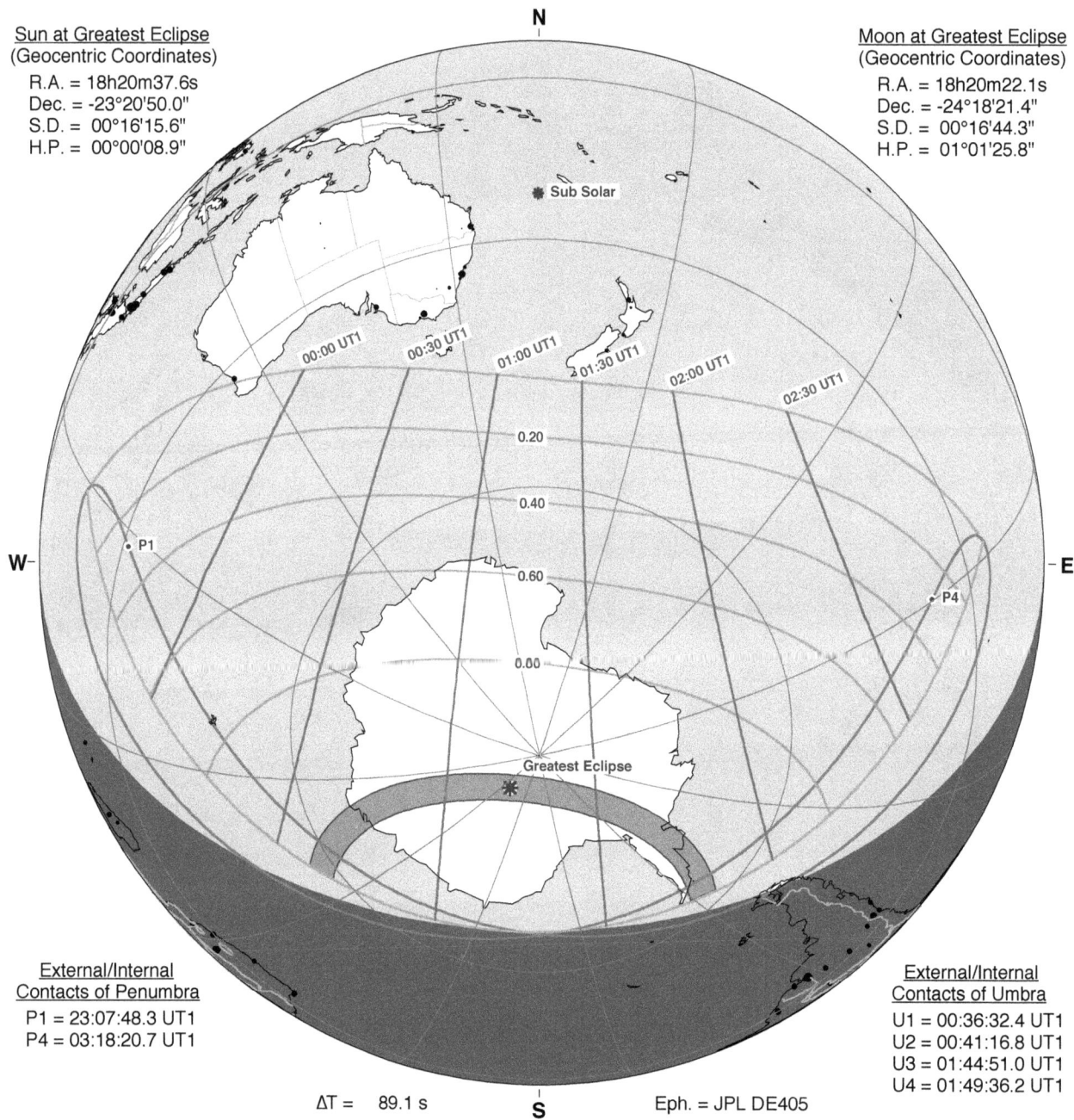

N

Sub Solar

00:00 UT1 00:30 UT1 01:00 UT1 01:30 UT1 02:00 UT1 02:30 UT1

0.20

0.40

W — — E

P1

0.60

P4

0.00

Greatest Eclipse

External/Internal
Contacts of Penumbra
P1 = 23:07:48.3 UT1
P4 = 03:18:20.7 UT1

External/Internal
Contacts of Umbra
U1 = 00:36:32.4 UT1
U2 = 00:41:16.8 UT1
U3 = 01:44:51.0 UT1
U4 = 01:49:36.2 UT1

ΔT = 89.1 s

S

Eph. = JPL DE405

Circumstances at Greatest Eclipse: 01:13:06.1 UT1		Circumstances at Greatest Duration: 01:13:13.9 UT1	
Lat. = 84°51.1'S	Sun Alt. = 19.4°	Lat. = 84°53.7'S	Sun Alt. = 19.4°
Long. = 021°42.3'E	Sun Azm. = 141.4°	Long. = 021°02.3'E	Sun Azm. = 142.0°
Path Width = 354.8 km	Duration = 01m49.9s	Path Width = 354.8 km	Duration = 01m49.9s

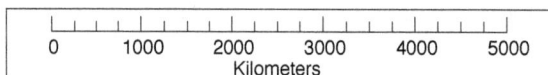

©2016 F. Espenak
www.EclipseWise.com

0 1000 2000 3000 4000 5000
Kilometers

Partial Solar Eclipse of 2058 May 22

Greatest Eclipse = 10:39:25.5 TD   (= 10:37:56.2 UT1)

Eclipse Magnitude = 0.4141   Saros Series = 119
Gamma = -1.3194   Saros Member = 68 of 71

Sun at Greatest Eclipse
(Geocentric Coordinates)
R.A. = 03h58m00.8s
Dec. = +20°28'40.9"
S.D. = 00°15'48.1"
H.P. = 00°00'08.7"

Moon at Greatest Eclipse
(Geocentric Coordinates)
R.A. = 03h59m32.2s
Dec. = +19°18'44.2"
S.D. = 00°15'09.0"
H.P. = 00°55'36.1"

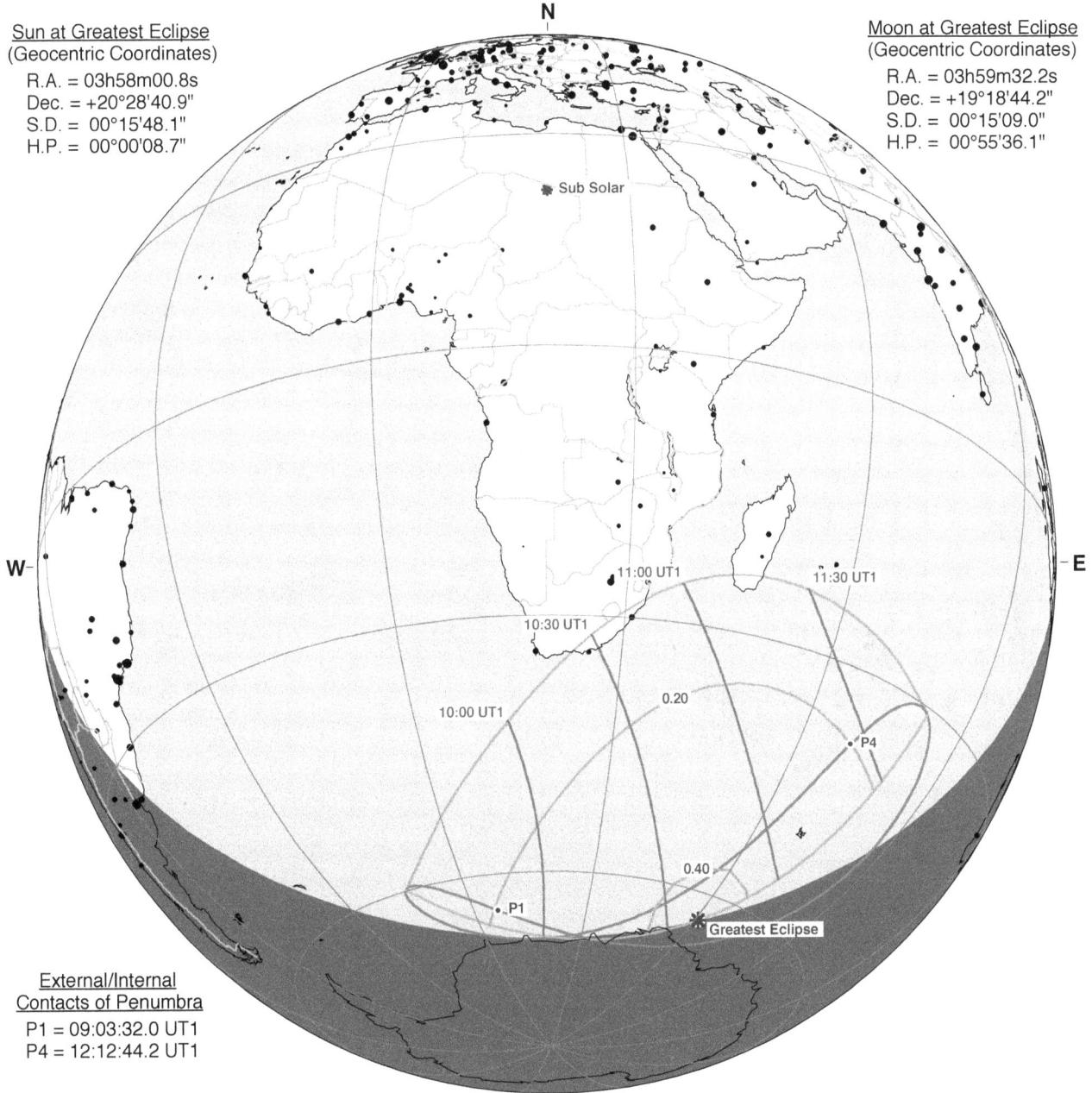

N

Sub Solar

W

E

11:00 UT1
11:30 UT1
10:30 UT1
0.20
10:00 UT1
P4
0.40
P1
Greatest Eclipse

External/Internal
Contacts of Penumbra
P1 = 09:03:32.0 UT1
P4 = 12:12:44.2 UT1

ΔT = 89.3 s   S   Eph. = JPL DE405

Circumstances at Greatest Eclipse: 10:37:56.2 UT1
Lat. = 63°31.8'S   Sun Alt. = 0.0°
Long. = 061°03.2'E   Sun Azm. = 321.7°

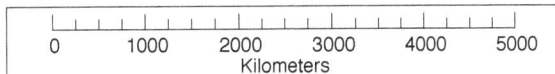

0   1000   2000   3000   4000   5000
Kilometers

©2016 F. Espenak
www.EclipseWise.com

Partial Solar Eclipse of 2058 Jun 21

Greatest Eclipse = 00:19:34.6 TD (= 00:18:05.3 UT1)

Eclipse Magnitude = 0.1261	Saros Series = 157
Gamma = 1.4869	Saros Member = 1 of 70

Sun at Greatest Eclipse
(Geocentric Coordinates)
R.A. = 05h59m41.6s
Dec. = +23°25'56.0"
S.D. = 00°15'44.3"
H.P. = 00°00'08.7"

Moon at Greatest Eclipse
(Geocentric Coordinates)
R.A. = 05h59m06.9s
Dec. = +24°46'21.8"
S.D. = 00°14'50.9"
H.P. = 00°54'29.6"

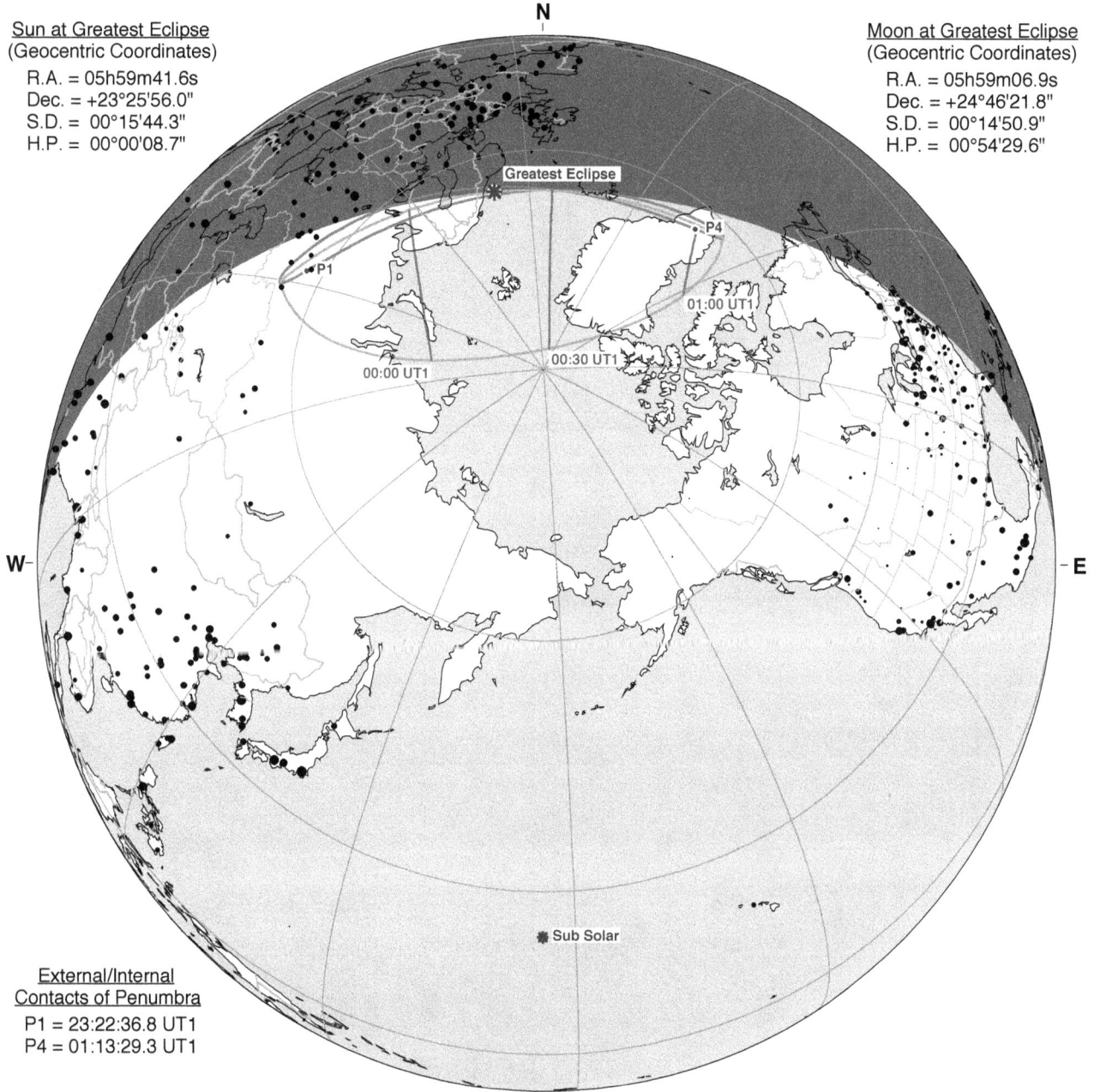

N

Greatest Eclipse

P4

P1

01:00 UT1

00:00 UT1

00:30 UT1

W

E

External/Internal
Contacts of Penumbra
P1 = 23:22:36.8 UT1
P4 = 01:13:29.3 UT1

Sub Solar

ΔT = 89.4 s

S

Eph. = JPL DE405

Circumstances at Greatest Eclipse: 00:18:05.3 UT1

Lat. = 65°57.0'N	Sun Alt. = 0.0°
Long. = 009°47.8'E	Sun Azm. = 12.7°

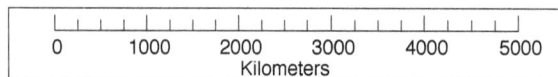

0 1000 2000 3000 4000 5000
Kilometers

©2016 F. Espenak
www.EclipseWise.com

Partial Solar Eclipse of 2058 Nov 16

Greatest Eclipse = 03:23:07.3 TD (= 03:21:37.7 UT1)

Eclipse Magnitude = 0.7644   Saros Series = 124
Gamma = 1.1224   Saros Member = 57 of 73

Sun at Greatest Eclipse
(Geocentric Coordinates)
R.A. = 15h26m32.8s
Dec. = -18°46'09.8"
S.D. = 00°16'10.2"
H.P. = 00°00'08.9"

Moon at Greatest Eclipse
(Geocentric Coordinates)
R.A. = 15h28m05.5s
Dec. = -17°45'10.1"
S.D. = 00°15'47.0"
H.P. = 00°57'55.4"

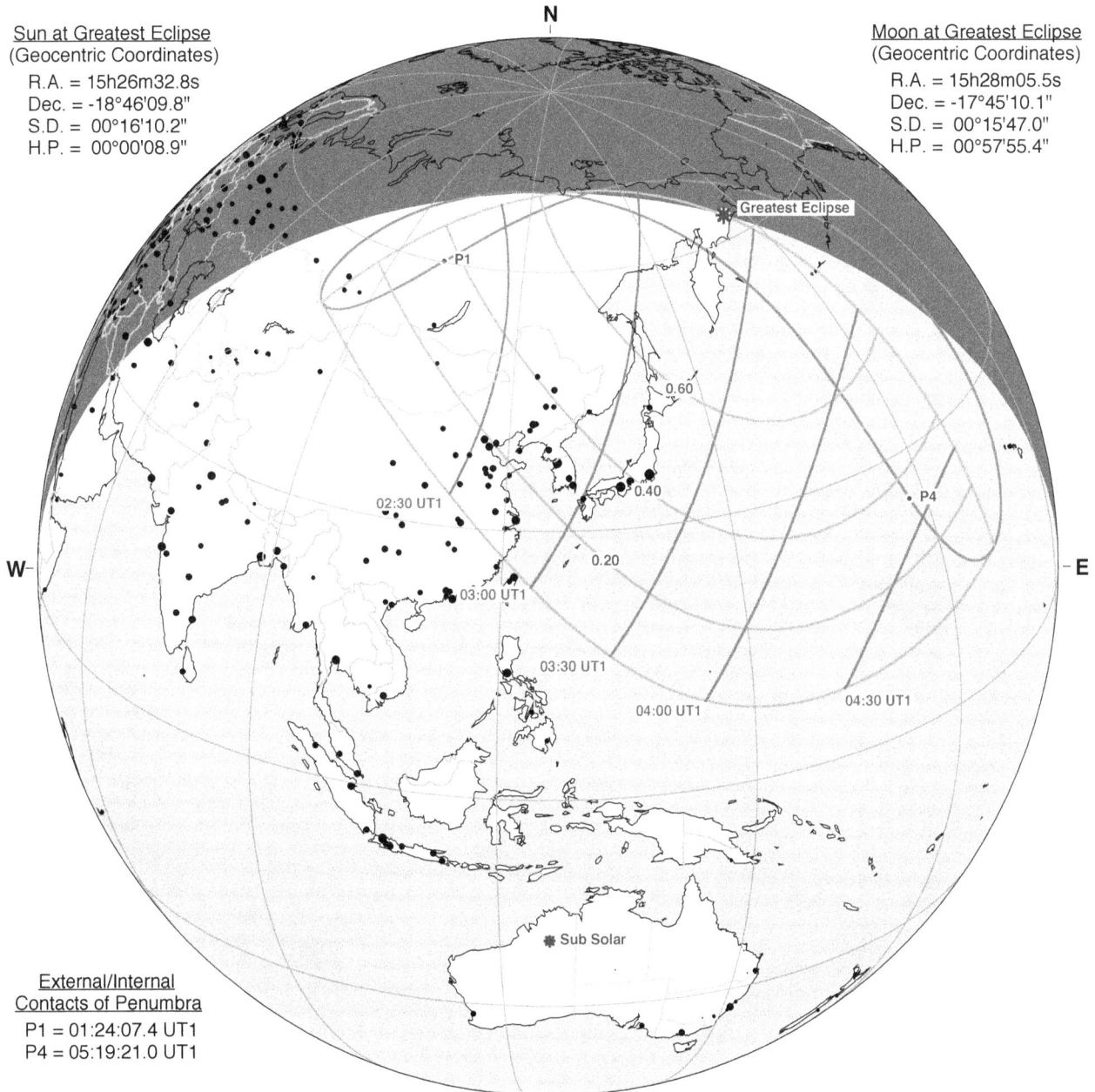

N

Greatest Eclipse

P1

0.60

02:30 UT1

0.40

P4

W

0.20

03:00 UT1

E

03:30 UT1

04:00 UT1   04:30 UT1

Sub Solar

External/Internal
Contacts of Penumbra
P1 = 01:24:07.4 UT1
P4 = 05:19:21.0 UT1

ΔT =   89.7 s   S   Eph. = JPL DE405

Circumstances at Greatest Eclipse: 03:21:37.7 UT1
Lat. = 62°54.0'N   Sun Alt. =   0.0°
Long. = 174°05.5'E   Sun Azm. = 225.1°

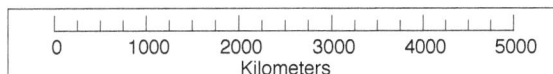

0	1000	2000	3000	4000	5000

Kilometers

©2016 F. Espenak
www.EclipseWise.com

Total Solar Eclipse of 2059 May 11

Greatest Eclipse = 19:22:15.6 TD (= 19:20:45.6 UT1)

Eclipse Magnitude = 1.0242	Saros Series = 129
Gamma = -0.5080	Saros Member = 54 of 80

Sun at Greatest Eclipse
(Geocentric Coordinates)
R.A. = 03h14m47.9s
Dec. = +18°02'08.6"
S.D. = 00°15'50.2"
H.P. = 00°00'08.7"

Moon at Greatest Eclipse
(Geocentric Coordinates)
R.A. = 03h15m32.3s
Dec. = +17°34'20.5"
S.D. = 00°15'59.6"
H.P. = 00°58'41.8"

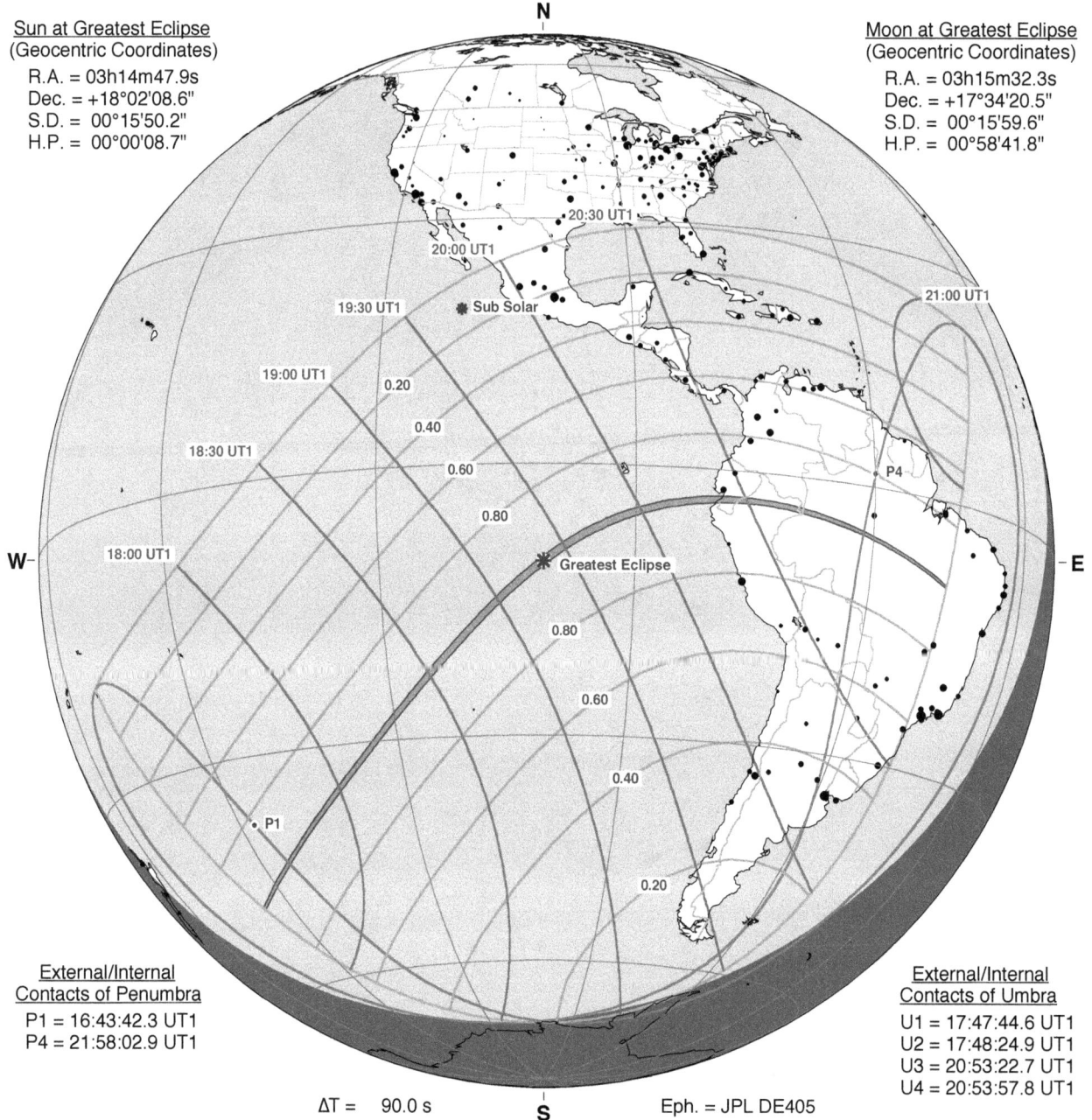

N

20:30 UT1
20:00 UT1
19:30 UT1
Sub Solar
19:00 UT1
0.20
18:30 UT1
0.40
0.60
18:00 UT1
0.80
W
Greatest Eclipse
E
21:00 UT1
P4
0.80
0.60
P1
0.40
0.20
S

External/Internal Contacts of Penumbra
P1 = 16:43:42.3 UT1
P4 = 21:58:02.9 UT1

External/Internal Contacts of Umbra
U1 = 17:47:44.6 UT1
U2 = 17:48:24.9 UT1
U3 = 20:53:22.7 UT1
U4 = 20:53:57.8 UT1

ΔT = 90.0 s   Eph. = JPL DE405

Circumstances at Greatest Eclipse: 19:20:45.6 UT1

Lat. = 10°43.0'S	Sun Alt. = 59.4°
Long. = 100°29.3'W	Sun Azm. = 339.9°
Path Width = 94.6 km	Duration = 02m23.4s

Circumstances at Greatest Duration: 19:22:42.0 UT1

Lat. = 10°21.8'S	Sun Alt. = 59.4°
Long. = 099°59.3'W	Sun Azm. = 338.0°
Path Width = 94.8 km	Duration = 02m23.4s

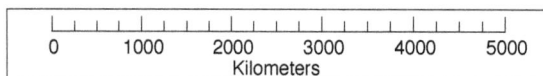

0 1000 2000 3000 4000 5000
Kilometers

©2016 F. Espenak
www.EclipseWise.com

Annular Solar Eclipse of 2059 Nov 05

Greatest Eclipse = 09:18:14.6 TD   (= 09:16:44.3 UT1)

Eclipse Magnitude = 0.9417   Saros Series = 134
Gamma = 0.4454   Saros Member = 46 of 71

Sun at Greatest Eclipse
(Geocentric Coordinates)
R.A. = 14h42m02.6s
Dec. = -15°43'28.3"
S.D. = 00°16'07.6"
H.P. = 00°00'08.9"

Moon at Greatest Eclipse
(Geocentric Coordinates)
R.A. = 14h42m42.7s
Dec. = -15°21'02.7"
S.D. = 00°14'58.8"
H.P. = 00°54'58.7"

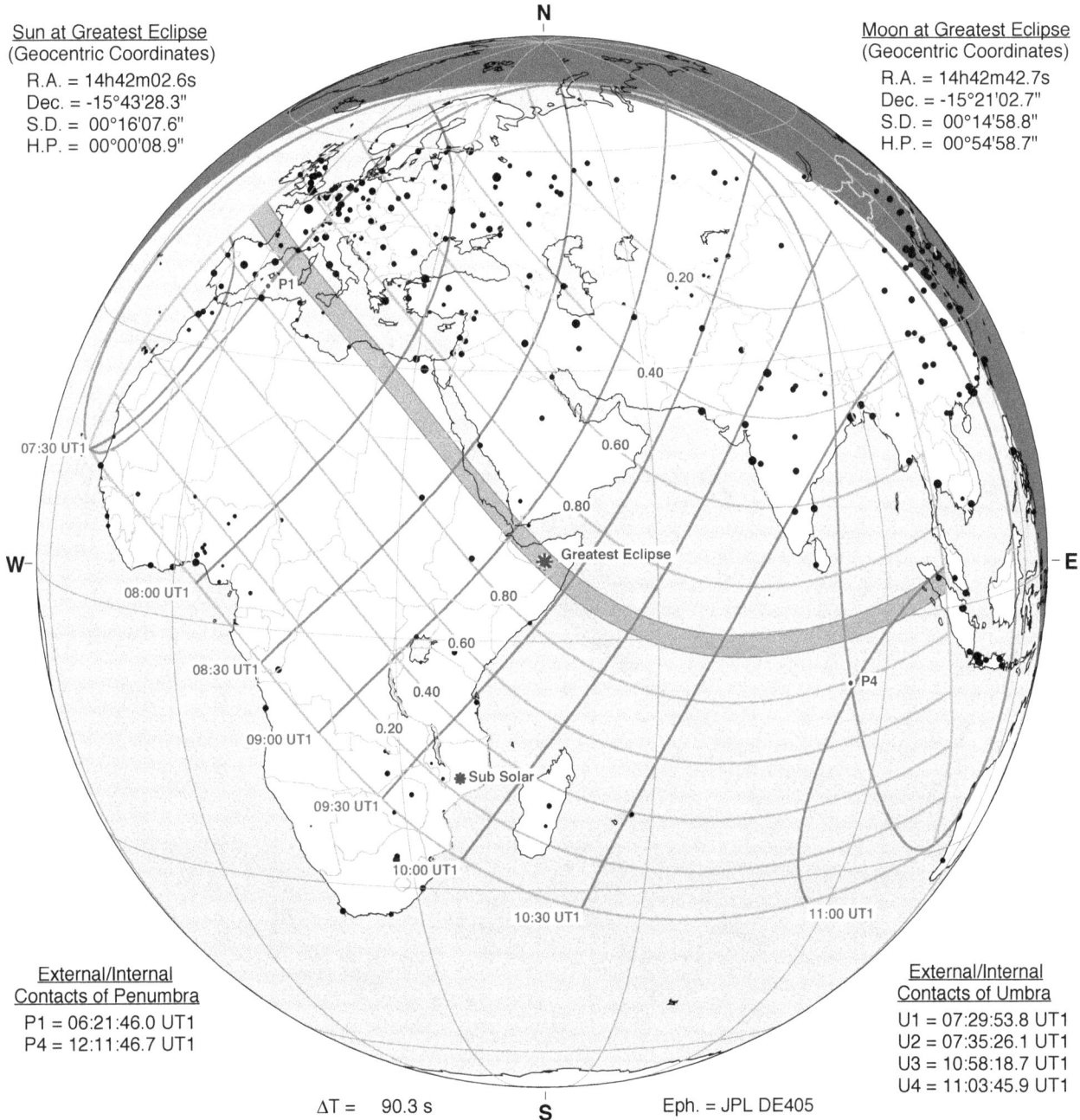

N

07:30 UT1

W

E

0.20

0.40

0.60

0.80

Greatest Eclipse

0.80

0.60

0.40

0.20

P1

08:00 UT1

08:30 UT1

09:00 UT1

09:30 UT1

10:00 UT1

10:30 UT1

11:00 UT1

P4

Sub Solar

External/Internal
Contacts of Penumbra
P1 = 06:21:46.0 UT1
P4 = 12:11:46.7 UT1

External/Internal
Contacts of Umbra
U1 = 07:29:53.8 UT1
U2 = 07:35:26.1 UT1
U3 = 10:58:18.7 UT1
U4 = 11:03:45.9 UT1

ΔT =   90.3 s   S   Eph. = JPL DE405

Circumstances at Greatest Eclipse: 09:16:44.3 UT1

Lat. = 08°44.6'N	Sun Alt. = 63.5°
Long. = 046°59.3'E	Sun Azm. = 202.7°
Path Width = 238.3 km	Duration = 06m59.6s

Circumstances at Greatest Duration: 09:31:43.2 UT1

Lat. = 06°02.0'N	Sun Alt. = 62.3°
Long. = 050°11.1'E	Sun Azm. = 217.9°
Path Width = 243.9 km	Duration = 07m01.5s

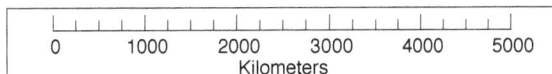

©2016 F. Espenak
www.EclipseWise.com

0   1000   2000   3000   4000   5000
Kilometers

Total Solar Eclipse of 2060 Apr 30

Greatest Eclipse = 10:09:59.8 TD   (= 10:08:29.2 UT1)

Eclipse Magnitude = 1.0660	Saros Series = 139
Gamma = 0.2422	Saros Member = 32 of 71

Sun at Greatest Eclipse
(Geocentric Coordinates)
R.A. = 02h33m38.4s
Dec. = +15°04'16.7"
S.D. = 00°15'52.6"
H.P. = 00°00'08.7"

Moon at Greatest Eclipse
(Geocentric Coordinates)
R.A. = 02h33m13.6s
Dec. = +15°17'46.8"
S.D. = 00°16'38.8"
H.P. = 01°01'05.8"

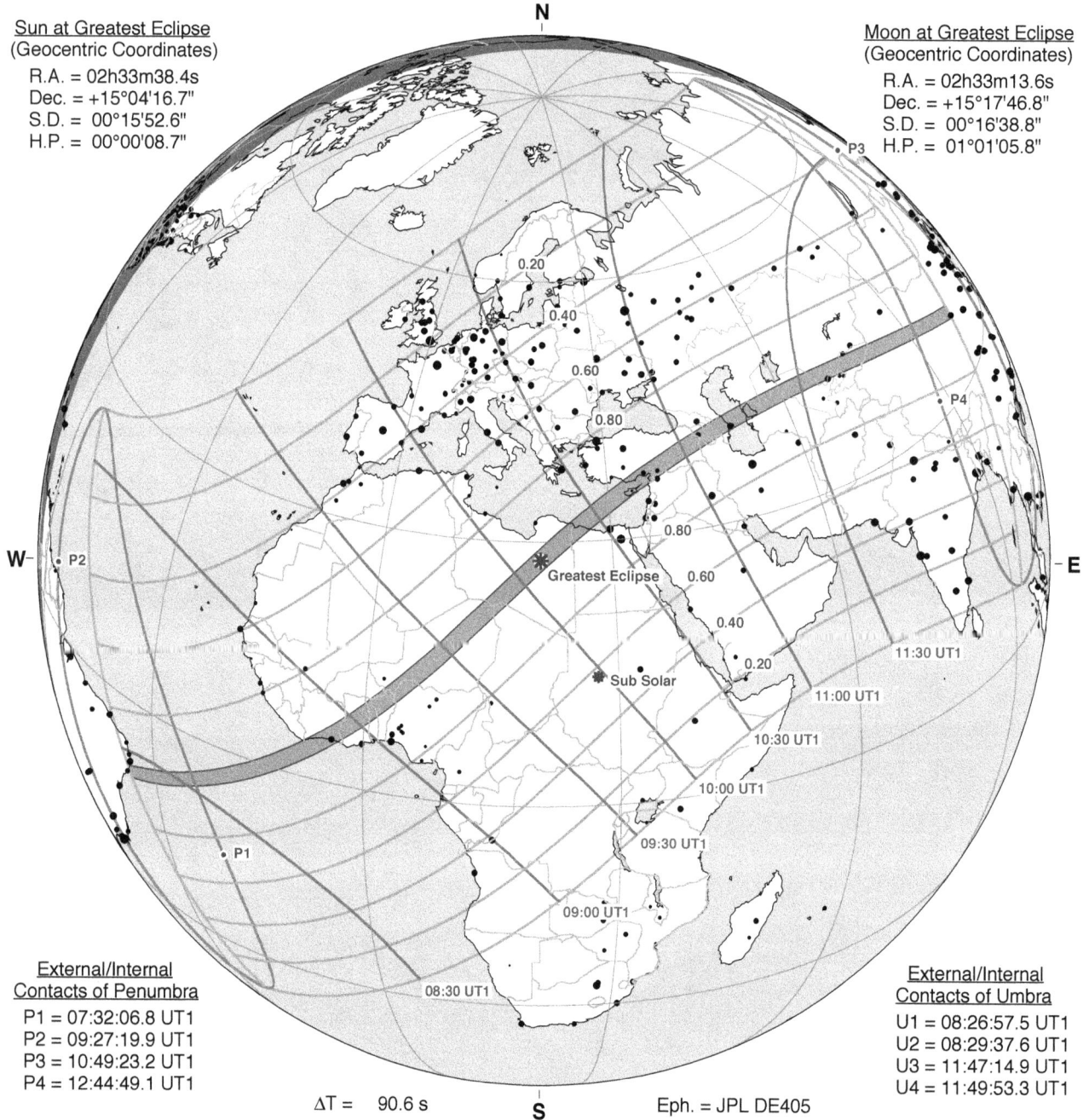

External/Internal Contacts of Penumbra
P1 = 07:32:06.8 UT1
P2 = 09:27:19.9 UT1
P3 = 10:49:23.2 UT1
P4 = 12:44:49.1 UT1

External/Internal Contacts of Umbra
U1 = 08:26:57.5 UT1
U2 = 08:29:37.6 UT1
U3 = 11:47:14.9 UT1
U4 = 11:49:53.3 UT1

ΔT =   90.6 s

Eph. = JPL DE405

Circumstances at Greatest Eclipse: 10:08:29.2 UT1

Lat. = 27°57.4'N	Sun Alt. = 75.8°
Long. = 020°49.0'E	Sun Azm. = 154.1°
Path Width = 222.0 km	Duration = 05m15.2s

Circumstances at Greatest Duration: 10:13:24.7 UT1

Lat. = 29°09.3'N	Sun Alt. = 75.5°
Long. = 022°23.1'E	Sun Azm. = 166.1°
Path Width = 221.3 km	Duration = 05m15.5s

0   1000   2000   3000   4000   5000
Kilometers

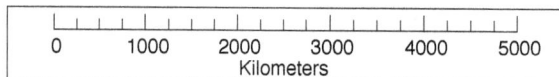

©2016 F. Espenak
www.EclipseWise.com

Annular Solar Eclipse of 2060 Oct 24

Greatest Eclipse = 09:24:10.4 TD   (= 09:22:39.4 UT1)

Eclipse Magnitude = 0.9277   Saros Series = 144

Gamma = -0.2625   Saros Member = 19 of 70

Sun at Greatest Eclipse
(Geocentric Coordinates)
R.A. = 13h58m17.5s
Dec. = -12°04'28.2"
S.D. = 00°16'04.8"
H.P. = 00°00'08.8"

Moon at Greatest Eclipse
(Geocentric Coordinates)
R.A. = 13h57m52.2s
Dec. = -12°17'09.7"
S.D. = 00°14'42.1"
H.P. = 00°53'57.3"

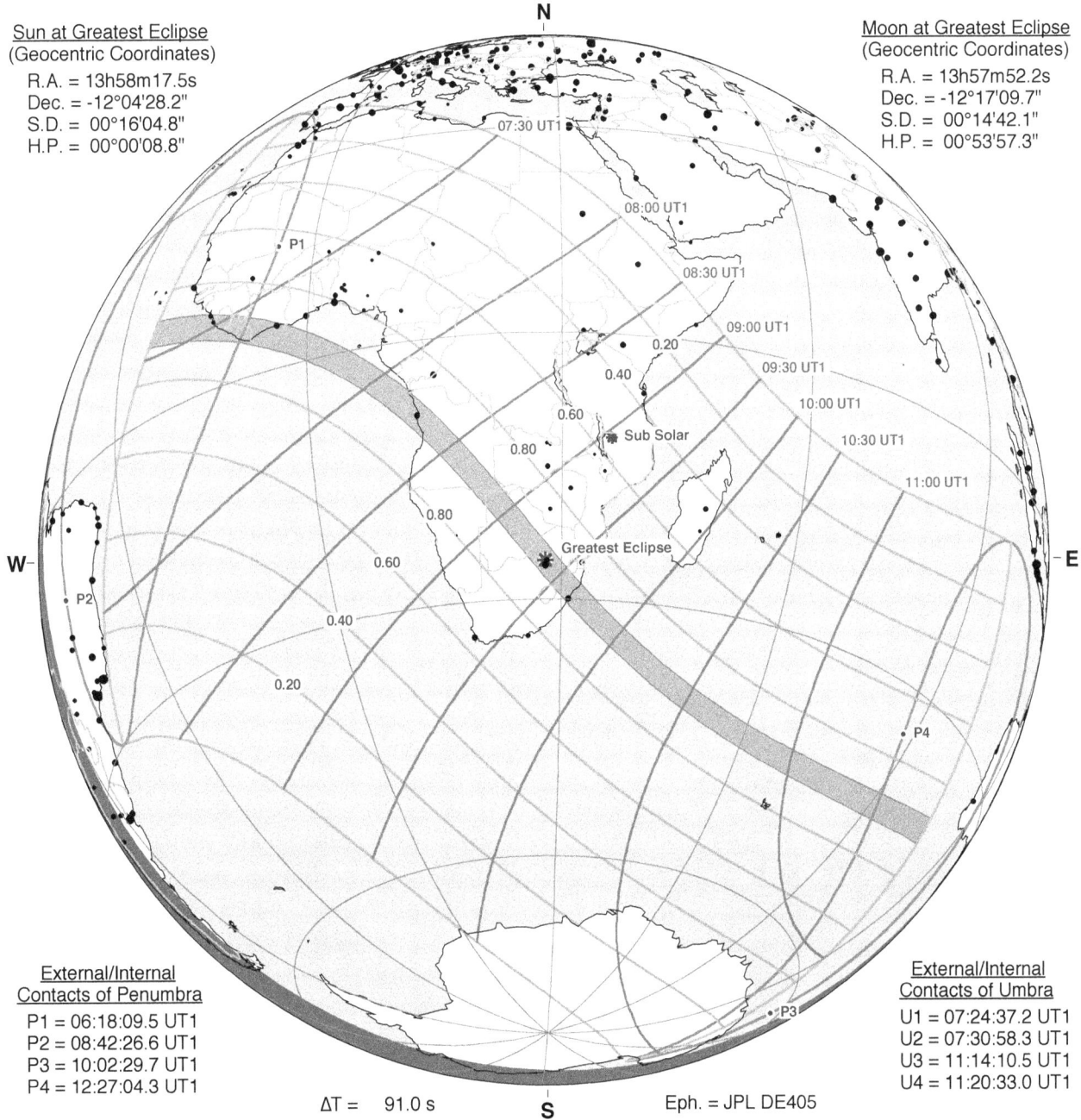

External/Internal
Contacts of Penumbra
P1 = 06:18:09.5 UT1
P2 = 08:42:26.6 UT1
P3 = 10:02:29.7 UT1
P4 = 12:27:04.3 UT1

External/Internal
Contacts of Umbra
U1 = 07:24:37.2 UT1
U2 = 07:30:58.3 UT1
U3 = 11:14:10.5 UT1
U4 = 11:20:33.0 UT1

ΔT = 91.0 s   Eph. = JPL DE405

Circumstances at Greatest Eclipse: 09:22:39.4 UT1
Lat. = 25°47.2'S   Sun Alt. = 74.6°
Long. = 028°01.1'E   Sun Azm. = 28.1°
Path Width = 280.5 km   Duration = 08m05.8s

Circumstances at Greatest Duration: 09:39:59.5 UT1
Lat. = 29°42.0'S   Sun Alt. = 72.4°
Long. = 032°03.8'E   Sun Azm. = 356.6°
Path Width = 278.5 km   Duration = 08m08.0s

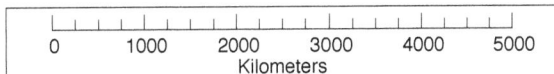

©2016 F. Espenak
www.EclipseWise.com

Total Solar Eclipse of 2061 Apr 20

Greatest Eclipse = 02:56:49.1 TD (= 02:55:17.8 UT1)

Eclipse Magnitude = 1.0476   Saros Series = 149
Gamma = 0.9578   Saros Member = 23 of 71

Sun at Greatest Eclipse
(Geocentric Coordinates)
R.A. = 01h53m47.8s
Dec. = +11°39'59.8"
S.D. = 00°15'55.3"
H.P. = 00°00'08.8"

Moon at Greatest Eclipse
(Geocentric Coordinates)
R.A. = 01h52m03.2s
Dec. = +12°32'19.1"
S.D. = 00°16'36.4"
H.P. = 01°00'56.9"

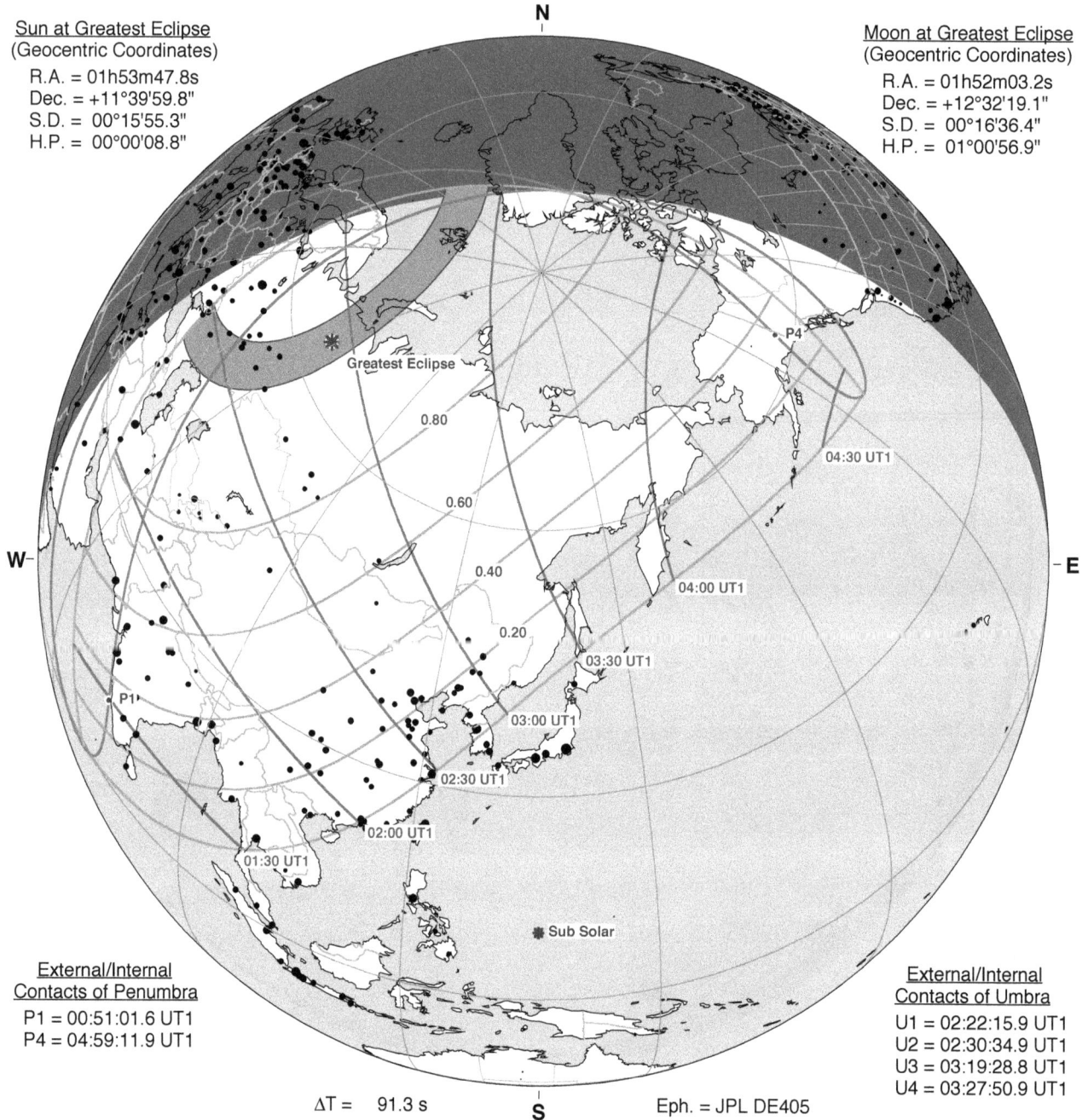

N

Greatest Eclipse

0.80

0.60

0.40

0.20

P4

04:30 UT1

04:00 UT1

03:30 UT1

03:00 UT1

02:30 UT1

02:00 UT1

01:30 UT1

P1

W —

— E

Sub Solar

External/Internal
Contacts of Penumbra
P1 = 00:51:01.6 UT1
P4 = 04:59:11.9 UT1

External/Internal
Contacts of Umbra
U1 = 02:22:15.9 UT1
U2 = 02:30:34.9 UT1
U3 = 03:19:28.8 UT1
U4 = 03:27:50.9 UT1

ΔT = 91.3 s

S

Eph. = JPL DE405

Circumstances at Greatest Eclipse: 02:55:17.8 UT1

Lat. = 64°32.2'N	Sun Alt. = 16.2°
Long. = 059°03.3'E	Sun Azm. = 96.8°
Path Width = 558.9 km	Duration = 02m37.4s

Circumstances at Greatest Duration: 02:55:31.2 UT1

Lat. = 64°39.5'N	Sun Alt. = 16.2°
Long. = 059°01.8'E	Sun Azm. = 96.9°
Path Width = 558.2 km	Duration = 02m37.4s

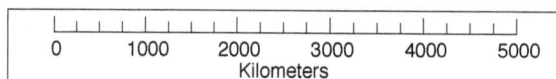

©2016 F. Espenak
www.EclipseWise.com

Annular Solar Eclipse of 2061 Oct 13

Greatest Eclipse = 10:32:09.7 TD (= 10:30:38.0 UT1)

Eclipse Magnitude = 0.9469   Saros Series = 154

Gamma = -0.9639   Saros Member = 9 of 71

Sun at Greatest Eclipse
(Geocentric Coordinates)
R.A. = 13h16m11.1s
Dec. = -08°03'03.6"
S.D. = 00°16'01.7"
H.P. = 00°00'08.8"

Moon at Greatest Eclipse
(Geocentric Coordinates)
R.A. = 13h14m30.5s
Dec. = -08°50'16.1"
S.D. = 00°15'07.5"
H.P. = 00°55'30.4"

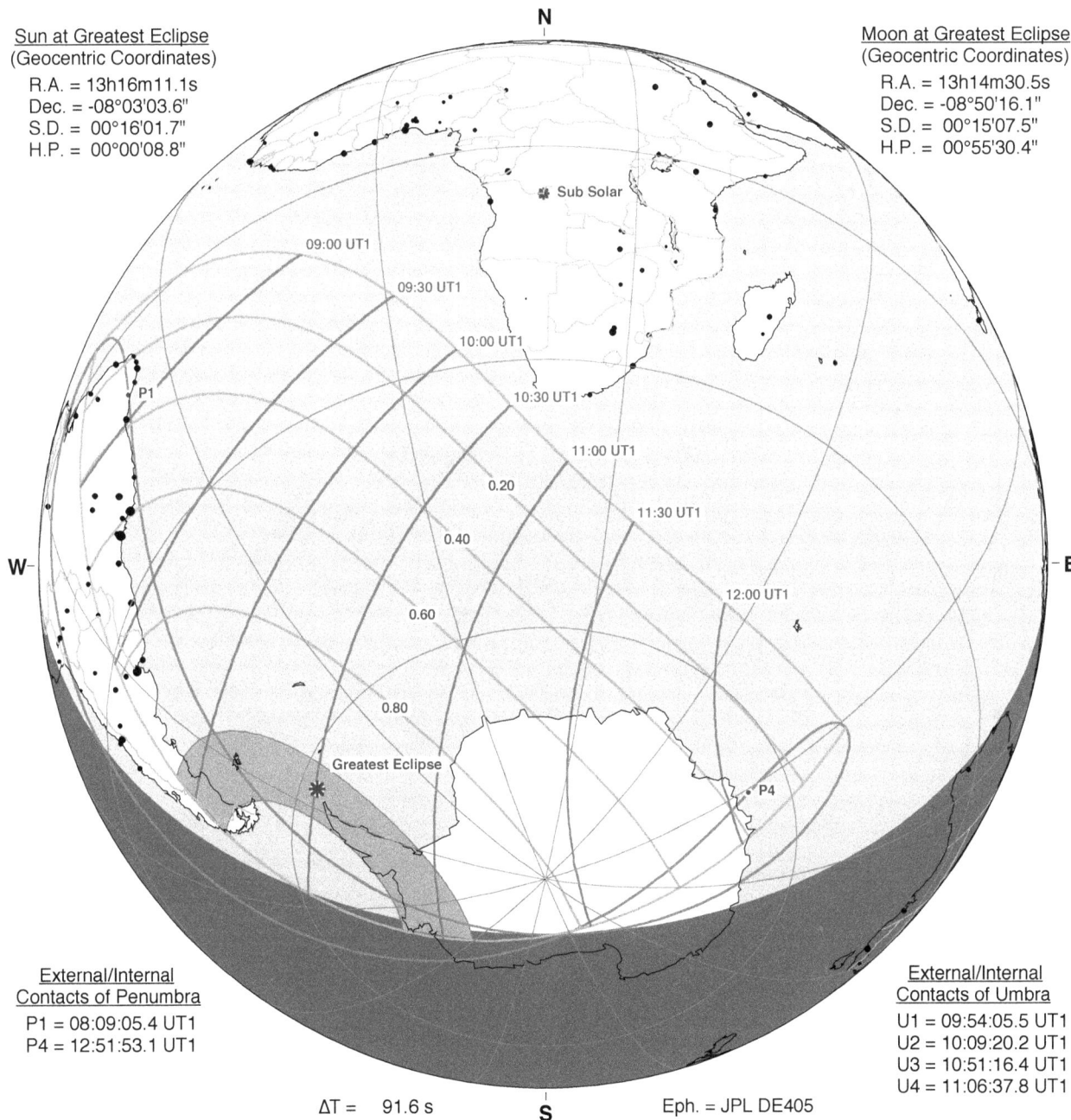

N

Sub Solar

09:00 UT1
09:30 UT1
10:00 UT1
10:30 UT1
11:00 UT1
11:30 UT1
12:00 UT1

P1

0.20
0.40
0.60
0.80

W

E

Greatest Eclipse

P4

External/Internal Contacts of Penumbra
P1 = 08:09:05.4 UT1
P4 = 12:51:53.1 UT1

External/Internal Contacts of Umbra
U1 = 09:54:05.5 UT1
U2 = 10:09:20.2 UT1
U3 = 10:51:16.4 UT1
U4 = 11:06:37.8 UT1

ΔT = 91.6 s

S

Eph. = JPL DE405

Circumstances at Greatest Eclipse: 10:30:38.0 UT1
Lat. = 62°08.2'S   Sun Alt. = 14.9°
Long. = 054°29.2'W   Sun Azm. = 79.0°
Path Width = 742.6 km   Duration = 03m41.0s

Circumstances at Greatest Duration: 10:33:54.1 UT1
Lat. = 63°45.1'S   Sun Alt. = 14.7°
Long. = 054°50.1'W   Sun Azm. = 78.1°
Path Width = 728.7 km   Duration = 03m41.0s

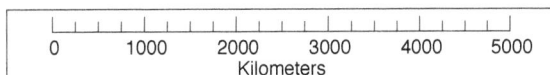

0   1000   2000   3000   4000   5000
Kilometers

©2016 F. Espenak
www.EclipseWise.com

157

Partial Solar Eclipse of 2062 Mar 11

Greatest Eclipse = 04:26:16.2 TD   (= 04:24:44.3 UT1)

Eclipse Magnitude = 0.9331   Saros Series = 121
Gamma = -1.0238   Saros Member = 63 of 71

Sun at Greatest Eclipse
(Geocentric Coordinates)
R.A. = 23h26m28.0s
Dec. = -03°36'57.3"
S.D. = 00°16'06.2"
H.P. = 00°00'08.9"

Moon at Greatest Eclipse
(Geocentric Coordinates)
R.A. = 23h28m20.0s
Dec. = -04°27'39.9"
S.D. = 00°15'26.8"
H.P. = 00°56'41.5"

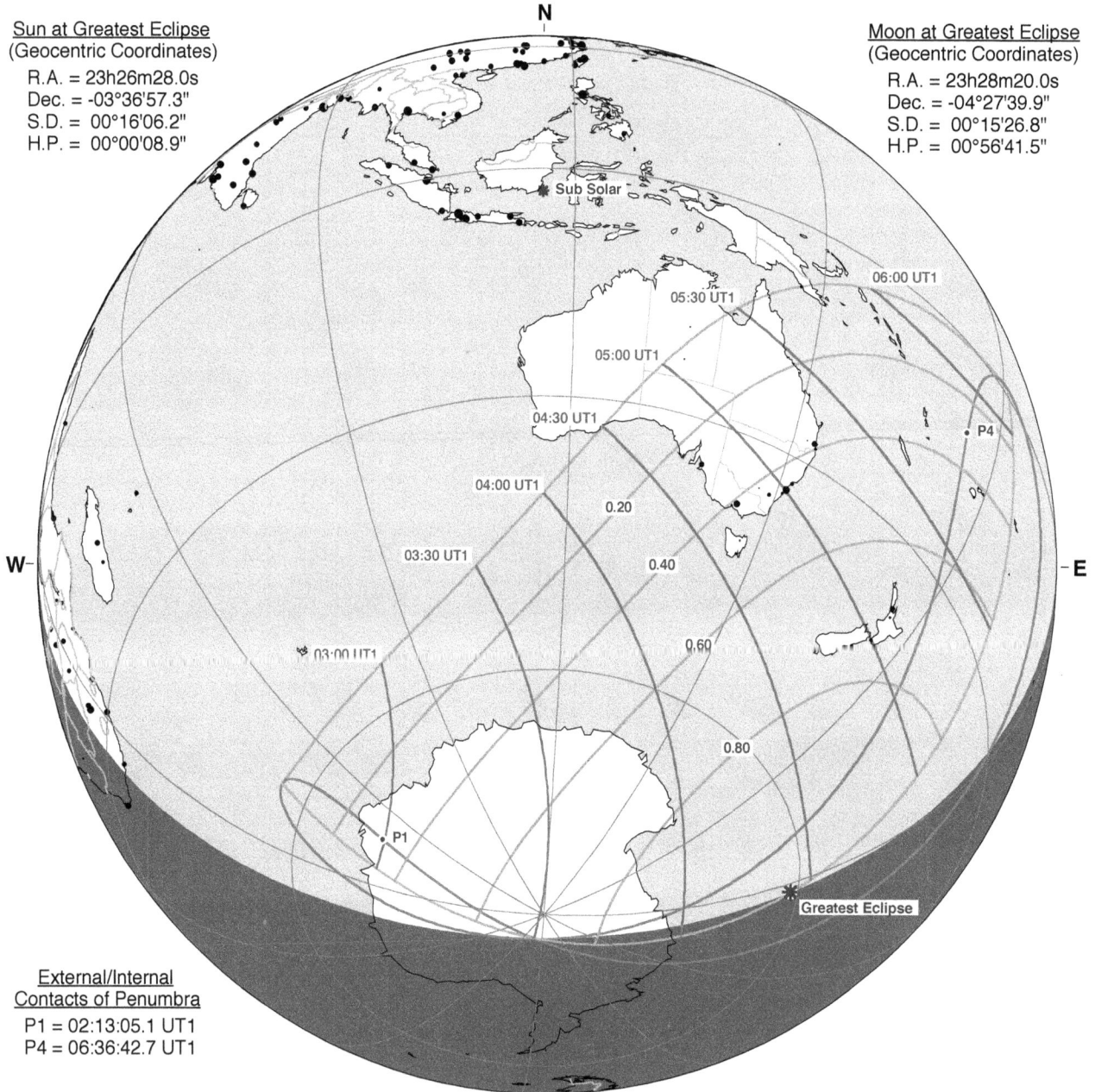

N

Sub Solar

06:00 UT1
05:30 UT1
05:00 UT1
04:30 UT1
P4
04:00 UT1
0.20
03:30 UT1
0.40
W
E
03:00 UT1
0.60
0.80
P1
Greatest Eclipse

**External/Internal
Contacts of Penumbra**
P1 = 02:13:05.1 UT1
P4 = 06:36:42.7 UT1

ΔT =   91.9 s   S   Eph. = JPL DE405

Circumstances at Greatest Eclipse: 04:24:44.3 UT1
Lat. = 60°58.3'S   Sun Alt. =   0.0°
Long. = 147°14.6'W   Sun Azm. = 262.5°

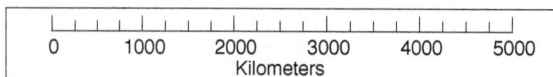

0   1000   2000   3000   4000   5000
Kilometers

Partial Solar Eclipse of 2062 Sep 03

Greatest Eclipse = 08:54:27.4 TD   (= 08:52:55.1 UT1)

Eclipse Magnitude = 0.9749   Saros Series = 126
Gamma = 1.0192   Saros Member = 50 of 72

Sun at Greatest Eclipse
(Geocentric Coordinates)
R.A. = 10h50m30.3s
Dec. = +07°22'28.5"
S.D. = 00°15'51.2"
H.P. = 00°00'08.7"

Moon at Greatest Eclipse
(Geocentric Coordinates)
R.A. = 10h52m25.5s
Dec. = +08°16'29.0"
S.D. = 00°16'22.2"
H.P. = 01°00'04.6"

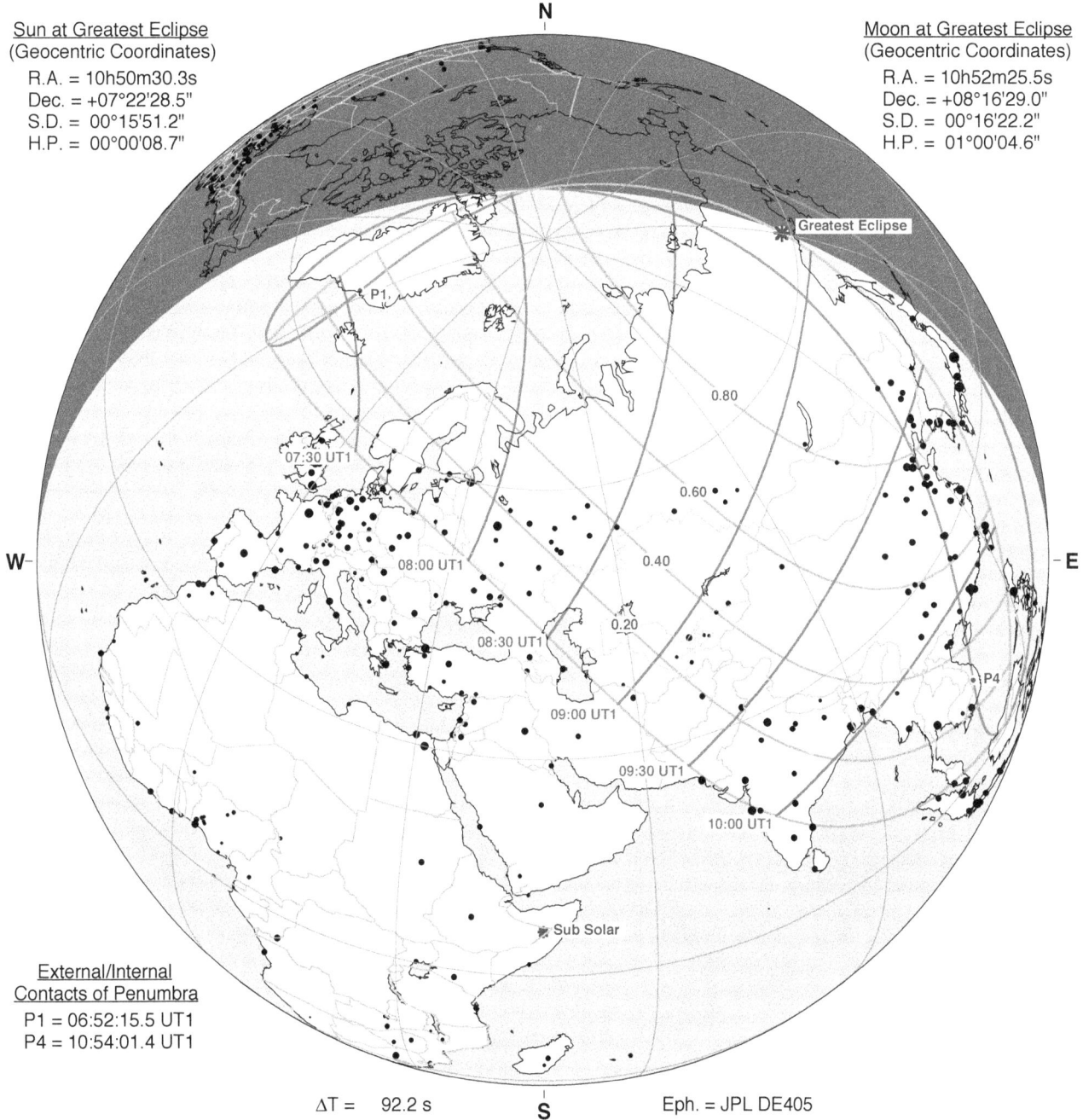

N

P1

07:30 UT1

0.80

0.60

08:00 UT1

0.40

W

E

0.20

08:30 UT1

09:00 UT1

P4

09:30 UT1

10:00 UT1

Greatest Eclipse

Sub Solar

External/Internal
Contacts of Penumbra
P1 = 06:52:15.5 UT1
P4 = 10:54:01.4 UT1

ΔT = 92.2 s

S

Eph. = JPL DE405

Circumstances at Greatest Eclipse: 08:52:55.1 UT1
Lat. = 61°18.3'N   Sun Alt. =   0.0°
Long. = 150°12.6'E   Sun Azm. = 285.5°

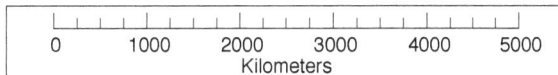

0   1000   2000   3000   4000   5000
Kilometers

©2016 F. Espenak
www.EclipseWise.com

159

Annular Solar Eclipse of 2063 Feb 28

Greatest Eclipse = 07:43:30.0 TD (= 07:41:57.4 UT1)

Eclipse Magnitude = 0.9293   Saros Series = 131

Gamma = -0.3360   Saros Member = 53 of 70

Sun at Greatest Eclipse
(Geocentric Coordinates)
R.A. = 22h45m11.8s
Dec. = -07°54'42.4"
S.D. = 00°16'08.9"
H.P. = 00°00'08.9"

Moon at Greatest Eclipse
(Geocentric Coordinates)
R.A. = 22h45m46.2s
Dec. = -08°10'47.1"
S.D. = 00°14'47.6"
H.P. = 00°54'17.7"

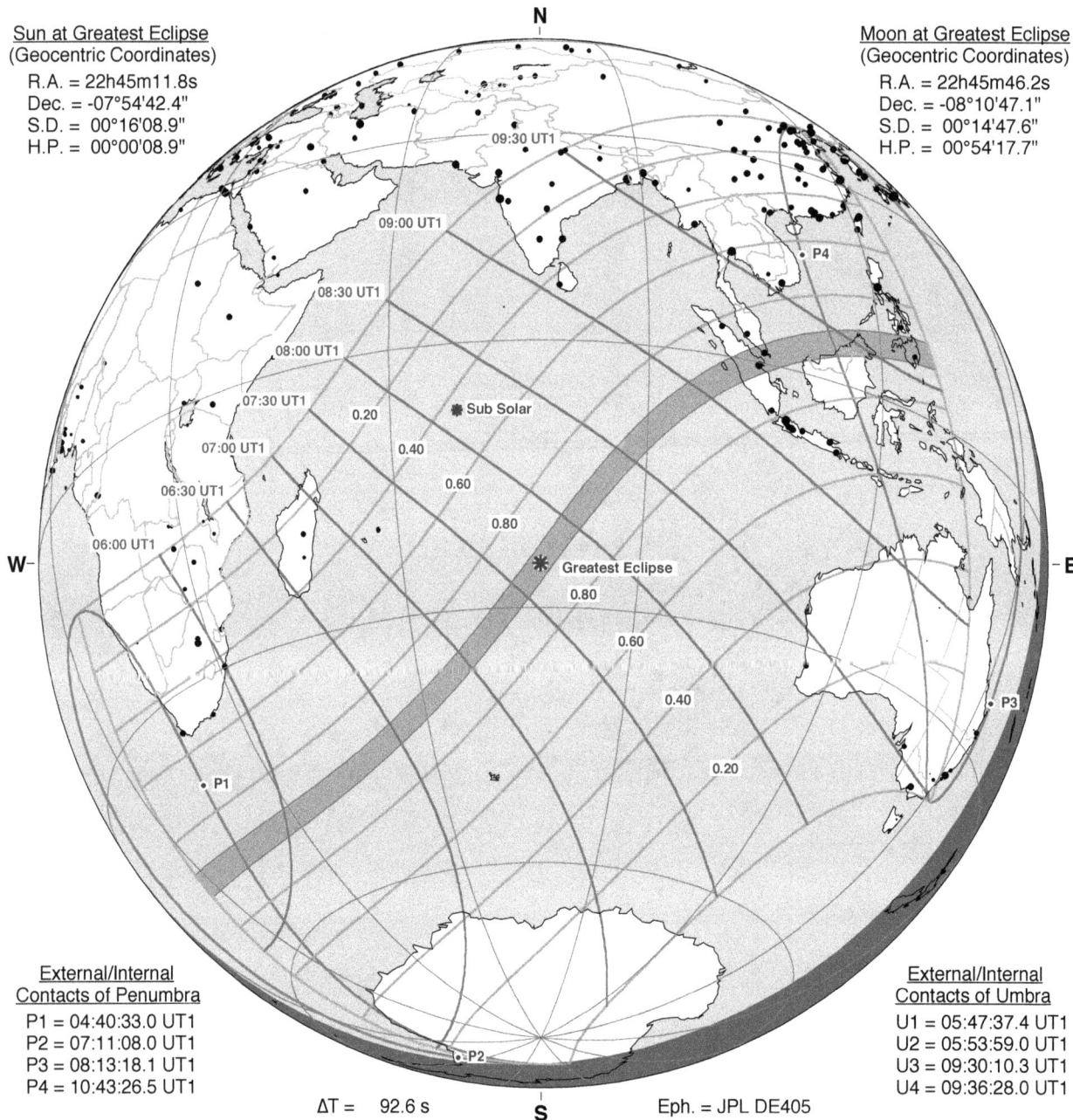

External/Internal Contacts of Penumbra
P1 = 04:40:33.0 UT1
P2 = 07:11:08.0 UT1
P3 = 08:13:18.1 UT1
P4 = 10:43:26.5 UT1

External/Internal Contacts of Umbra
U1 = 05:47:37.4 UT1
U2 = 05:53:59.0 UT1
U3 = 09:30:10.3 UT1
U4 = 09:36:28.0 UT1

ΔT = 92.6 s

Eph. = JPL DE405

Circumstances at Greatest Eclipse: 07:41:57.4 UT1

Lat. = 25°13.9'S	Sun Alt. = 70.2°
Long. = 077°37.8'E	Sun Azm. = 329.4°
Path Width = 279.6 km	Duration = 07m41.2s

Circumstances at Greatest Duration: 07:27:17.6 UT1

Lat. = 28°52.6'S	Sun Alt. = 68.8°
Long. = 074°24.5'E	Sun Azm. = 351.4°
Path Width = 276.8 km	Duration = 07m42.5s

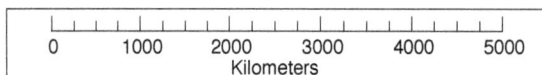

©2016 F. Espenak
www.EclipseWise.com

Total Solar Eclipse of 2063 Aug 24

Greatest Eclipse = 01:22:10.6 TD   (= 01:20:37.6 UT1)

Eclipse Magnitude = 1.0750   Saros Series = 136
Gamma = 0.2771   Saros Member = 40 of 71

Sun at Greatest Eclipse
(Geocentric Coordinates)
R.A. = 10h12m03.7s
Dec. = +11°07'34.9"
S.D. = 00°15'48.9"
H.P. = 00°00'08.7"

Moon at Greatest Eclipse
(Geocentric Coordinates)
R.A. = 10h12m34.5s
Dec. = +11°22'46.8"
S.D. = 00°16'43.4"
H.P. = 01°01'22.6"

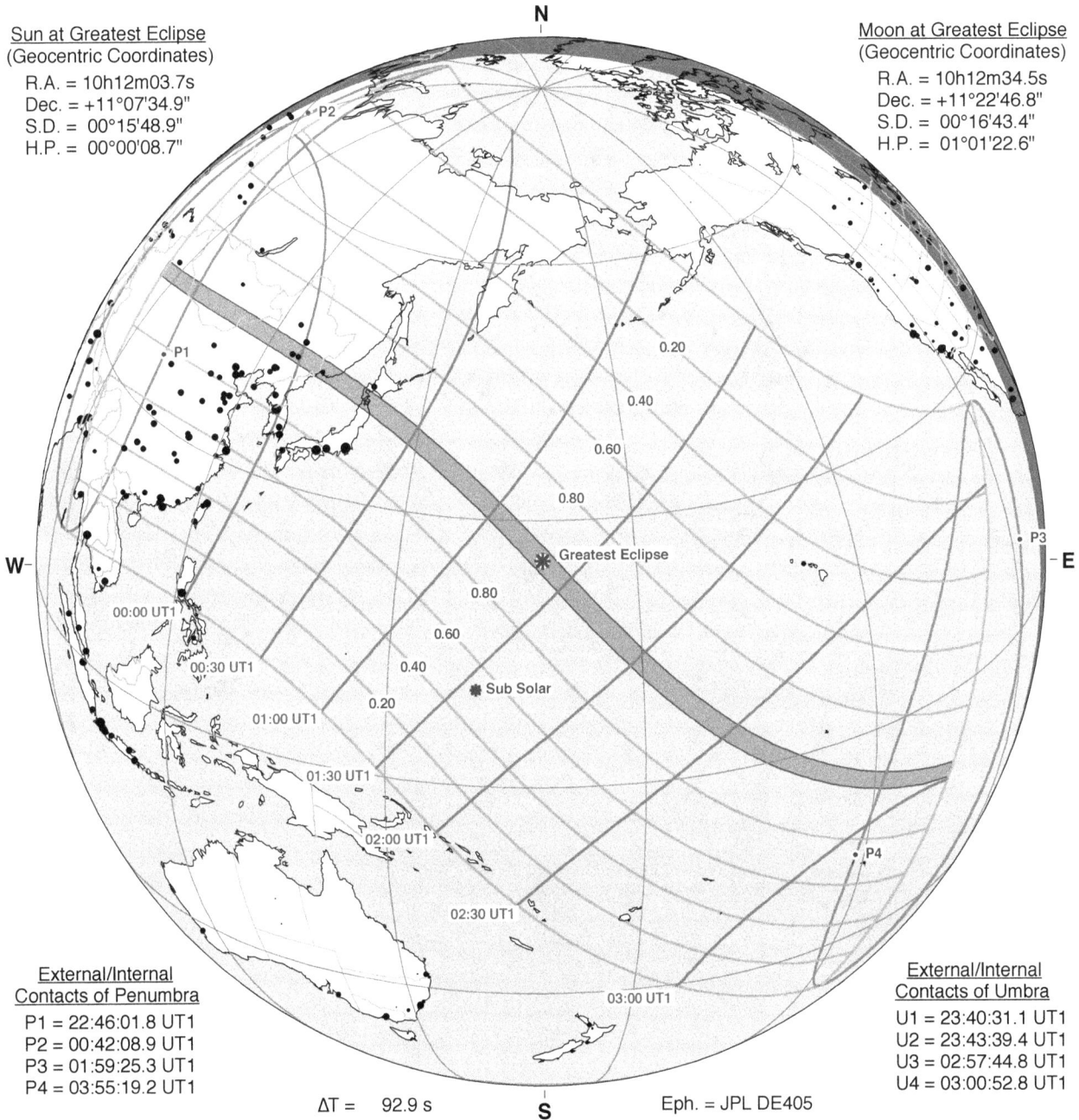

N

P2

P1

0.20
0.40
0.60
0.80

W

Greatest Eclipse

0.80
0.60
0.40
0.20

Sub Solar

P3
E

00:00 UT1
00:30 UT1
01:00 UT1
01:30 UT1
02:00 UT1
02:30 UT1
03:00 UT1

P4

External/Internal
Contacts of Penumbra
P1 = 22:46:01.8 UT1
P2 = 00:42:08.9 UT1
P3 = 01:59:25.3 UT1
P4 = 03:55:19.2 UT1

ΔT = 92.9 s

S

Eph. = JPL DE405

External/Internal
Contacts of Umbra
U1 = 23:40:31.1 UT1
U2 = 23:43:39.4 UT1
U3 = 02:57:44.8 UT1
U4 = 03:00:52.8 UT1

Circumstances at Greatest Eclipse: 01:20:37.6 UT1
Lat. = 25°33.7'N   Sun Alt. = 73.8°
Long. = 168°19.2'E   Sun Azm. = 208.6°
Path Width = 252.2 km   Duration = 05m49.1s

Circumstances at Greatest Duration: 01:15:57.6 UT1
Lat. = 26°49.2'N   Sun Alt. = 73.5°
Long. = 166°56.6'E   Sun Azm. = 198.7°
Path Width = 251.3 km   Duration = 05m49.3s

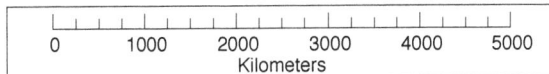

0   1000   2000   3000   4000   5000
Kilometers

Annular Solar Eclipse of 2064 Feb 17

Greatest Eclipse = 07:00:23.3 TD   (= 06:58:50.1 UT1)

Eclipse Magnitude = 0.9262   Saros Series = 141
Gamma = 0.3597   Saros Member = 26 of 70

Sun at Greatest Eclipse
(Geocentric Coordinates)
R.A. = 22h02m13.8s
Dec. = -12°01'37.5"
S.D. = 00°16'11.3"
H.P. = 00°00'08.9"

Moon at Greatest Eclipse
(Geocentric Coordinates)
R.A. = 22h01m38.9s
Dec. = -11°44'08.3"
S.D. = 00°14'47.1"
H.P. = 00°54'15.6"

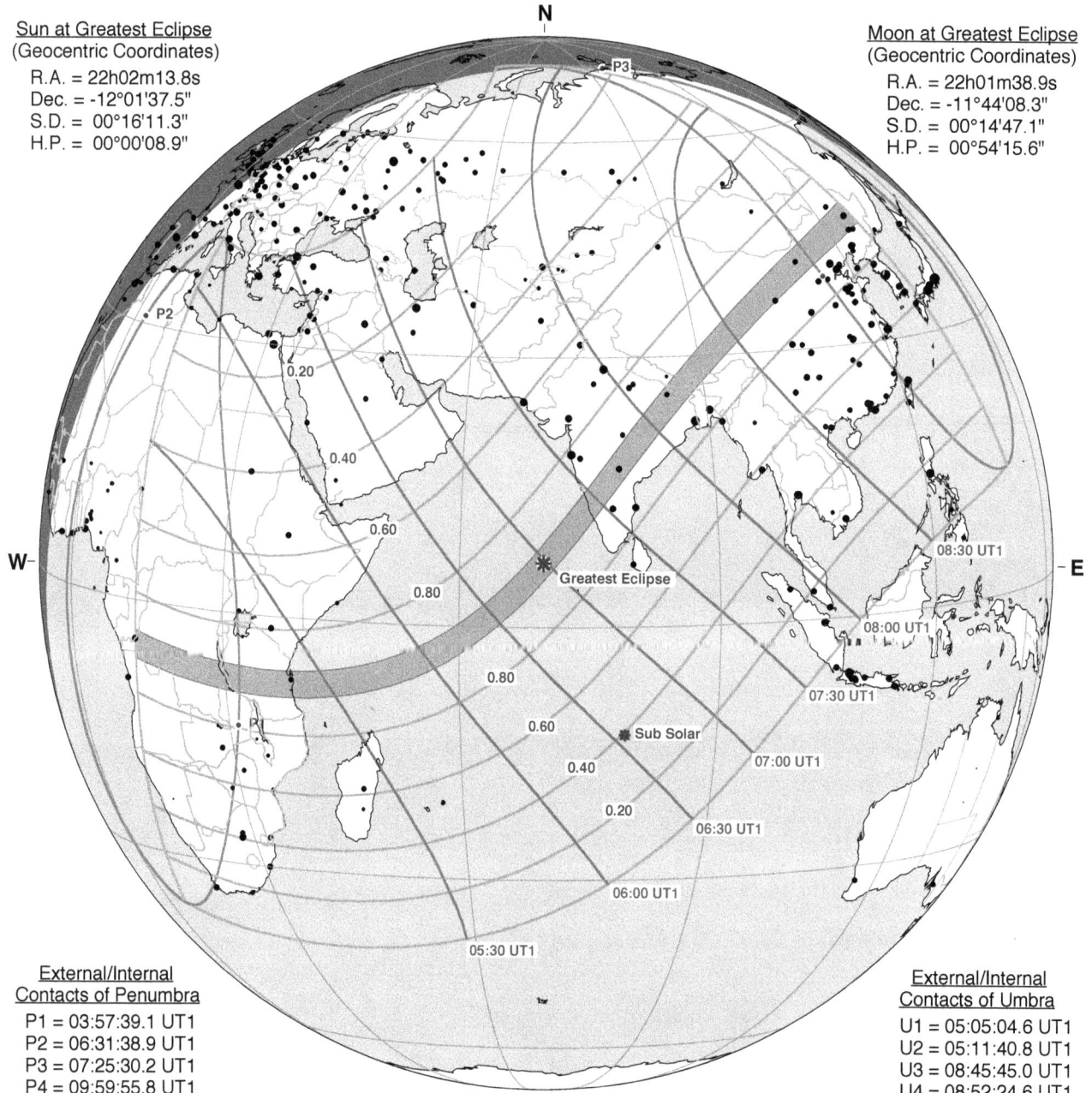

External/Internal Contacts of Penumbra
P1 = 03:57:39.1 UT1
P2 = 06:31:38.9 UT1
P3 = 07:25:30.2 UT1
P4 = 09:59:55.8 UT1

External/Internal Contacts of Umbra
U1 = 05:05:04.6 UT1
U2 = 05:11:40.8 UT1
U3 = 08:45:45.0 UT1
U4 = 08:52:24.6 UT1

ΔT = 93.2 s

Eph. = JPL DE405

Circumstances at Greatest Eclipse: 06:58:50.1 UT1
Lat. = 07°02.0'N   Sun Alt. = 68.9°
Long. = 069°37.3'E   Sun Azm. = 154.4°
Path Width = 294.6 km Duration = 08m56.4s

Circumstances at Greatest Duration: 06:43:05.7 UT1
Lat. = 03°50.2'N   Sun Alt. = 67.4°
Long. = 066°34.3'E   Sun Azm. = 134.9°
Path Width = 300.9 km Duration = 08m58.7s

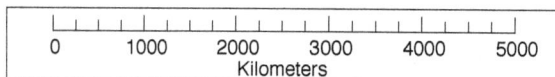

©2016 F. Espenak
www.EclipseWise.com

Total Solar Eclipse of 2064 Aug 12

Greatest Eclipse = 17:46:06.3 TD (= 17:44:32.7 UT1)

Eclipse Magnitude = 1.0495   Saros Series = 146

Gamma = -0.4652   Saros Member = 30 of 76

Sun at Greatest Eclipse
(Geocentric Coordinates)
R.A. = 09h32m49.7s
Dec. = +14°33'07.1"
S.D. = 00°15'47.0"
H.P. = 00°00'08.7"

Moon at Greatest Eclipse
(Geocentric Coordinates)
R.A. = 09h32m02.7s
Dec. = +14°07'45.3"
S.D. = 00°16'19.3"
H.P. = 00°59'54.2"

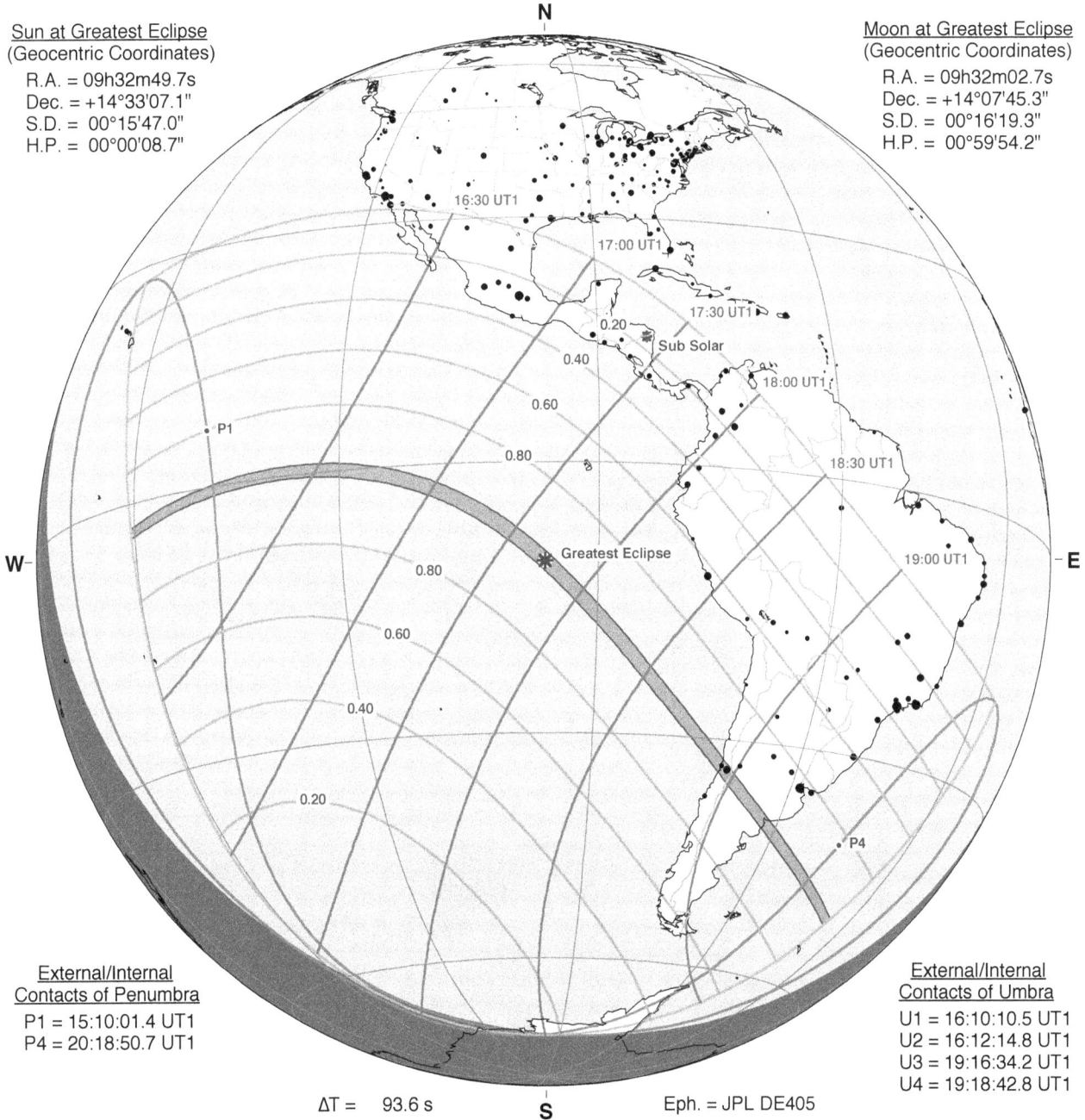

N

16:30 UT1

17:00 UT1

17:30 UT1

0.20

0.40   Sub Solar

0.60   18:00 UT1

P1   0.80   18:30 UT1

W   0.80   Greatest Eclipse   19:00 UT1   E

0.60

0.40

0.20

P4

S

External/Internal
Contacts of Penumbra
P1 = 15:10:01.4 UT1
P4 = 20:18:50.7 UT1

External/Internal
Contacts of Umbra
U1 = 16:10:10.5 UT1
U2 = 16:12:14.8 UT1
U3 = 19:16:34.2 UT1
U4 = 19:18:42.8 UT1

ΔT = 93.6 s   Eph. = JPL DE405

Circumstances at Greatest Eclipse: 17:44:32.7 UT1

Lat. = 10°55.8'S	Sun Alt. = 62.2°
Long. = 096°05.9'W	Sun Azm. = 23.8°
Path Width = 183.7 km	Duration = 04m27.7s

Circumstances at Greatest Duration: 17:40:44.1 UT1

Lat. = 10°06.0'S	Sun Alt. = 62.1°
Long. = 097°04.6'W	Sun Azm. = 28.0°
Path Width = 184.6 km	Duration = 04m27.8s

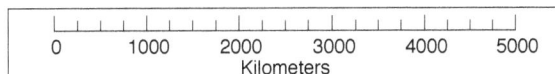

©2016 F. Espenak
www.EclipseWise.com

0   1000   2000   3000   4000   5000
Kilometers

163

Partial Solar Eclipse of 2065 Feb 05

Greatest Eclipse = 09:52:25.5 TD (= 09:50:51.6 UT1)

Eclipse Magnitude = 0.9123   Saros Series = 151
Gamma = 1.0336   Saros Member = 17 of 72

Sun at Greatest Eclipse
(Geocentric Coordinates)
R.A. = 21h18m22.7s
Dec. = -15°41'30.6"
S.D. = 00°16'13.3"
H.P. = 00°00'08.9"

Moon at Greatest Eclipse
(Geocentric Coordinates)
R.A. = 21h16m47.2s
Dec. = -14°47'52.5"
S.D. = 00°15'25.7"
H.P. = 00°56'37.5"

N

Greatest Eclipse

P4

0.80

0.60

11:00 UT1

W — E

P1

0.40

10:30 UT1

0.20

10:00 UT1

09:30 UT1

08:30 UT1

09:00 UT1

Sub Solar

External/Internal
Contacts of Penumbra
P1 = 07:39:11.3 UT1
P4 = 12:02:17.3 UT1

ΔT = 93.9 s   S   Eph. = JPL DE405

Circumstances at Greatest Eclipse: 09:50:51.6 UT1
Lat. = 62°09.7'N   Sun Alt. = 0.0°
Long. = 022°01.5'W   Sun Azm. = 125.4°

0   1000   2000   3000   4000   5000
Kilometers

©2016 F. Espenak
www.EclipseWise.com

Partial Solar Eclipse of 2065 Jul 03

Greatest Eclipse = 17:33:52.5 TD (= 17:32:18.3 UT1)

Eclipse Magnitude = 0.1639   Saros Series = 118
Gamma = 1.4619   Saros Member = 71 of 72

Sun at Greatest Eclipse
(Geocentric Coordinates)
R.A. = 06h53m43.9s
Dec. = +22°51'26.7"
S.D. = 00°15'43.9"
H.P. = 00°00'08.6"

Moon at Greatest Eclipse
(Geocentric Coordinates)
R.A. = 06h54m50.6s
Dec. = +24°10'43.8"
S.D. = 00°15'05.3"
H.P. = 00°55'22.6"

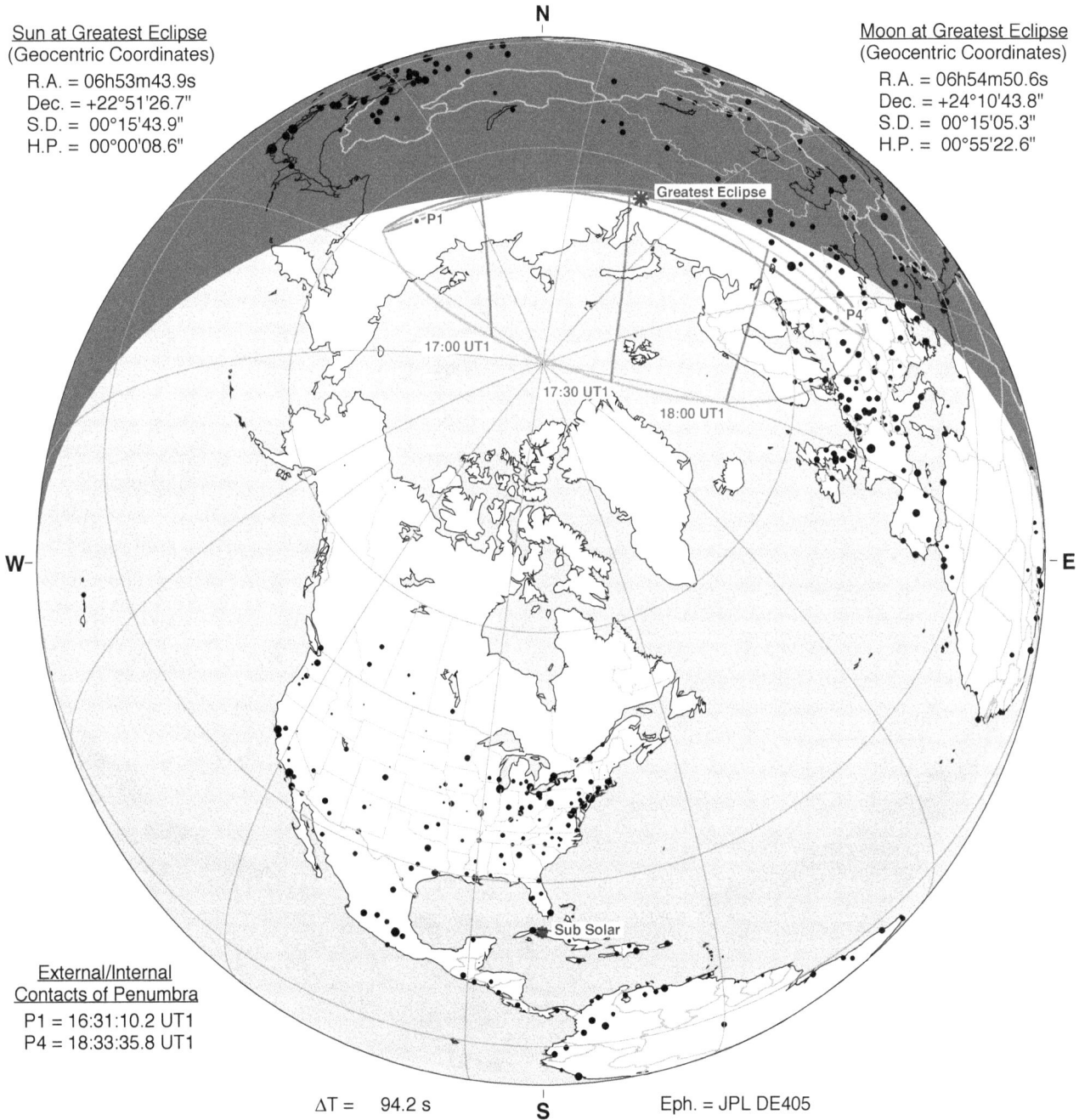

N

Greatest Eclipse

P1

P4

17:00 UT1

17:30 UT1

18:00 UT1

W

E

Sub Solar

External/Internal
Contacts of Penumbra
P1 = 16:31:10.2 UT1
P4 = 18:33:35.8 UT1

ΔT = 94.2 s   S   Eph. = JPL DE405

Circumstances at Greatest Eclipse: 17:32:18.3 UT1
Lat. = 64°49.3'N   Sun Alt. = 0.0°
Long. = 071°43.9'E   Sun Azm. = 335.9°

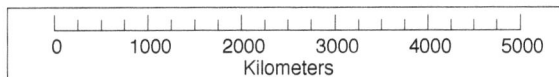

0   1000   2000   3000   4000   5000
Kilometers

©2016 F. Espenak
www.EclipseWise.com

Partial Solar Eclipse of 2065 Aug 02

Greatest Eclipse = 05:34:16.6 TD   (= 05:32:42.3 UT1)

Eclipse Magnitude = 0.4903   Saros Series = 156
Gamma = -1.2758   Saros Member = 4 of 69

Sun at Greatest Eclipse
(Geocentric Coordinates)
R.A. = 08h51m52.4s
Dec. = +17°35'43.5"
S.D. = 00°15'45.5"
H.P. = 00°00'08.7"

Moon at Greatest Eclipse
(Geocentric Coordinates)
R.A. = 08h50m03.4s
Dec. = +16°28'16.4"
S.D. = 00°15'28.9"
H.P. = 00°56'49.3"

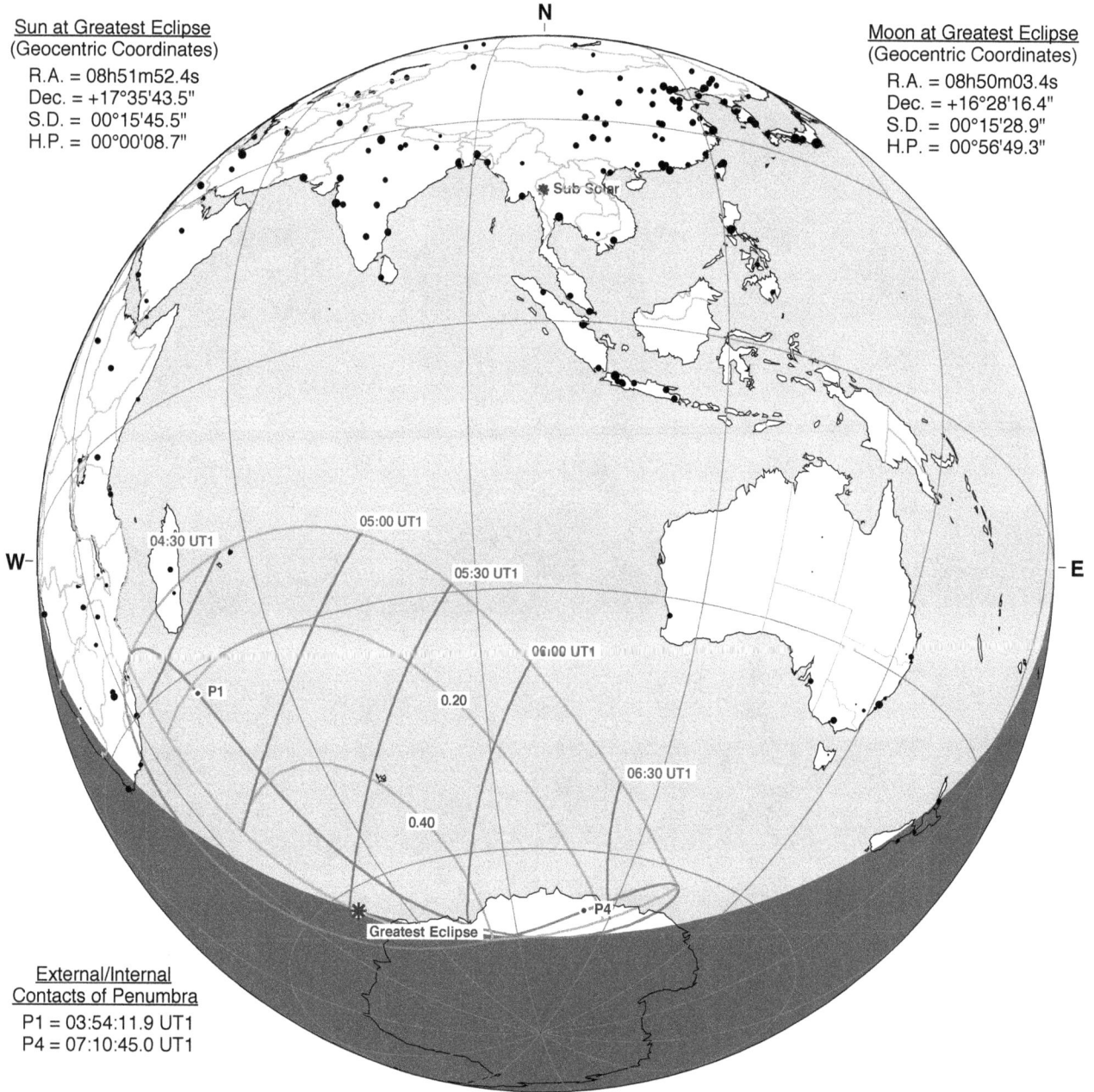

N

W

E

04:30 UT1
05:00 UT1
05:30 UT1
06:00 UT1
06:30 UT1

P1

0.20
0.40

P4

Greatest Eclipse

Sub Solar

External/Internal Contacts of Penumbra
P1 = 03:54:11.9 UT1
P4 = 07:10:45.0 UT1

ΔT = 94.3 s

S

Eph. = JPL DE405

Circumstances at Greatest Eclipse: 05:32:42.3 UT1
Lat. = 62°42.7'S   Sun Alt. = 0.0°
Long. = 046°24.5'E   Sun Azm. = 48.7°

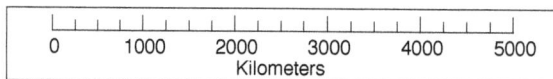

0   1000   2000   3000   4000   5000
Kilometers

Partial Solar Eclipse of 2065 Dec 27

Greatest Eclipse = 08:39:55.7 TD   (= 08:38:21.2 UT1)

Eclipse Magnitude = 0.8769   Saros Series = 123
Gamma = -1.0688   Saros Member = 56 of 70

Sun at Greatest Eclipse
(Geocentric Coordinates)
R.A. = 18h26m44.9s
Dec. = -23°17'20.3"
S.D. = 00°16'15.7"
H.P. = 00°00'08.9"

Moon at Greatest Eclipse
(Geocentric Coordinates)
R.A. = 18h27m25.5s
Dec. = -24°21'42.8"
S.D. = 00°16'37.3"
H.P. = 01°01'00.2"

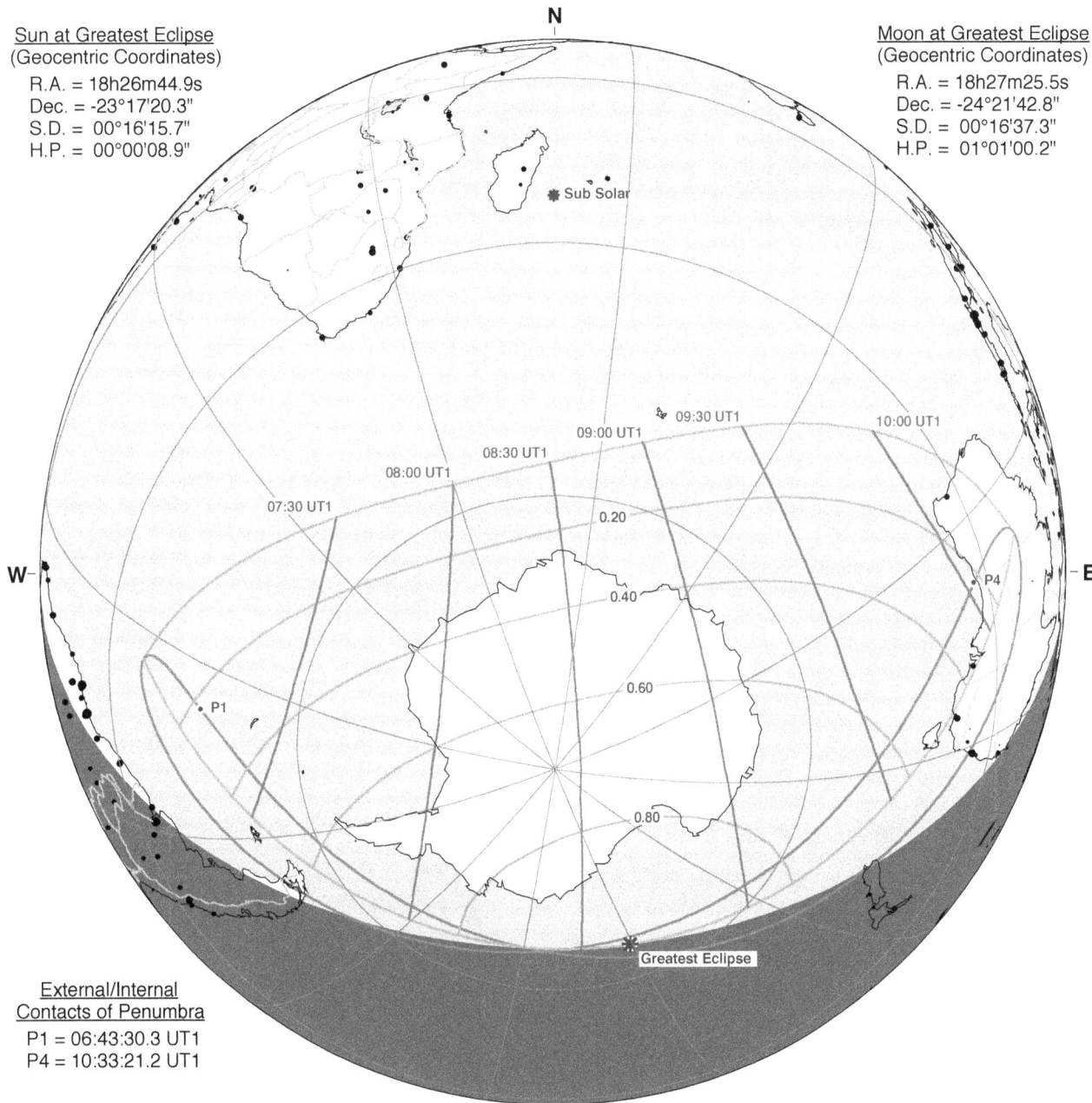

N

Sub Solar

09:30 UT1

09:00 UT1   10:00 UT1

08:30 UT1

08:00 UT1

07:30 UT1

0.20

W   E

0.40

0.60

P4

P1

0.80

Greatest Eclipse

External/Internal
Contacts of Penumbra
P1 = 06:43:30.3 UT1
P4 = 10:33:21.2 UT1

ΔT = 94.6 s   S   Eph. = JPL DE405

Circumstances at Greatest Eclipse: 08:38:21.2 UT1
Lat. = 65°23.8'S   Sun Alt. = 0.0°
Long. = 149°17.5'W   Sun Azm. = 198.3°

| 0 | 1000 | 2000 | 3000 | 4000 | 5000 |
Kilometers

©2016 F. Espenak
www.EclipseWise.com

Annular Solar Eclipse of 2066 Jun 22

Greatest Eclipse = 19:25:47.7 TD (= 19:24:12.8 UT1)

Eclipse Magnitude = 0.9435	Saros Series = 128
Gamma = 0.7330	Saros Member = 61 of 73

Sun at Greatest Eclipse
(Geocentric Coordinates)
R.A. = 06h07m28.7s
Dec. = +23°25'11.2"
S.D. = 00°15'44.2"
H.P. = 00°00'08.7"

Moon at Greatest Eclipse
(Geocentric Coordinates)
R.A. = 06h07m48.1s
Dec. = +24°04'22.4"
S.D. = 00°14'42.0"
H.P. = 00°53'57.0"

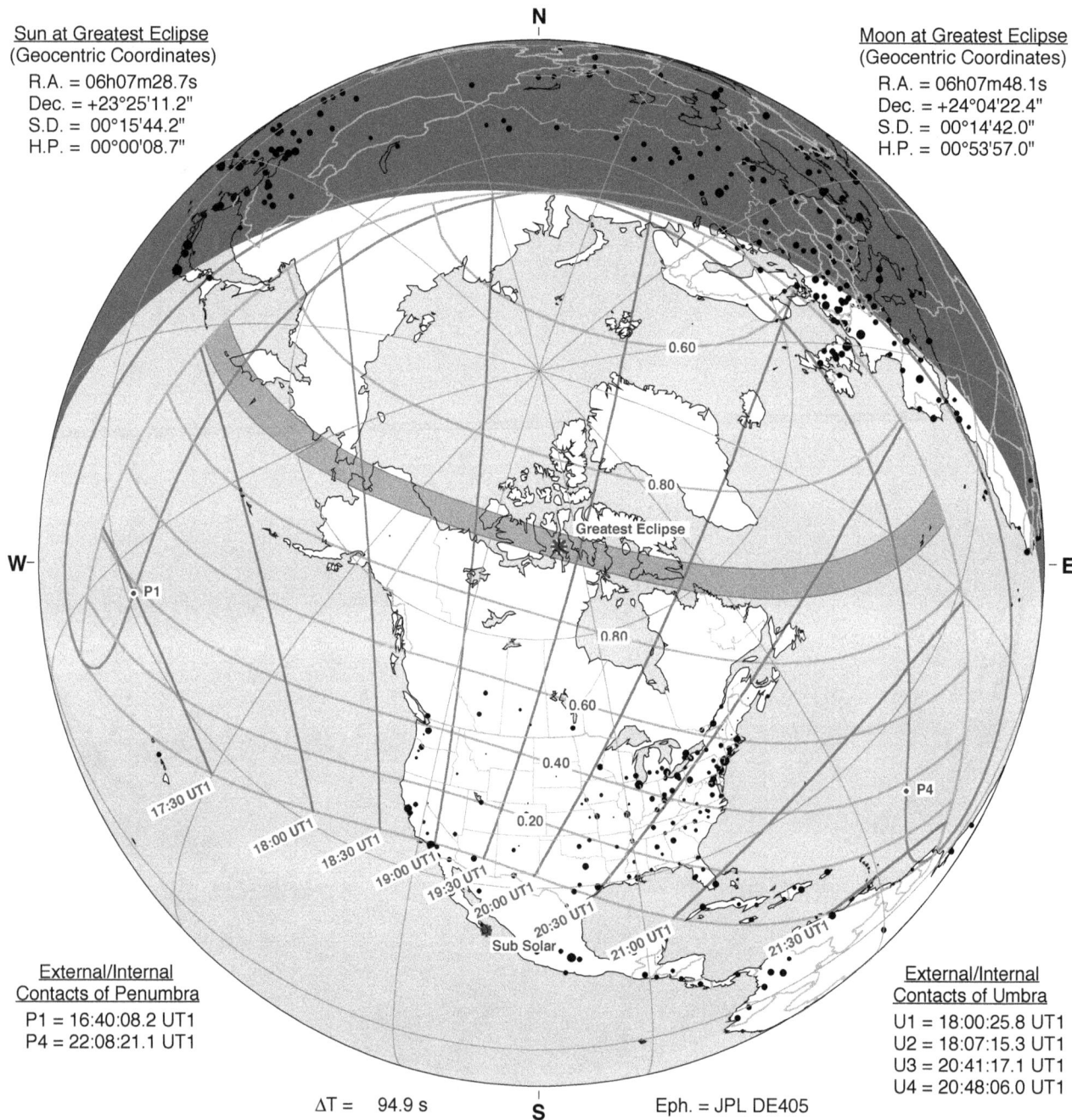

External/Internal Contacts of Penumbra
P1 = 16:40:08.2 UT1
P4 = 22:08:21.1 UT1

External/Internal Contacts of Umbra
U1 = 18:00:25.8 UT1
U2 = 18:07:15.3 UT1
U3 = 20:41:17.1 UT1
U4 = 20:48:06.0 UT1

ΔT = 94.9 s

Eph. = JPL DE405

Circumstances at Greatest Eclipse: 19:24:12.8 UT1
Lat. = 70°07.9'N
Long. = 096°29.9'W
Path Width = 308.5 km
Sun Alt. = 42.6°
Sun Azm. = 197.5°
Duration = 04m39.8s

Circumstances at Greatest Duration: 19:21:23.9 UT1
Lat. = 70°31.1'N
Long. = 099°39.3'W
Path Width = 308.3 km
Sun Alt. = 42.5°
Sun Azm. = 192.6°
Duration = 04m39.8s

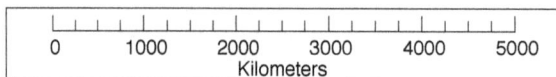

©2016 F. Espenak
www.EclipseWise.com

Total Solar Eclipse of 2066 Dec 17

Greatest Eclipse = 00:23:39.9 TD   (= 00:22:04.7 UT1)

Eclipse Magnitude = 1.0416   Saros Series = 133
Gamma = -0.4043   Saros Member = 48 of 72

Sun at Greatest Eclipse
(Geocentric Coordinates)
R.A. = 17h39m46.4s
Dec. = -23°20'56.0"
S.D. = 00°16'15.1"
H.P. = 00°00'08.9"

Moon at Greatest Eclipse
(Geocentric Coordinates)
R.A. = 17h39m53.3s
Dec. = -23°45'32.9"
S.D. = 00°16'39.9"
H.P. = 01°01'09.6"

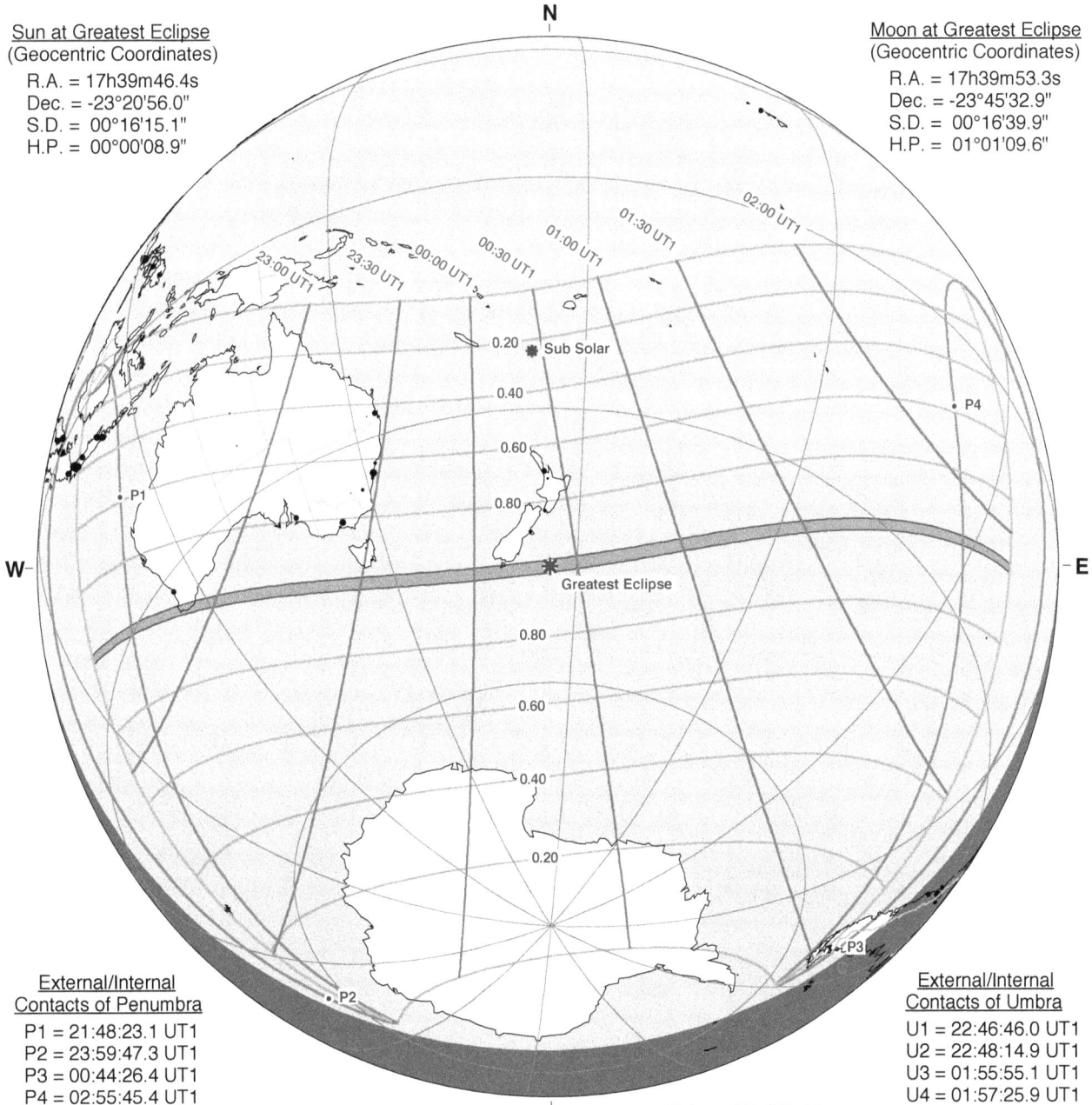

N

W

E

S

0.20 Sub Solar
0.40
0.60
0.80
Greatest Eclipse
0.80
0.60
0.40
0.20

P1
P2
P3
P4

02:00 UT1
01:30 UT1
01:00 UT1
00:30 UT1
00:00 UT1
23:30 UT1
23:00 UT1

External/Internal
Contacts of Penumbra
P1 = 21:48:23.1 UT1
P2 = 23:59:47.3 UT1
P3 = 00:44:26.4 UT1
P4 = 02:55:45.4 UT1

External/Internal
Contacts of Umbra
U1 = 22:46:46.0 UT1
U2 = 22:48:14.9 UT1
U3 = 01:55:55.1 UT1
U4 = 01:57:25.9 UT1

ΔT = 95.3 s   Eph. = JPL DE405

Circumstances at Greatest Eclipse: 00:22:04.7 UT1

Lat. = 47°22.1'S	Sun Alt. = 65.9°
Long. = 175°38.4'E	Sun Azm. = 355.1°
Path Width = 152.2 km	Duration = 03m14.5s

Circumstances at Greatest Duration: 00:21:56.4 UT1

Lat. = 47°22.5'S	Sun Alt. = 65.9°
Long. = 175°33.7'E	Sun Azm. = 355.3°
Path Width = 152.2 km	Duration = 03m14.5s

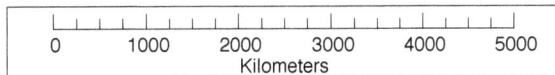

0   1000   2000   3000   4000   5000
Kilometers

©2016 F. Espenak
www.EclipseWise.com

169

Appendix D

Central Solar Eclipse Maps: 2001 to 2100

Key to Central Solar Eclipse Maps

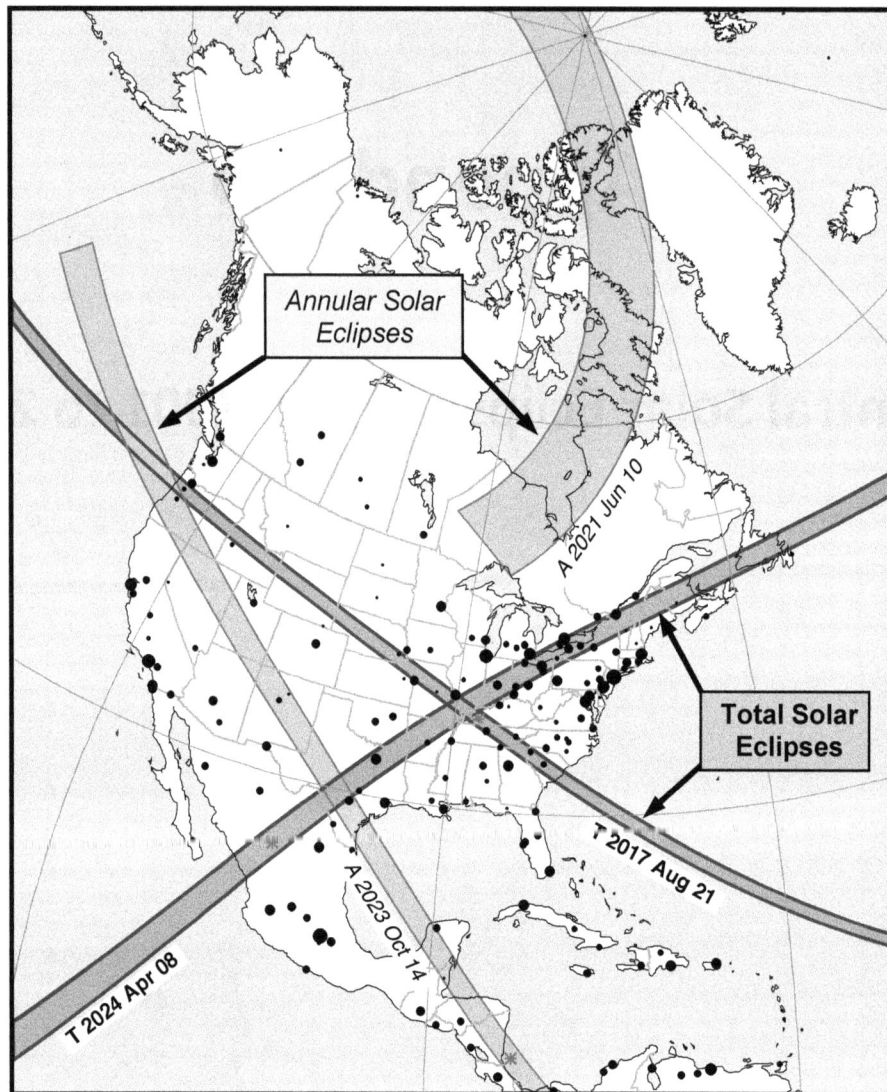

Total Eclipses – the path is darkly shaded (blue in Color & Deluxe Editions); the date appears in bold and is preceded by the letter **T**

Annular Eclipses – the path is lightly shaded (orange in Color & Deluxe Editions); the date appears in italics and is preceded by the letter **A**

Hybrid Eclipses – the path is darkly shaded (or blue) along the total sections and lightly shaded (or orange) along the annular sections; the date appears in bold-italics and is preceded by the letter **H**

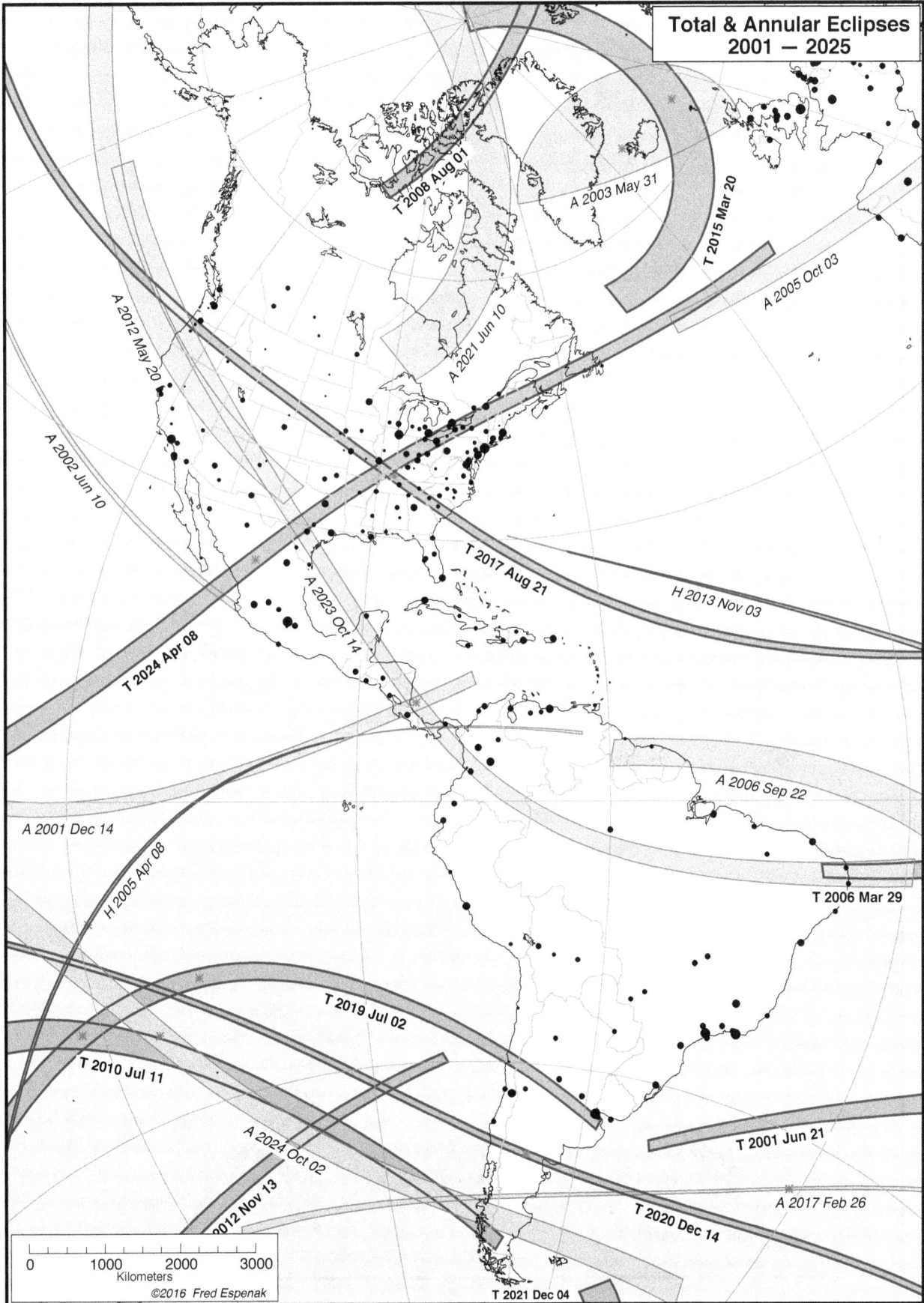

Total & Annular Eclipses
2001 — 2025

T 2008 Aug 01
A 2003 May 31
T 2015 Mar 20
A 2005 Oct 03
A 2012 May 20
A 2021 Jun 10
A 2002 Jun 10
T 2017 Aug 21
H 2013 Nov 03
A 2023 Oct 14
T 2024 Apr 08
A 2006 Sep 22
A 2001 Dec 14
H 2005 Apr 08
T 2006 Mar 29
T 2019 Jul 02
T 2010 Jul 11
A 2024 Oct 02
2012 Nov 13
T 2001 Jun 21
A 2017 Feb 26
T 2020 Dec 14
T 2021 Dec 04

0   1000   2000   3000
Kilometers
©2016 Fred Espenak

Total & Annular Eclipses
2001 — 2025

A 2021 Jun 10
T 2024 Apr 08
A 2003 May 31
T 2015 Mar 20
T 2008 Aug 01
T 2009 Jul 22
A 2019 Dec 26
A 2020 Jun 21
A 2005 Oct 03
T 2006 Mar 29
H 2013 Nov 03
A 2010 Jan 15
T 2001 Jun 21
A 2016 Sep 01
A 2006 Sep 22
A 2017 Feb 26
T 2020 Dec 14
T 2002 Dec 04
A 2009 Jan 26

0   1000   2000   3000
Kilometers
©2016 Fred Espenak

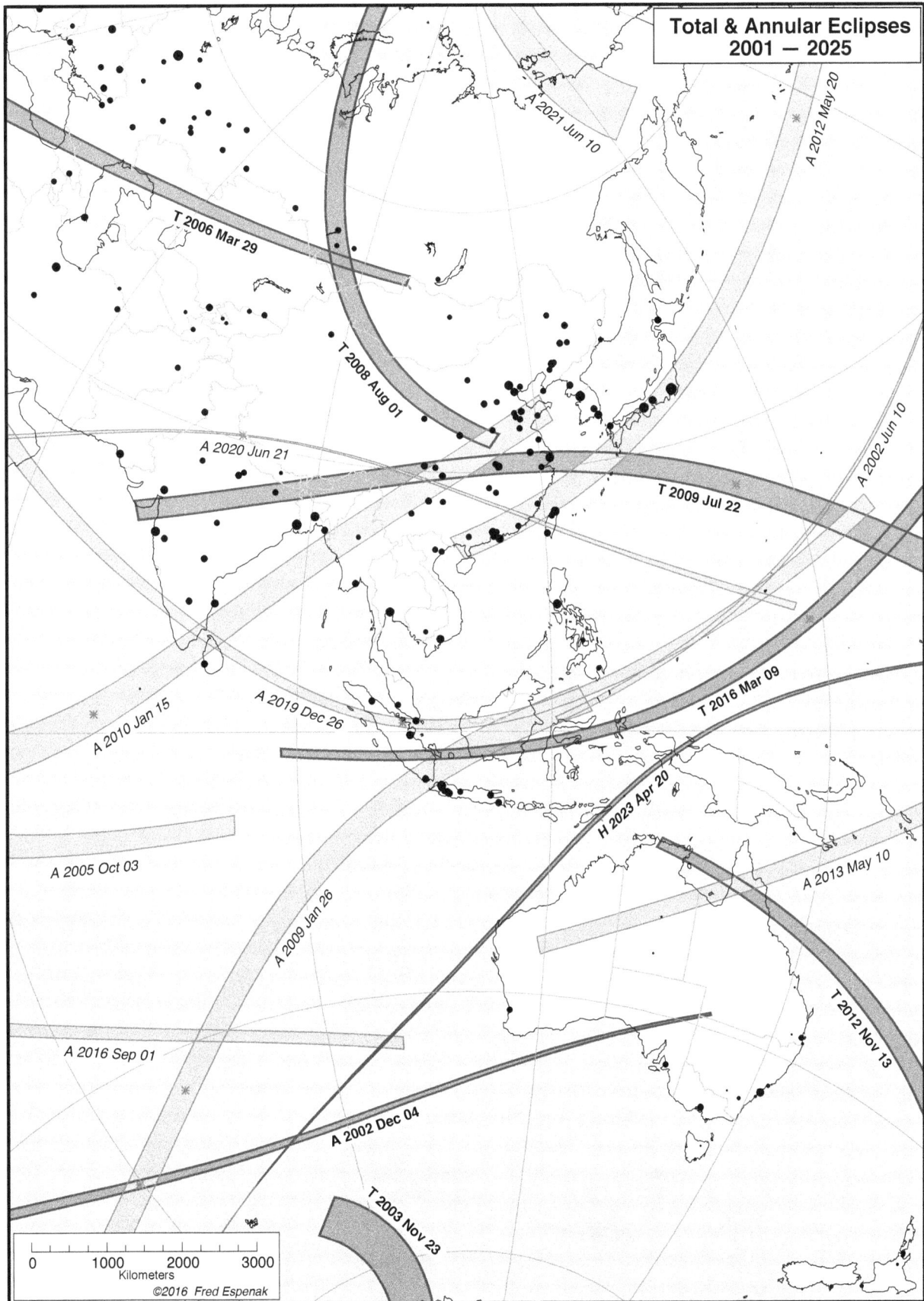

**Total & Annular Eclipses
2001 — 2025**

A 2021 Jun 10

A 2012 May 20

T 2006 Mar 29

T 2008 Aug 01

A 2020 Jun 21

T 2009 Jul 22

A 2002 Jun 10

A 2010 Jan 15

A 2019 Dec 26

T 2016 Mar 09

A 2005 Oct 03

A 2009 Jan 26

H 2023 Apr 20

A 2013 May 10

A 2016 Sep 01

T 2012 Nov 13

A 2002 Dec 04

T 2003 Nov 23

| 0 | 1000 | 2000 | 3000 |
Kilometers

©2016 Fred Espenak

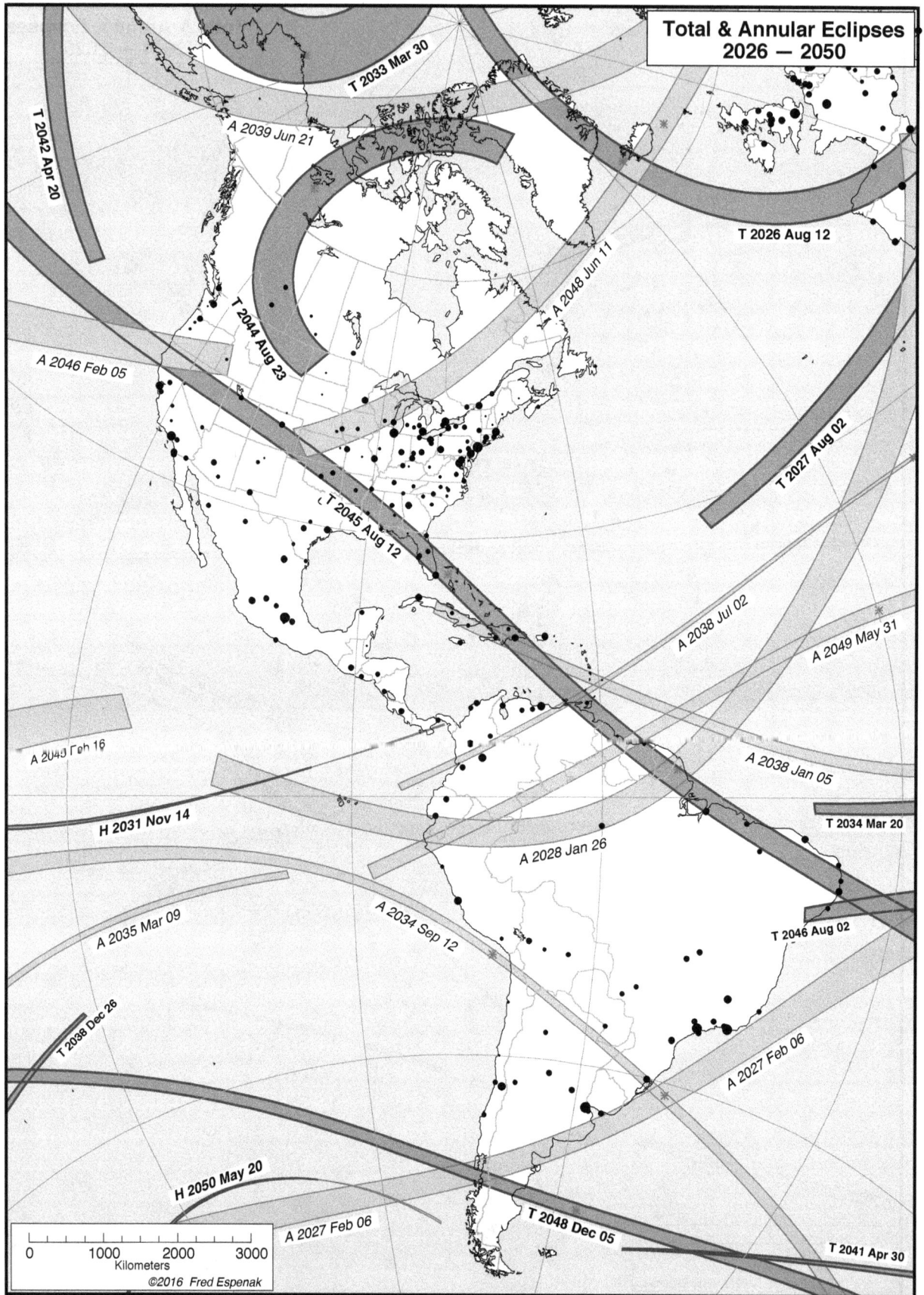

Total & Annular Eclipses
2026 — 2050

T 2033 Mar 30
A 2039 Jun 21
T 2042 Apr 20
T 2026 Aug 12
A 2048 Jun 11
T 2044 Aug 23
A 2046 Feb 05
T 2027 Aug 02
T 2045 Aug 12
A 2038 Jul 02
A 2049 May 31
A 2049 Feb 16
A 2038 Jan 05
H 2031 Nov 14
T 2034 Mar 20
A 2028 Jan 26
A 2034 Sep 12
T 2046 Aug 02
A 2035 Mar 09
A 2027 Feb 06
T 2038 Dec 26
H 2050 May 20
A 2027 Feb 06
T 2048 Dec 05
T 2041 Apr 30

0 1000 2000 3000
Kilometers
©2016 Fred Espenak

Total & Annular Eclipses
2026 — 2050

T 2044 Aug 23
T 2033 Mar 30
A 2041 O
A 2039 Jun 21
A 2048 Jun 11
A 2030 Jun 01
T 2035 Sep 02
T 2026 Aug 12
A 2028 Jan 26
T 2027 Aug 02
A 2038 Jul 02
A 2049 May 31
T 2034 Mar 20
H 2049 Nov 25
A 2038 Jan 05
A 2027 Feb 06
A 2031 May 21
T 2046 Aug 02
T 2045 Aug 12
T 2041 Apr 30
T 2030 Nov 25
T 2048 Dec 05
A 2034 Sep 12
A 2032 May 09

0   1000   2000   3000
Kilometers
©2016 Fred Espenak

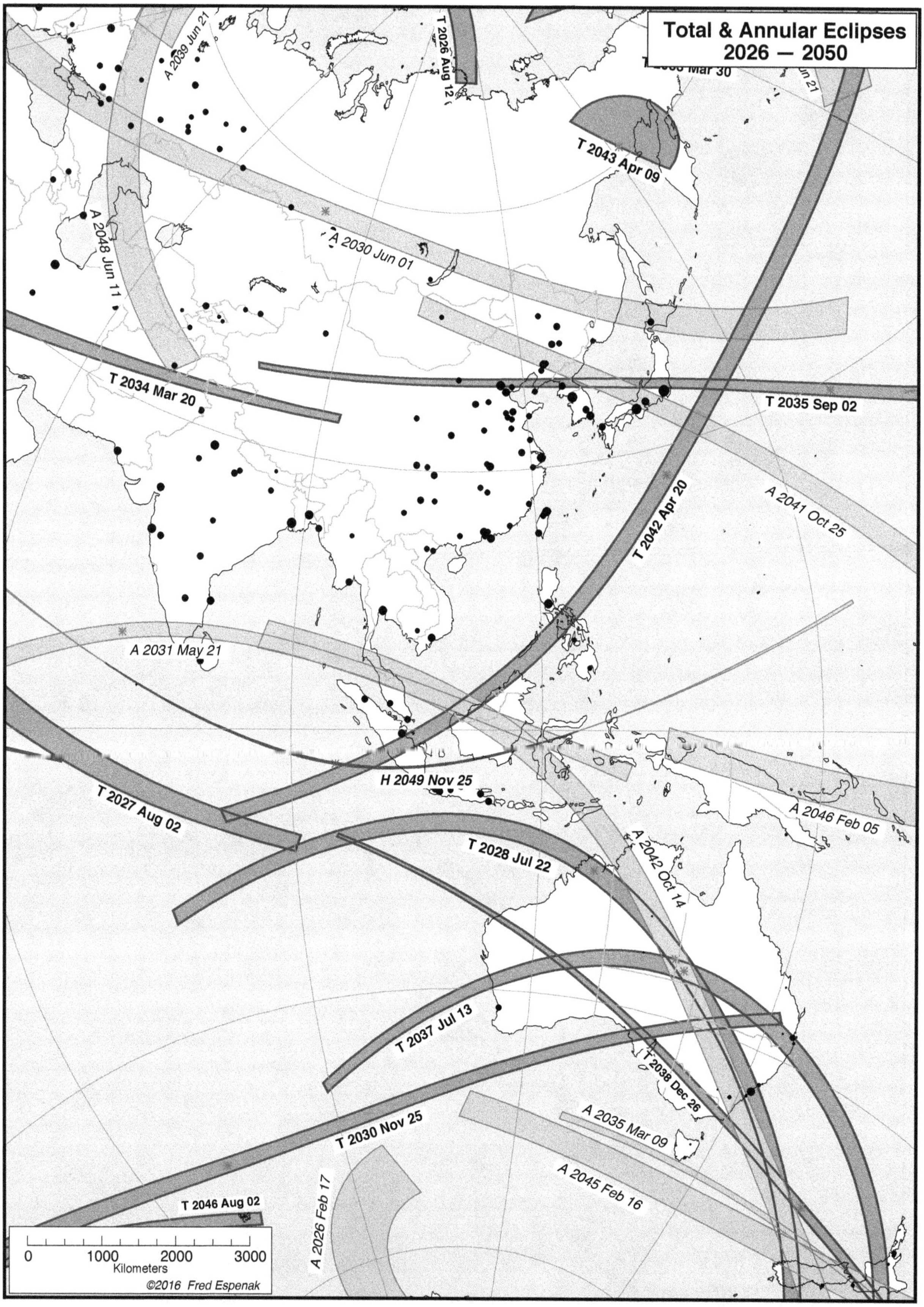

Total & Annular Eclipses
2026 — 2050

A 2039 Jun 21
T 2026 Aug 12
T 2043 Apr 09
A 2048 Jun 11
A 2030 Jun 01
T 2034 Mar 20
T 2035 Sep 02
A 2041 Oct 25
T 2042 Apr 20
A 2031 May 21
H 2049 Nov 25
A 2046 Feb 05
T 2027 Aug 02
T 2028 Jul 22
A 2042 Oct 14
T 2037 Jul 13
T 2038 Dec 26
T 2030 Nov 25
A 2035 Mar 09
A 2026 Feb 17
A 2045 Feb 16
T 2046 Aug 02

0 1000 2000 3000
Kilometers
©2016 Fred Espenak

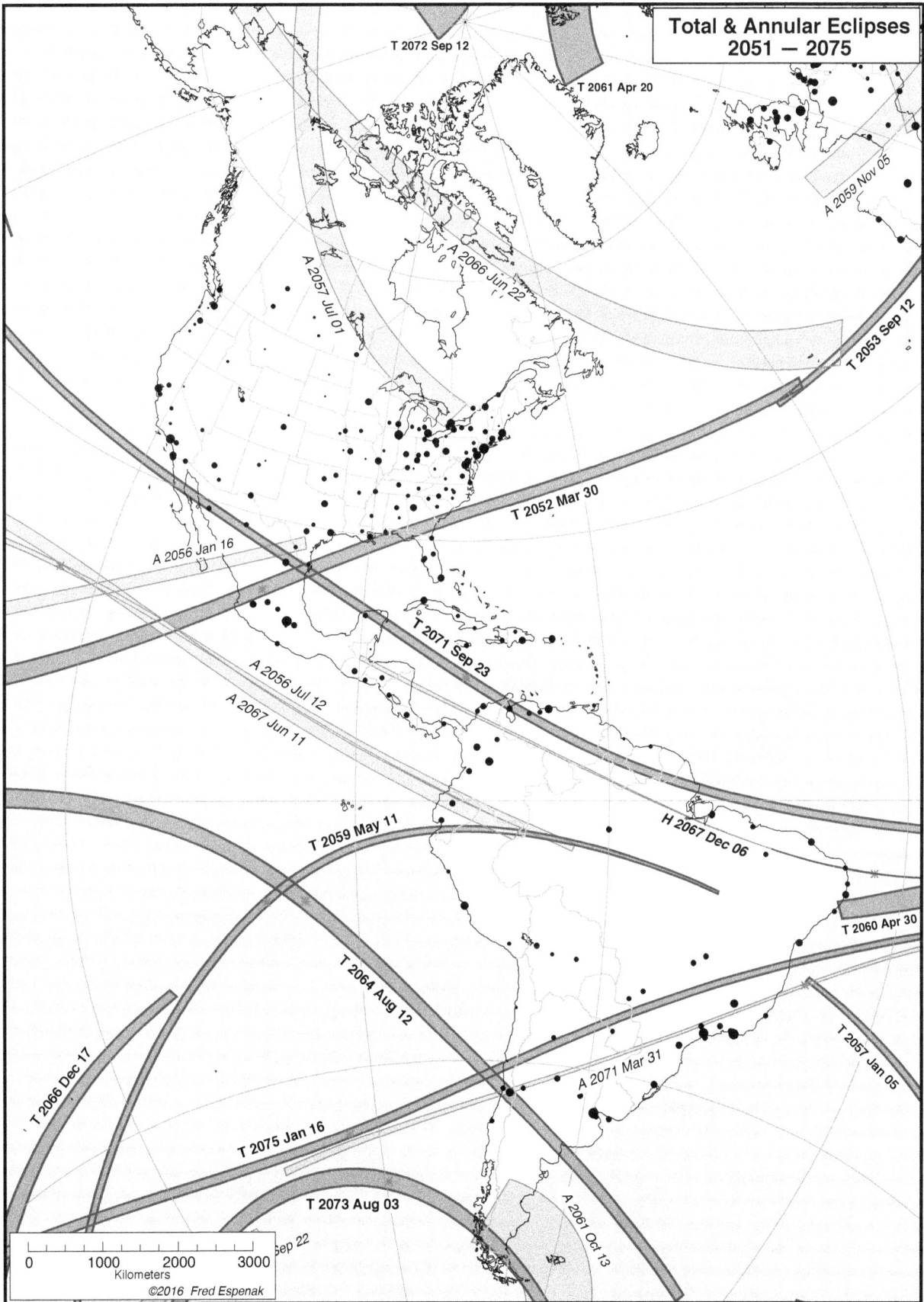

Total & Annular Eclipses
2051 — 2075

T 2072 Sep 12
T 2061 Apr 20
A 2059 Nov 05
A 2057 Jul 01
A 2066 Jun 22
T 2053 Sep 12
T 2052 Mar 30
A 2056 Jan 16
T 2071 Sep 23
A 2056 Jul 12
A 2067 Jun 11
H 2067 Dec 06
T 2059 May 11
T 2060 Apr 30
T 2064 Aug 12
T 2057 Jan 05
A 2071 Mar 31
T 2066 Dec 17
T 2075 Jan 16
A 2061 Oct 13
T 2073 Aug 03
Sep 22

0 1000 2000 3000
Kilometers
©2016 Fred Espenak

179

Total & Annular Eclipses 2051 — 2075

T 2072 Sep 12

T 2061 Apr 20

A 2057 Jul 01

A 2075 Jul 13

T 2063 Aug 24

A 2066 Jun 22

T 2052 Mar 30

T 2060 Apr 30

T 2053 Sep 12

T 2060 Apr 30

A 2059 Nov 05

T 2053 Sep 12

A 2074 Jan 27

T 2071 Sep 23

H 2067 Dec 06

A 2071 Mar 31

A 2064 Feb 17

A 2060 Oct 24

T 2055 Jul 24

A 2070 Oct 04

T 2057 Jan 05

A 2053 Mar 20

A 2063 Feb 28

0   1000   2000   3000
Kilometers

©2016 Fred Espenak

**Total & Annular Eclipses
2051 — 2075**

T 2072 Sep 12

A 2057 Jul 01

A 2066 Jun 22

A 2075 Jul 13

T 2061 Apr 20

T 2060 Apr 30

T 2063 Aug 24

T 2070 Apr 11

A 2056 Jan 16

A 2064 Feb 17

A 2074 Jan 27

A 2074 Jul 24

T 2053 Sep 12

A 2059 Nov 05

A 2063 Feb 28

A 2053 Mar 20

A 2052 Sep 22

T 2057 Jan 05

T 2068 May 31

T 2066 Dec 17

A 2070 Oct 04

A 2060 Oct 24

T 2055 Jul 24

0	1000	2000	3000

Kilometers

©2016 Fred Espenak

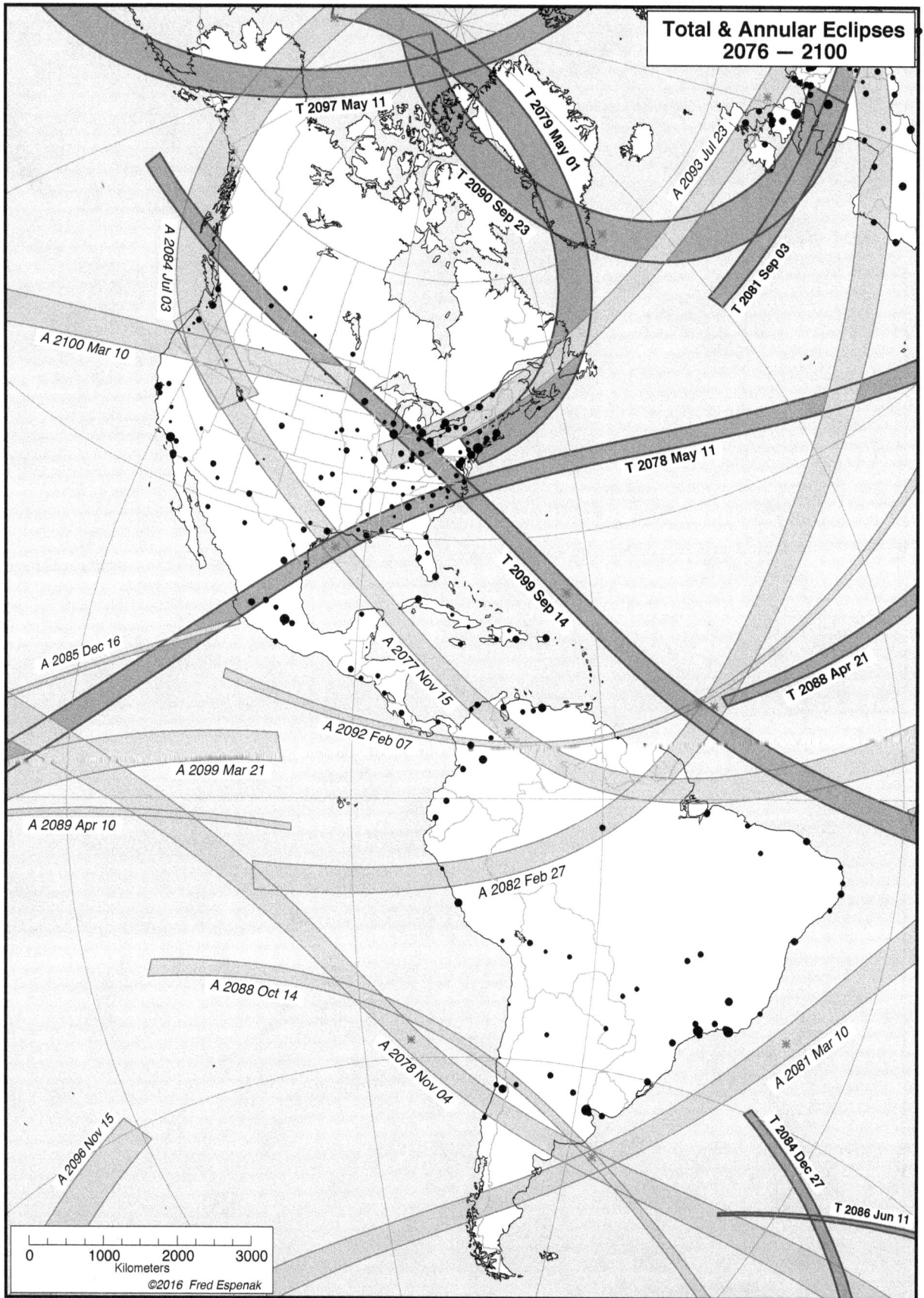

Total & Annular Eclipses
2076 — 2100

T 2097 May 11
T 2079 May 01
T 2090 Sep 23
A 2093 Jul 23
A 2084 Jul 03
T 2081 Sep 03
A 2100 Mar 10
T 2078 May 11
A 2085 Dec 16
T 2099 Sep 14
A 2077 Nov 15
T 2088 Apr 21
A 2092 Feb 07
A 2099 Mar 21
A 2089 Apr 10
A 2088 Oct 14
A 2082 Feb 27
A 2081 Mar 10
A 2078 Nov 04
T 2084 Dec 27
A 2096 Nov 15
T 2086 Jun 11

0   1000   2000   3000
Kilometers
©2016 Fred Espenak

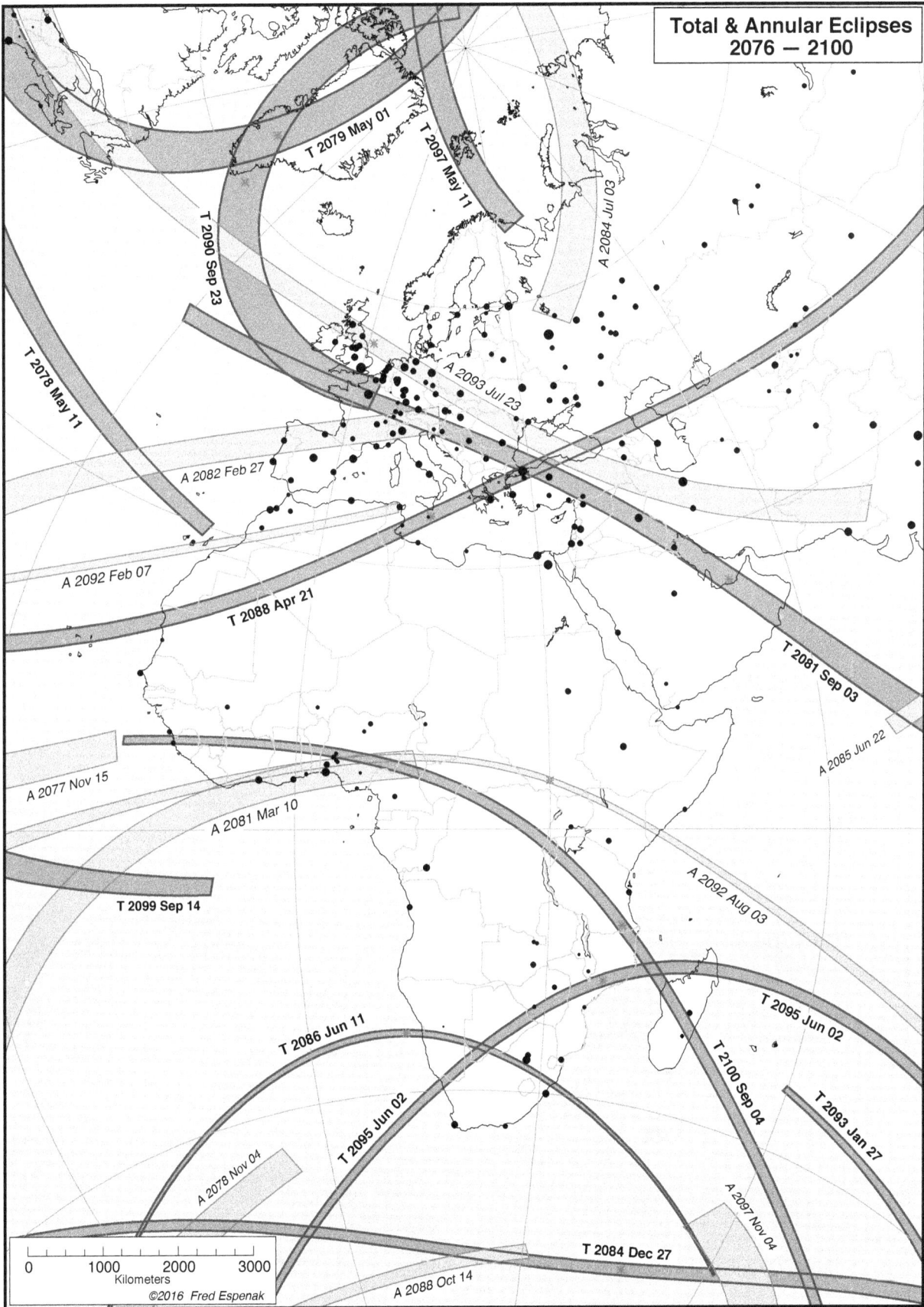

Total & Annular Eclipses
2076 — 2100

T 2079 May 01
T 2097 May 11
T 2090 Sep 23
A 2084 Jul 03
T 2078 May 11
A 2093 Jul 23
A 2082 Feb 27
A 2092 Feb 07
T 2088 Apr 21
T 2081 Sep 03
A 2085 Jun 22
A 2077 Nov 15
A 2081 Mar 10
T 2099 Sep 14
A 2092 Aug 03
T 2095 Jun 02
T 2086 Jun 11
T 2100 Sep 04
T 2093 Jan 27
T 2095 Jun 02
A 2078 Nov 04
A 2097 Nov 04
T 2084 Dec 27
A 2088 Oct 14

0   1000   2000   3000
Kilometers
©2016 Fred Espenak

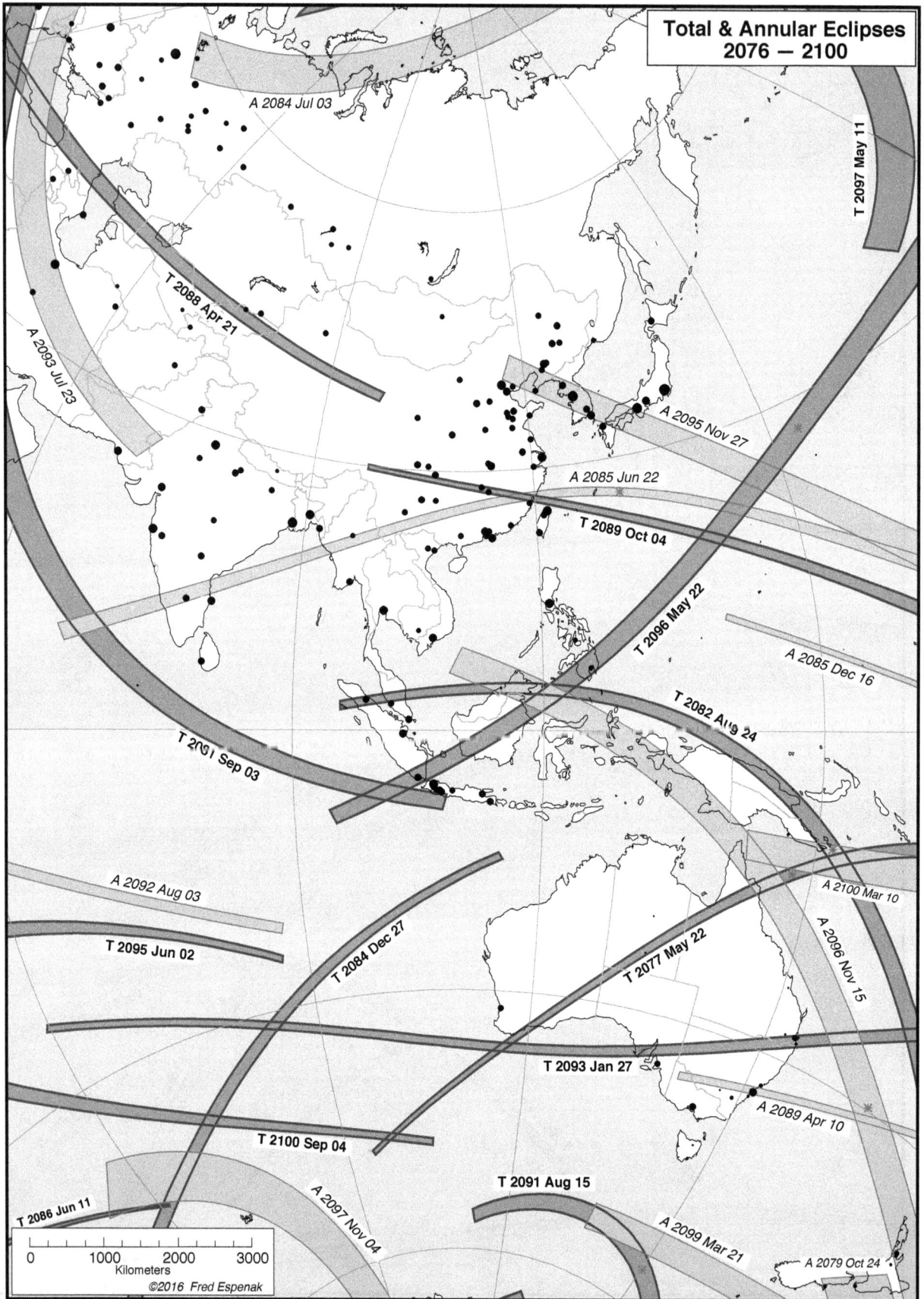

Total & Annular Eclipses
2076 — 2100

A 2084 Jul 03

T 2097 May 11

T 2088 Apr 21

A 2093 Jul 23

A 2095 Nov 27

A 2085 Jun 22

T 2089 Oct 04

T 2096 May 22

A 2085 Dec 16

T 2082 Aug 24

T 2091 Sep 03

A 2092 Aug 03

A 2100 Mar 10

T 2095 Jun 02

T 2084 Dec 27

T 2077 May 22

A 2096 Nov 15

T 2093 Jan 27

A 2089 Apr 10

T 2100 Sep 04

T 2091 Aug 15

T 2086 Jun 11

A 2097 Nov 04

A 2099 Mar 21

A 2079 Oct 24

0   1000   2000   3000
Kilometers

©2016 Fred Espenak

Astropixels Publications

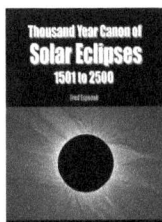

Thousand Year Canon of Solar Eclipses 1501 to 2500 (Fred Espenak) contains maps and data for each of the 2,389 solar eclipses occurring over the ten-century period centered on the present era. A comprehensive catalog lists the essential characteristics of each eclipse while a series of global maps show the exact geographic extent of each eclipse.

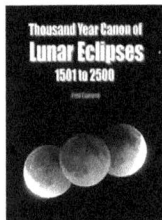

The Thousand Year Canon of Lunar Eclipses 1501 to 2500 (Fred Espenak) contains diagrams, maps and data for each of the 2,424 lunar eclipses occurring over the ten-century period centered on the present era. A comprehensive catalog lists the essential characteristics of each eclipse while a series of diagrams and maps illustrate the Moon-shadow geometry and geographic regions of visibility of each eclipse.

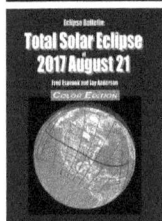

Eclipse Bulletin: Total Solar Eclipse of 2017 August 21 (Fred Espenak & Jay Anderson) is the ultimate guide to this highly anticipated event. The bulletin is a treasure trove of facts on every conceivable aspect of the eclipse. The exact details about the path of totality can be found in a series of tables containing geographic coordinates, times, altitudes, and more. Detailed maps plot the total eclipse path across the USA. Local circumstances for over 1000 cities provide times of each phase of the eclipse plus the eclipse magnitude, duration and Sun's altitude.

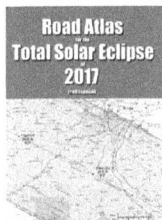

Road Atlas for the Total Solar Eclipse of 2017 (Fred Espenak) contains a comprehensive series of 37 high-resolution maps of the path of totality across the USA. The large scale (1:700,000 or 1 inch = 11 miles) shows both major and minor roads, towns and cities, rivers, lakes, parks, national forests, wilderness areas and mountain ranges. The duration of totality is plotted in 20-second steps, making it easy to estimate the length of the total eclipse from any location in the eclipse path.

TOTAL Eclipse or Bust! A Family Road Trip (Patricia Espenak & Fred Espenak) is a book for the entire family. The story follows a typical family on a road trip to see the 2017 total eclipse of the Sun. Along the way the children learn all about the how and why of eclipses in a friendly and an uncomplicated way. The book also provides basic information about how to view a total solar eclipse and where to go for America's great eclipse on August 21, 2017.

Atlas of Central Solar Eclipses in the USA (Fred Espenak) contains of a series of 499 global maps showing the track of every total and annular solar eclipse across the USA from 1001 through 3000. It is accompanied by a catalog that lists the major characteristics of each eclipse. A set of 20 detailed maps, each covering a 50-year period and centered on the lower 48 states, shows the path of every total and annular eclipse. The maps include state boundaries and major cities. These maps also cover southern Canada and northern Mexico.

The 2017 eclipse publications have been recognized by *Sky & Telescope magazine* as **Hot Products for 2016**. For complete details and ordering information on the above *Astropixels Publications*, visit:

astropixels.com/pubs/

www.ingramcontent.com/pod-product-compliance
Lightning Source LLC
Chambersburg PA
CBHW061754210326
41518CB00036B/2376